The Randlords

Geoffrey Wheatcroft

A TOUCHSTONE BOOK
Published by Simon & Schuster, Inc.
NEW YORK

First Touchstone Edition, 1987

Published by Simon & Schuster, Inc.
Simon & Schuster Building
Rockefeller Center
1230 Avenue of the Americas
New York, New York 10020

Published by arrangement with Atheneum Publishers

TOUCHSTONE and colophon are registered trademarks
of Simon & Schuster, Inc.

Maps drawn by Richard Natkiel
Manufactured in the United States of America

10 9 8 7 6 5 4 3 2 1 Pbk.

Library of Congress Cataloging-in-Publication Data

Wheatcroft, Geoffrey.
   The Randlords.

   (A Touchstone book)
   Bibliography: p.
   Includes index.
   1. Gold industry—South Africa—History—19th
century. 2. Diamond industry and trade—South Africa—
History—19th century. 3. Capitalists and financiers—
South Africa—History—19th century. 4. South Africa—
Economic conditions—To 1918. I. Title.
[HD9536.S62W44   1985]   338.2'0968   87-4513

ISBN 0-671-63993-5 Pbk.

The author and publisher gratefully acknowledge permission to reproduce
copyright photographs from the following sources: *Star,* Johannesburg, for the
portraits of Paul Kruger, Cecil Rhodes and Friedrich Eckstein; BBC Hulton
Picture Library for the portrait of Alfred Beit; De Beers for the portraits of
Ernest and Harry Oppenheimer; and Africa Collection, Johannesburg Public
Library for all the rest.

In Memory of My Mother

Contents

Illustrations

Friedrich Eckstein
Alois Nellmapius
Edward Lippert
Headgear
Outside the goldmine shaft
Stamp battery
In a drive
Drilling in the stopes
Adolf Goerz
Abe Bailey
Ernest Openheimer
Harry Oppenheimer

Just as the power of gold to shape mighty destinies was the theme of the old saga which Wagner utilized immortally in his *Ring*, so in the story of the Witwatersrand have we a homily – albeit one from real life – on the power of gold.

Hedley A. Chilvers, *Out of the Crucible* (1931)

Happily for the Witwatersrand its immense possibilities have attracted a class of capitalist . . . certainly not excelled in any other branch of finance.

G.A. Denny, *The Deep Level Mines of the Rand* (1902)

I could a tale or two unfold, if I liked . . . of certain South African 'financiers', who are really the greatest rascals that Nature ever spawned, and who would sell their souls if a purchaser could be found for such vile trash.

Louis Cohen, *Reminiscences of Kimberley* (1911)

The men who dug for gold and diamonds and were lucky enough to find what they sought appeared to be unlucky in everything else. So far as I could judge they were the unhappiest men on earth. A camarilla of spite, envy and hatred engulfed them fiercely on all sides immediately they grew rich.

P. Tennyson Cole, *Vanity Varnished* (1931)

In 1885 the burghers of the lovely, fever-stricken De Kaap valley came to their old President announcing the discovery of a vast auriferous Eldorado . . . Oom Paul remained silent and was lost in deep silence for many minutes before replying as follows: 'Do not talk to me of gold, the element which brings more dissension, misfortunes and unexpected plagues in its trail than benefits. Pray to God, as I am doing, that the curse connected with its coming may not overshadow our dear land . . . Pray and implore Him Who has stood by us that He will continue to do so, for I tell you today that every ounce of gold taken from the bowels of our soil will yet have to be weighed up with rivers of tears.'

Leo Weinthal, *Memories, Mines and Millions* (1931)

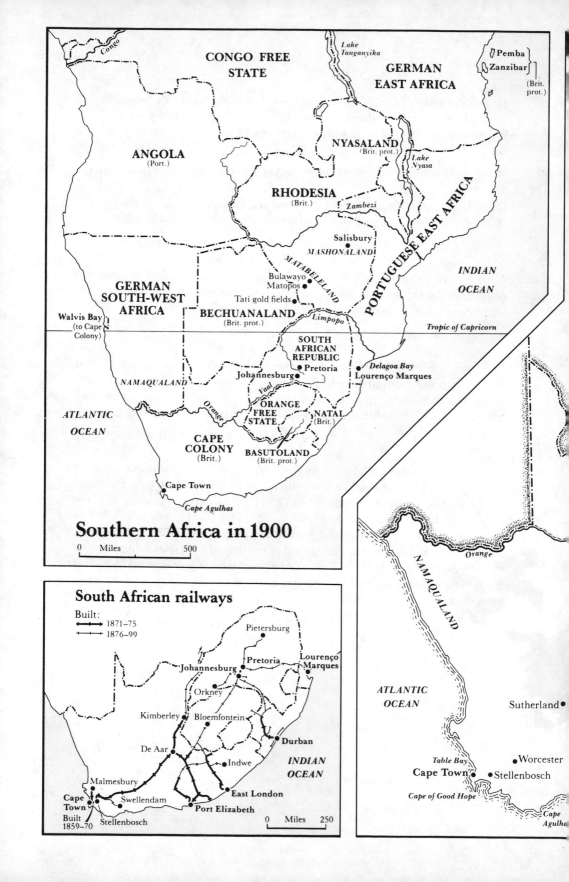

CONGO FREE
STATE

Congo

ANGOLA
(Port.)

GERMAN
EAST AFRICA

Lake
Tanganyika

Pemba
Zanzibar

(Brit.
prot.)

NYASALAND
(Brit. prot.)

Lake
Nyasa

RHODESIA
(Brit.)

Zambezi

Salisbury
MASHONALAND

MATABELELAND

Bulawayo
Matopos

Tati gold fields

INDIAN
OCEAN

GERMAN
SOUTH-WEST
AFRICA

BECHUANALAND
(Brit. prot.)

Limpopo

Walvis Bay
(to Cape
Colony)

Tropic of Capricorn

NAMAQUALAND

SOUTH
AFRICAN
REPUBLIC

Pretoria

Delagoa Bay
Lourenço Marques

Johannesburg

Vaal

Orange

ATLANTIC
OCEAN

ORANGE
FREE
STATE

NATAL
(Brit.)

CAPE
COLONY
(Brit.)

BASUTOLAND
(Brit. prot.)

Cape Town

Cape Agulhas

Southern Africa in 1900

0   Miles   500

South African railways

Built:
■■■■ 1871–75
+++ 1876–99

Pietersburg

Johannesburg

Pretoria

Lourenço
Marques

Orkney

Kimberley

Bloemfontein

Durban

De Aar

Indwe

INDIAN
OCEAN

Malmesbury

Swellendam

East London

Cape
Town
Built
1859–70

Stellenbosch

Port Elizabeth

0   Miles   250

Orange

NAMAQUALAND

ATLANTIC
OCEAN

Sutherland

Table Bay

Cape Town

Worcester

Stellenbosch

Cape of Good Hope

Cape
Agulhas

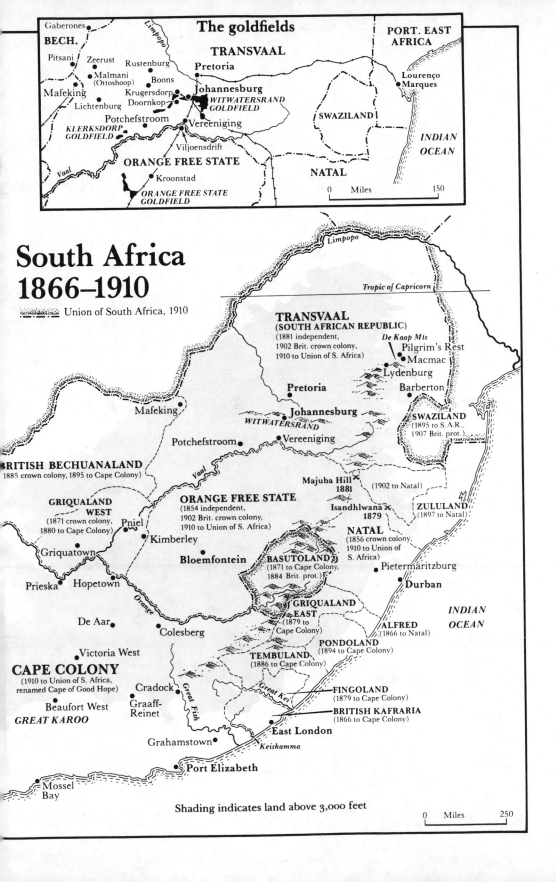

The goldfields

PORT. EAST AFRICA

Gaberones
BECH.
Pitsani Zeerust Rustenburg
Malmani
(Ottoshoop) Boons
Mafeking Krugersdorp
Lichtenburg Doornkop
Potchefstroom
KLERKSDORP GOLDFIELD
Viljoensdrift

TRANSVAAL
Pretoria
Johannesburg
WITWATERSRAND GOLDFIELD
Vereeniging

Lourenço Marques

SWAZILAND

INDIAN OCEAN

ORANGE FREE STATE
Kroonstad
ORANGE FREE STATE GOLDFIELD

NATAL

0   Miles   150

South Africa
1866–1910

Union of South Africa, 1910

Limpopo

Tropic of Capricorn

**TRANSVAAL
(SOUTH AFRICAN REPUBLIC)**
(1881 independent,
1902 Brit. crown colony,
1910 to Union of S. Africa)

De Kaap Mts
Pilgrim's Rest
Macmac
Lydenburg
Barberton

Pretoria

Mafeking

Johannesburg
WITWATERSRAND
Vereeniging

SWAZILAND
(1895 to S.A.R.,
1907 Brit. prot.)

Potchefstroom

Vaal

BRITISH BECHUANALAND
(1885 crown colony, 1895 to Cape Colony)

**GRIQUALAND
WEST**
(1871 crown colony,
1880 to Cape Colony)
Pniel
Kimberley
Griquatown

ORANGE FREE STATE
(1854 independent,
1902 Brit. crown colony,
1910 to Union of S. Africa)

Majuba Hill
1881   (1902 to Natal)

Isandhlwana
1879

ZULULAND
(1897 to Natal)

NATAL
(1856 crown colony,
1910 to Union of
S. Africa)

Pietermaritzburg

Bloemfontein

BASUTOLAND
(1871 to Cape Colony,
1884 Brit. prot.)

Durban

Prieska Hopetown

Orange

De Aar
Colesberg

**GRIQUALAND
EAST**
(1879 to
Cape Colony)

ALFRED
(1866 to Natal)

*INDIAN
OCEAN*

PONDOLAND
(1894 to Cape Colony)

Victoria West

TEMBULAND
(1886 to Cape Colony)

CAPE COLONY
(1910 to Union of S. Africa,
renamed Cape of Good Hope)

Cradock
Graaff-Reinet

Beaufort West
GREAT KAROO

Great Fish

Great Kei

FINGOLAND
(1879 to Cape Colony)

BRITISH KAFRARIA
(1866 to Cape Colony)

East London

Grahamstown
Keiskamma

Port Elizabeth

Mossel
Bay

Shading indicates land above 3,000 feet

0   Miles   250

Preface

When I want to read a book I write one, Disraeli said, and that has been the best answer I could think of when asked why this book came to be written. Some years ago I became fascinated by the story of the mining industry and the mining magnates of South Africa, a country with which I had no connection and which I had never then visited. There seemed to be no entirely satisfactory book on the subject. For one thing, more even than with most historical subjects, there was a yawning gap between the popular book and the academic; if one wanted to be sharp, between the uncritical and the unreadable. So I set off in the hope of bridging that gap, having managed to convey a little of my rapt interest to others who made the writing of this book possible.

As I proceded, I found, like a character in a thriller, that I was on to something bigger. The Randlords were not just a group of extraordinary men (though they were that, and human personality is the salt of history), but their story, or rather the story of their mines, is the central story of South Africa for the last hundred years. A poor and unknown country became rich, famous and powerful all because of the mines. Governments were subverted and overthrown and bolstered up, because of gold. Every one of the social and ethnic groups in the country has seen its history changed by mining. There has been a direct connexion between gold and the political systems of the country, the relationships between rulers and ruled, the continuing success of white supremacy. All that South Africa has been and has become relates directly to this one industry. A mighty book could be written on such a mighty theme. This is not that.

What has emerged from my efforts is a modest little book with a good deal to be modest about. All that I have been able to do is pick out some themes here and there and highlight them, while trying to keep a continuous narrative flowing. The book is at any rate written if not for what Nabokov called that limp and amorphous creature, the general reader, then at least for the non-specialist. Some time ago a disgruntled author whose book I had reviewed complained that it should have been entrusted to someone who 'understood the purpose of academic history'. His point was well taken. I do not know what that purpose is and hope that the day never comes when historians write only for an esoteric professional côterie.

This book has taken longer to appear than I once hoped and if a small thing may be compared with a great I might use Johnson's words: it was 'written with little assistance of the learned, and without any patronage of the great, not in the soft obscurities of retirement, or under the shelter of academic bowers, but amid inconvenience and distraction' – in this case, the inconvenience and distraction of earning a living as a freelance journalist. On the whole I am glad it was so. There are snares entangled with the assistance of the learned: at an Oxford college where I impudently sought help I was asked if my book would be critical of Rhodes, whose Trust is still an important source of income. And also with the patronage of the great: a good few of the books on the mining magnates have patently been written with the financial backing of the Randlords' heirs, the present-day mining companies, who would not be human (or corporate) if they did not look to a friendly portrait of their antecedents. There is something to be said for ploughing a lonely furrow.

On the other hand, it is a pleasant duty to thank the many people who did help me. I list them in the hope that those whom I have forgotten will forgive me and that those who have forgotten why I thank them will not mind.

First I must thank the respective librarians, archivists and staffs of the British Library, the London Library, the Newspaper Library of the British Library at Colindale, Rhodes House Library at Oxford, and the Islington Public Library; the South African Library, the Johannesburg Public Library and especially the Africana Section, Kimberley Public Library, the Jagger Library at the University of Cape Town, the Cory Library at Rhodes University, Grahamstown, the Cullen Library at the University of the Witwatersrand, the National English Literary Museum and Documentation Centre at Grahamstown, the De Beers Library in Johannesburg, the Cape Archives, the Transvaal Archives, and the Barlow Rand Archives.

I should also like to thank the booksellers who have provided me with many rare and useful books: Keegans's Bookshop of Cape Town, Frank R. Thorold of Johannesburg, Headgear Books of Croydon, McBlain Books of Hamden, Connecticut, and especially Mr Paul Mills of Clarke's Bookshop in Cape Town.

Three visits to South Africa were made possible by obliging editors: Mr Alexander Chancellor, then the editor of the *Spectator*, for whom I once worked and who gave me a generous period of paid leave which I rewarded by not returning to his employ; Mr Nicholas Parsons, then the editor of the *Business Traveller;* and Mr J.W.M. Thompson, then the editor of the *Sunday Telegraph.*

In Southern Africa I enjoyed hospitality and other kindnesses from Mrs Mary Hope, Mr J.D.F. Jones and Miss Polly Hope, Mr and Mrs John Platter, and Mr and Mrs Michael Hornsby; the Hon. Mrs Richard Acton (Miss Judith Todd), Professor and Mrs Hermann Giliomee, Mr and Mrs Ian Greig, especially Doreen Greig, *vade mecum* to the Randlords' Parktown, Mr Michael Holman, Mrs Kiddie, Mr Otto Krause, Mr Tertius Myburgh, Mrs Chlöe Rolfes, Mr Timothy and the Hon. Mrs Sheehy, Mr Peter van Blommenstein, and Professor Charles van Onselen.

Much of the book was written in the houses of friends: the Hon. Kieran and Mrs Guinness of Dunlewey, Mr and Mrs Nicholas Johnston of Pieve di Compito, Mr and Mrs David Leeming of Boxted, and Mr Panagiotis Theodoracopulos of Milton-under-Wychwood.

Others to whom I am grateful are Mr Rodney Blumer, Mrs Eric Christiansen, Mr J.G. Cluff, Mr John Cowan, Mr Michael Davie, Mr Donat Gallagher, Miss Mary Gordon, Mr Colin Haycraft, Mr Terence Kilmartin, Mr Andrew MacDonald, Professor Lord McGregor of Durris, Mr Shiva Naipaul, Mr Colin Newbury, the Provost of the Queen's College, Oxford (Lord Blake), Mr Brian Roberts, Mr Kenneth Rose, and Mr and Mrs Penry Williams, especially June Williams, cataloguer of Rhodes's papers in Rhodes House Library. The book has benefited from the attention which different drafts were given by three friends: Professor Norman Stone, Mr Alan Watkins and Mr Richard West, who introduced me in the first place to the immortal Louis Cohen. Mrs Ferdinand Mount twice turned a hideous manuscript into a handsome typescript and Mrs Lawson typed a final draft in record time.

Since the first edition was published several errors and omissions have been brought to my attention and corrected in the paperback edition. For this, I am most grateful to Professor Shlomo Avineri, Dr A. M. Davey, Mr Ian Fraser, Mrs Betty Gray, Lord Neidpath, Mrs S. K. Rumble and the Hon. John Skeffington.

Authors often thank their friends and literary connexions; less often, I notice, their bank managers. That is all the more reason for my saying that without the sympathy and steel nerves of, successively, Mr G.G. Bartholomew and Mr A.R. Butler of Messrs Coutts & Co. this book would not have been written.

Those can at least see the book which they helped and encouraged me to write. The late Sir Hugh Fraser MP shared my fascination with

the Randlords (though he disliked them as my title) ; we used to meet
to talk about Barnato over lunch and I am sad that Hugh cannot see
my pages on Barney.

Nor can Stephen Koss, whose early death robs those who knew him
only by his work of a distinguished American scholar of recent English
history, and those of us fortunate enough to have known him personally of the most generous of colleagues.

Nor can another. On my way to the airport, we met in Soho to talk
and drink away the afternoon as ever. Then I left for my flight to
Johannesburg, where ten days later I heard that he had been killed in
a stupid accident, at the age of thirty-seven. A light went out in the
lives of many people, not least of me who write the name of Michael
Cornelius Dempsey, my beloved friend.

Even on its modest level this book raises questions about history
which are interesting in themselves. Moralism is always a thorny
problem. History should record the past and those who made it rather
than pass judgment on them. But that is easier said than done. No one
can write of, say, the great exterminations of peoples in our time in the
same tone of voice as of, say, the Investiture Controversy or the
Exclusion Crisis. South Africa is a living question, exciting stronger
passions than almost any other, and no one who writes about its history
can fail to know what the great C. W. de Kiewiet meant : 'Most of these
pages tell of South Africa in the Seventies or Eighties of the last century ;
and yet they are also about today.' At all events I have tried hard to
avoid moralising and neither to idolize nor to traduce the Randlords, as
different generations and schools of historians have tended to do, one or
the other. There is a moral here, I think, but it is left for the reader to
draw.

A book about South Africa is a book about race. In one sense all
racial distinctions are absurd. In another sense it would be still more
absurd in this of all settings to pretend that they do not exist. Racial
categorization in South Africa a century ago was rougher and readier
than it has since become under the pedantic nicety of 'apartness'. But,
in any case, one comes up against the problem of what Evelyn Waugh
called 'the kaleidoscopic change of euphemism' – and not only
euphemism – in this field. White (or rather pinkish) men called
themselves white and called black men black, sometimes. The
universal word for the indigenous African peoples of the sub-continent
was 'Kaffir', in origin an Islamic word for an unbeliever, picked up by
Europeans, and not a century ago as offensive as it has since come to
be regarded. In a more robust, or possibly honest, age South African
mining shares were known on the London stock market as 'Kaffirs'.

At Kimberley during the 1870s J.B. Currey was shocked to hear for the first time blacks called 'niggers', a word doubtless brought by American diggers. Black mine workers were also often called 'boys' or 'natives' (an idiotic usage by whites who prided themselves on being South African-born). For reasons not far to seek black South Africans nowadays dislike the name Bantu as much as black Americans dislike Negro, but these are, in fact, as respectable ethnographic terms as Celt or Slav.

The kaleidoscope produces many ironies. Until the late nineteenth century 'Africander' or 'Afrikaner' often meant mulatto or mixed-race, the people who are more recently called 'Coloured' (which in the United States, until not long ago, was the polite name for a black or 'negro'). By a further irony, until the middle of the twentieth century 'racialism' and the 'race question' in the South African context meant the conflict between Briton and Boer, English and Afrikaner (in its later sense). And so it goes on. Influenced by my contemporary sources I, too, may have been somewhat rough and ready in this matter, but intend no offence and hope that none is taken.

A book about mining is a book about money. Sharp inflation is disagreeable for everyone, but, as Mr Geoffrey Blainey has said, it harms the historian especially because it breaks one of his most useful measuring sticks. It is, however, easy enough to remember that throughout most of the period with which this book deals, the happy age of the gold standard, currencies and prices were more or less stable. Prices from the last quarter of the nineteenth century may be multiplied by twenty-five to get a very rough impression in today's terms. Until 1914 a troy ounce of gold was worth fractionally less than 85s. (And it is depressing for an English writer still in his thirties to realize that he must explain to younger readers what 'shillings' and 'pence' were: 12d or pennies = 1s or shilling; 20s = £1; sometimes larger sums were expressed in shillings, e.g. 85s, which is £4 5s or £4 25p.) Throughout that period a pound sterling was worth United States $4.20, so that there were $17.85 to a troy ounce of gold.

GEOFFREY WHEATCROFT

Prologue
The Gold-Reef City 1895

Spring comes to South Africa in September. Rain clouds break and water the uplands which stretch from the dusty Karoo inland of the Cape to the Transvaal high veld. The rains are eagerly awaited. This parched country has always needed water to grow grass for cattle to graze and crops for food.

By 1895 rain was needed for something else, quite new, not only for the farmlands but to lubricate an industrial city which in less than ten years had sprouted up on the veld of the southern Transvaal like an exotic mushroom: Johannesburg, the gold-reef city. The townspeople of Johannesburg needed water for all the necessities of life. No less important, water was needed in huge quantities on the gold-mines which were the city's reason for being. The new town stood at 6,000 feet on top of a ridge, far from any river; in the absence of rain, water was desperately scarce. By the end of September 1895 the rains had not come.

This city was set in the strange land of South Africa, far from what was the civilized world of Europe and North America. Until very recently those northern countries had scarcely heard of South Africa, utterly remote and obscure. South Africa was not indeed a 'country' in any ordinary sense, scarcely even a geographical entity. The map showed a patchwork of different political units: the two British colonies of the Cape, where diamonds had been discovered a quarter-century before, and Natal; two inland Dutch republics, the Orange Free State and the South African Republic or the Transvaal, where the gold lay; and one or two African protectorates or kingdoms. There were fewer African territories than there had been twenty years before. In the 1880s the scramble for Africa had got into its stride, one kingdom after another was gobbled up, and in South Africa the pace had been forced by the discovery of vast mineral wealth in this hitherto poverty-stricken land. The pace by now was hot indeed as the paramount power in the subcontinent, Great Britain, sparred with African kingdoms and Boer republics alike.

The new gold-reef city was like nowhere else on earth. 'Monte Carlo superimposed on Sodom and Gomorrah' was how it struck one visitor from the Cape, a fantastical and gaudy place dropped as though by accident into a faraway pastoral backwater. It was not even certain how the city had got its name. Fifty miles north of the reef was Pretoria, the sleepy little capital city of the Republic, from where the old President, Stephanus Johannes Paulus Kruger, had watched the mine town spring up. He was tactfully told that it had been named after him when there were in fact at least two others with more likely eponymic claims. However it may be, the name remained, adapted from time to time as Joey's, Judasberg, Jewhannesburg or Jo'burg. The Zulus do better: they call it simply Igoli, the city of gold.

On this lofty ridge twenty years earlier a few dozen men had tended cattle and sheep. Then prospectors followed graziers and, after mounting excitement and expectancy, the Main Reef of the Witwatersrand had been discovered in 1886. At first no one could know that much the greatest gold-field on earth had been found, but what was known was enough to trigger off a gold-rush and within a few years to raise a strange man-made landscape along the line of the Reef, with the golden city at its centre. By 1895 this 'Randscape' stretched for nearly forty miles along the outcrop of the gold-reef. The Reef is a vast sheet of ore which runs underground in a long curve. Along the crest of the Rand (as its name was usually abbreviated) it breaks surface. When it was first discovered, mines were cut straight down into this outcrop, each one staked out and given its name, sometimes after the old Boer farms that had been there such as Langlaagte or Vogelstruis; or after the miners who had pitched camp, Robinson, Ferreira, Simmer & Jack; or fanciful, City & Suburban, Jubilees, Jumpers, Spes Bona.

Mining scenery is unmistakeable. In the centre of each mine stood the headgear, the great triangular metal box whose winding gear lowered miners below and hauled up ore. Near the headgear were the machinery and plant which treated the ore, the stamp batteries like giant pestles and mortars and the buildings where crushed ore was chemically processed. All around were the tailing sites or slagheaps, great man-made hills in unlikely ochre colours. In the centre of this line of mines was the city. Johannesburg had a rigidly fixed southern limit: the outcrop of the Reef and the mines. Immediately to the north was Marshall's Township (site of Marshallstown, now the financial centre of Johannesburg), with Jeppe's Township to the east, and Ferreira's and Fordburg Townships to the west.

In its early years Johannesburg had looked much like the mining camp that it was, with tents at first, then plain huts thrown together on the stands or building sites into which the townships were divided. In 1889 one newly arrived miner had felt 'sick at heart as I looked upon a mass of dusty galvanized iron buildings with a sprinkling of brick houses mostly in a state of construction'. By 1895 a more solid city had grown up around Market Square. To the south of Market Street with its horse-drawn trams were large, solid stone buildings: the Stock Exchange, banks, the offices of mining companies and the new Rand Club, the work of a young builder called Thomas Cullinan. Further north were smaller buildings; some of them were still built of corrugated iron, the essential material of early Johannesburg, while others were more substantial, two storeys in brick with their characteristic first-floor balcony. Further north still were the post office and the telephone exchange, and the curving line of the railway with its new station on the northern edge of town.

The town sprawled beyond its formal limits. To the north-west and to the east were what were called the locations or townships for the mass of black miners – 50,000 of them by 1895 – who had flocked to the Reef from as far away as Basutoland and Xhosa-speaking East Cape, from Zululand, from Mozambique and from the territories in the north which the British South Africa Company had recently seized. Conditions for these 'boys' were grim both on and off the mines. They came, nevertheless, in search of a cash wage to take back to their villages. If, that is, they took it back. All too much money earned at the mines was used to drown sorrows and all around Johannesburg you might find 'boys' lying dead on the veld from exposure and 'the effects of the vile liquids sold them by unscrupulous dealers'.

Not that the white men of the gold-reef city were distinguished by abstemiousness or any other conventional virtues. The population of the mining town was heterogenous and its way of life was hectic and dissipated. Dotted around the town were modest working-class suburbs, home to the skilled Cornish and Australian miners, 'hard-rock men' come to work as overseers and artisans in search of high wages. A rung above them were the Scottish and American mining engineers, all of them likewise earning high salaries, as well as bankers, lawyers and newspapermen – the men who would be found at the bar of the Rand Club. But although there were classes in Johannesburg and great economic differences, it was not a rigid caste society; far from it. Among this mêlée of English and Boers, Americans and Germans, Jews and Frenchmen, Italians and

Greeks – Uitlanders (outlanders or incomers) as the Boers called them – few snobbish distinctions were drawn. It was a place to which men would come with no question asked.

And women also. A ridiculous but haunting poem inspired by events at the end of 1895 included the line, 'There are girls in the gold-reef city', and so there were. A survey in October reckoned that in addition to some 25,000 white men living within three miles of Market Square, there were 14,000 women, of whom at least 1,000 were prostitutes. The town had ninety-seven brothels of various nationalities: thirty-six French, twenty German, five Russian and so on. Often there was little distinction between brothels and bars. A reporter observing Johannesburg by night found 'scenes of squalor, misery and vice'. White miners were served in their bars by young women who were not angels and who supplemented their income in the oldest way. When they were not drinking or whoring, many miners spent their time gambling. The men of Johannesburg would bet on anything: horses, prize fights, rat-killing dogs, cards, billiards, how many flies would land on a lump of sugar – and of course gold shares. Among the richest of the city play ran ferociously high with fortunes made and lost.

It is small wonder that this town upset fastidious visitors. They flinched from its brazenness, the vulgar glitter, the naked pursuit of quick riches. The great South African writer Olive Schreiner began a letter,

Hell ...

Dear Mr Merriman,

You will perceive from the superscription that I am still in Johannesburg.

Only a woman, she thought could see the place 'in all its full hideousness. It is the women who are the most terrible thing ... not the poor outcast women ... the apparently respectable women.' Another stern female voice was sounded by Flora Shaw, the gifted colonial correspondent of *The Times* in South Africa, in whose affairs she was soon to play a striking part. For her: 'Johannesburg at present has no politics. It is much too busy with material problems. It is hideous and detestable, luxury without order, sensual enjoyment without art, riches without refinement, display without dignity. Everything in fact which is most foreign to the principles alike of morality and taste by which decent life has been guided in every state of civilisation.' There was not a man there, she said, who knew a violin from a vegetable. Luxury without comfort, Miss Shaw might almost have said.

Johannesburg was built as a modern industrial and commercial city in which money and business efficiency came first. Characteristically, there were telephones long before there were main sewers. Everyday life was repellent in the drought. The rich could wash in bottled soda water, but even for them – let alone the poor – the arid, dusty town was horrible. Besides, the heat and the dust raised men's temperatures and aggravated both public and private passions.

As September gave way to October, heavy clouds, dense and gravid, hung over the town, but they did not break. On 24 October at the Wanderer's cricket ground an experiment took place. A dandyish man just turned thirty, Solomon Barnato Joel, fired rockets into the clouds to see what effect they would have. One of the rockets pierced a heavy cloud and caused a shower. People in Johannesburg thought that Solly Joel's fireworks were funny or silly. But from Krugersdorp devout Calvinist burghers sent a memorial protesting that the rocket-provoked shower was 'a defiance of God and would most likely bring down a visitation from the Almighty' – a foolish episode which epitomised the cultural chasm between Boer and Uitlander. Whether in response to Joel's rockets or to the burghers' prayers, on 6 November the weather broke at last and the rains came.

There were other clouds besides over Johannesburg.

When Miss Shaw wrote 'no politics', her pen ran away with her and her reporter's instinct deserted her for the moment. There were politics in Johannesburg all right, but of an idiosyncratic sort conditioned by the city's curious circumstances. Johannesburg's position high on the ridge suggested the image of an island, which is, in a sense, what it was. It belonged to a different generation from the countryside, a different century even, this city of trams, electricity, telephones, typewriters, set amid rustic backwardness. It had a quite different population from its surrounding sea. The island-city is a phenomenon with a long history. All over central and eastern Europe at the time were cultural, ethnic, linguistic islands; yet none of these was as odd as Johannesburg. For one thing there were two human seas around it. The larger sea was the black population of the Transvaal, getting on for a million people: Pedi, Sotho, Tswana, who had been living there for centuries. They were an underclass, conquered, visible but ignored, like the American Indians or the Irish peasantry in the eighteenth century. To the English radical J.A. Hobson visiting Johannesburg, they were 'the Kaffirs, who do the work but don't count'.

If the Kaffirs did not count, the other human sea certainly did. The Boers had been living in what they called the Transvaal for two generations. Their distant forebears were Dutch, German and Huguenot settlers who had come to the Cape of Good Hope in the late seventeenth and early eighteenth centuries. At the turn of the nineteenth century they had passed from Dutch to British rule and in the 1830s the more obdurate of them had left the British Cape Colony and trekked north. There they had founded 'republics'. The British had reluctantly recognized these republics, then they thought of absorbing them, and then finally and still more reluctantly they had been obliged to acknowledge their independence. The Orange Free State was politically and economically insignificant; not so the Transvaal, bordered on the west and north by the Limpopo, on the south by the Vaal, and the east by the northern range of the Drakensberg. There the Republic was cut off from the sea, to which, at Delagoa Bay, it looked yearningly.

It called itself a republic, a state, but it did not much resemble a European nation state. The burghers or citizens of the Republic and their families were no more than a few score thousand. They ruled the whole territory of the Transvaal only notionally. The Boers kept the blacks in submission by corralling them on locations in different parts of the country and, whenever the blacks rose up (as they had in early 1895), by violently subduing them. All the same, for much of the time and in much of the country the authority of the Republic did not really run. A Boer – 'farmer' in Dutch – paid little attention to the Republic or, indeed, to society outside his family. He had his farm, comprising a large pastoral acreage where cattle and sheep grazed and from which at infrequent intervals he drove his family and his produce into the nearest *dorp* or township. True colonial farmers, the Boers were suspicious of authority and disinclined to pay taxes. Their elected representatives met from time to time in the Volksraad, a legislative assembly, but the power, such as it was, in the land was the remarkable figure of Paul Kruger, seventy that October of 1895 and who had already been President of the South African Republic for twelve years.

To borrow Gardiner's phrase about Cromwell, Kruger was the greatest because the most typical of Boers, the perfect representative of 'this strange and formidable people', as Winston Churchill once called them. Just as the Boers seemed to outsiders absurd, so too did Kruger with his uncouth ways. He entirely lacked what passed for learning or letters in European eyes. With his great beard and his shabby-genteel clothes, sometimes covered with self-awarded decorations, he looked the part of the rude backwoodsman. Appearances were deceptive. Like his people, he had other qualities. The Boers shared the

colonial frontiersman's habitual combination of cultural backwardness and technical sophistication. Kruger had read no book in his life
except the Bible and, although he travelled far across the earth, he
died in the belief that it was flat. But he saw to it that his burghers
possessed the latest Mauser rifles and could use them better than any
infantry in Europe, as some of those who mocked the rude Boer were
soon to discover.

As a small boy, Kruger had come north on the Great Trek to escape
'the English Flag'. He had fought the blacks and he had fought the
British, the two enemies of his blood. Now he had a new foe. Kruger
had watched the finding of gold first on the east Transvaal escarpment then on the Rand with apprehension. Gold might bolster
the shaky finances of his Republic and pay for its defence. But
it also brought with it a horde of these Uitlanders who would disrupt the Transvaal's way of life more effectively than any Imperial
force.

Sitting on the *stoep* of his house in Pretoria, drinking coffee, smoking
his pipe and spitting every so often, he contemplated the new city on
the ridge fifty miles to the south. Kruger knew his enemy: they were in
plain view. Standing above all others in Johannesburg were the great
plutocrats who owned and controlled the city. 'The Randlords' the
London press had catchily named them: two dozen very powerful and
rich financiers who had most of them come to South Africa and learnt
their trade on the Kimberley diamond-fields in the 1870s and 1880s
before moving on to the Rand after 1886.

Most notable of the group – along with Kruger, the outstanding
figure in South Africa – was Cecil John Rhodes. He was forty-three in
1895, a man who seemed to hold most of the strings in the subcontinent and wanted to hold them all. Rhodes was a neurotic, a bundle of
energy, his boyish appearance now suddenly ageing behind the neat
moustache; a man of strange imaginings and with dreams revolving
in his impenetrable mind. He was a phenomenon in an age of
phenomena: diamond magnate, politician, empire-builder and now
gold-mine owner. Most of his time was spent at the Cape, where he
was Prime Minister of the Cape Colony; but his eyes were fixed on the
north, on his new territories in Zambesia, as well as on the
Transvaal. From Groote Schuur, the old Dutch farmhouse outside
Cape Town which was his home, he tried to make sense of the South
African scene. The finding of the Rand gold-fields had been a
complication as well as a boon for Rhodes. It brought him new riches
to add to those he had acquired at Kimberley. Whereas in 1888 he
had told his fellow legislators in Cape Town that the Colony – which

included the diamond-fields and their mines, newly amalgamated by
Rhodes and his associates – was the dominant state in South Africa,
by 1895 this was ceasing to be true.

In the struggle to amalgamate the diamond mines at Kimberley,
Rhodes had made friends and foes who had since moved on to the
gold-fields. By 1895 the most important financial group on the Rand –
more so than Rhodes's own company, Consolidated Goldfields of
South Africa – traded under the modest name of H. Eckstein & Co.,
after its first Johannesburg principal, but was always known as the
Corner House.* The company was controlled from London, where
the name of the firm was those of its animating senior partners:
Wernher, Beit & Co. The tall, bearded, former cavalry officer Julius
Wernher was based in London. His partner was a far less commanding figure, but Alfred Beit was in truth the genius of the partnership, in
fact the one great financial mind in the mining business. He had
worked closely with Rhodes in the last phase at Kimberley and was
now applying his remarkable gifts to the maximization of profits in the
gold mines, as he once had to the rationalization of the diamond
trade. Those who met the mousy, twitchy 'little Alfred' socially might
underestimate his ability; not those in the inner circle of mining,
especially his friend Rhodes, who owed much to him. The partnership
employed several able men: a younger Eckstein brother, J.B. Taylor,
Lionel Phillips, Ludwig Breitmeyer and Max Michaelis. None was in
the same division as Beit.

At Kimberley Rhodes's enemies had included the ferocious and
bullying J.B. Robinson, who missed out on the amalgamation
before moving early to the Rand, and also Barney Barnato, who was if
not the richest or the most powerful of the Randlords then the most
colourful. Having been unable to beat Rhodes and Beit, he had joined
them in the great De Beers Consolidated Company, which wrapped
up the diamond-fields. Now he had his own string of gold-mining
companies headed by 'Johnnies' – Johannesburg Consolidated
Investment – all of them enthusiastically followed by amateur investors. Barnato was despised by the City of London, and his operations
were watched with critical interest. That was partly because of his
personal brashness – the Cockney Jew who was building a gaudy
palace on Park Lane as his London home, rags to riches, no
manners or respect for the mighty – and partly it was a just reflection
on his conduct even in an age when standards of financial morality
were not high. Barnato directed his operation with the help of his

* Eckstein's offices stood on the corner of Simmonds and Commissioner Streets, and the name
was a pun – *Ecke* is 'corner' – easily understood in a town with so many German-speakers.

family, all, like him, from Aldgate: his brother and three Joel nephews, who included Solly the rocket-launcher.

From the outside these Randlords gave the impression of being a united, even tightly-knit group. So their enemies thought – and they had many enemies. As they had loomed up in the popular conscious-ness – by the 1890s the 'South African millionaire' was an instantly recognizable figure, to music hall-goer, Conservative politician and radical journalist alike – they had often been the target of obloquy. Their chiefest foe close at hand was Kruger, and he knew them better. The Randlords were not a united group. The great question of the day in the Transvaal was whether or not to support Kruger and his Republic, and this question quite divided them. On the one hand there were those who hoped for friendship with Kruger: Barnato was one, Robinson another, as were the Berlin-born brothers George and Leopold Albu. On the other were the largest mining houses – especially the Corner House and Goldfields – who increas-ingly chafed under the restrictions imposed by the government. The identity of these houses was not an accident. Those who were most vexed by 'Krugerism' and the policy of concessions which kept up the price of such essential materials as dynamite were those with the longest and deepest commitment to the mines. At first some had thought the Rand gold-fields no more than a flash in the pan – the well-worn phrase comes appropriately from gold prospecting – and their pessimism seemed to be borne out by early technical setbacks. The last few years before 1895 had answered that. Far from being merely a limited lode of gold near the surface, it had been discovered that the gold reef stretched far and away underground. It could be mined anywhere where it was physically possible to reach it. If, instead of mining along the edge of the outcrop, a man were to walk several hundred yards south of it and sink a shaft, he would strike gold once more. Those who had taken the plunge, literally as well as metaphorically, on what were called the deep levels were at once those with the longest-term stake in the mines and those who needed to count every penny of costs as they opened their new mines. They hoped to bend Kruger to their will but if they could not bend him then they must break him.

So the mine owners who had gambled on the deep levels and, above all, Rhodes and the Corner House partners, decided to take action. For some time Rhodes had not concealed from his intimates the desirability of replacing Kruger. He was in a territorially strong position to do so with his power bases in the Cape and in 'Rhodesia', the new northern territory owned by his Chartered Company. If

Kruger would not do the Randlords' bidding, he must go. Might there not be a popular rising of Uitlanders in Johannesburg? That could be inferred from listening to some rabble-rousing speeches as 1895 drew to its close. Young FitzPatrick of the Corner House spoke of a 'calculated, deliberate policy' by the 'pitiless autocrat' Kruger 'to crush the mining industry', and sarcastically proposed a toast to the Uitlanders' adopted country, 'which we have adopted, but which has declined to adopt us'.

Was a rising on the Rand enough, however? Might not outside intervention to topple Kruger be necessary also? Earlier in the year a strip of land in the British protectorate of Bechuanaland on the western border of the Transvaal had been granted by the Colonial Secretary in London, Joseph Chamberlain, to Rhodes's Chartered Company. On that strip in the last days of the year a group of armed horsemen assembled under the command of an associate of Rhodes's, Dr Leander Starr Jameson.

What was about to take place was no frivolous or minor historical episode. All those who lived through it were to understand that. Forty-five years later Winston Churchill would speak of it as the 'event which seems to me when I look back over the map of life to be the fountain of all ill'.

The rain clouds had burst; the political clouds were heavier and more oppressive than ever; Johannesburg was aquiver.

1
The Land

When the world was young, its dry surface was one great land mass surrounded by sea. At its centre was what became Africa. This super-continent began to break up some 175 million years ago, 15 million years before the first signs of life on Earth. All the other continents derived from it. Africa was formed in five long geological eras. Its surface was spattered by igneous eruptions and moulded by violent earth movements, by colossal forces which squeezed and pulled and folded it like a crumpled sheet of paper or a tablecloth, by crustal splitting which gave the eastern and southern half of Africa its geographical character. One shudder broke away the Arabian penin-sula, another pushed down the long crease of the Rift valley, another threw up the series of high plateaux which run from the Horn of Africa to the Cape.

In the course of the pre-Cambrian era, what is now the Transvaal – the land lying between the Vaal river to the south and the Limpopo to the north – was flooded by a shallow sea. This sea carried with it all kinds of matter including countless tiny particles of the element 'Au', atomic number 79: gold. The dense and heavy particles were mixed in with pebbles and quartzites. As the sea ebbed from the southern Transvaal, this mixture was left behind in layers, like the tidemarks on a beach. Further covering left the reefs deep down, where they would have remained hidden but for later fierce movements of the earth. Faulting was so violent that rock layers of great antiquity were forced up from the depths to protrude through the surface. Rock systems which had been quite flat were tilted at strange angles to the horizontal, as ridges were pushed up and valleys pulled down. Forty-five miles north of the Vaal one long ridge was thrown up to run for some sixty miles from east to west. Millions of years later Dutch-speaking farmers saw streams glistening on this upland and called it the ridge of white water, the Witwatersrand. High on that ridge or Rand the gold reefs come up at an angle and break surface.

Surface gold-reefs were beaten by wind and weather. The gold can be carried far away by streams and rivers, sinking to the bottom in alluvial deposits. The Sacramento river in California is a classic case. Or it may collect on the surface in eluvial deposits. Sometimes outside forces collect the gold particles and press them together to form nuggets. Freaks of geology produce small, intensely concentrated seams of gold like the four-mile-long Comstock Lode in Nevada. The Witwatersrand reefs are not like that. They hold few large pieces of gold. The very largest nugget found in South Africa was the Peacock, weighing 12 lb: it compares with the Welcome Stranger of 160 lb found at Ballarat in Australia in 1869.

Where the Rand reef comes to the surface it does not glisten and is only recognizable by its geological formation. The jagged edge of the gold-bearing reef is studded with pebbles. It reminded Dutch farmers of a Boer sweetmeat similarly studded with almonds called 'banket', the name which stuck for the reef. Deep down the layers left by the retreating sea are barely perceptible. Thousands of feet below the surface a miner sees caught on the rock face by the light of his helmet lamp a faint speckled band a few inches wide. That is the gold reef. The Rand is not distinguished by the quality of its ore, which is low; it is the quantity of ore which was to make it the most important goldfield on earth. Though the reefs are thin, they are vast in extent. These geological facts—poor quality ore in huge quantities—were pregnant with significance for the human history of South Africa.

Long after the formation of the gold reef, deposits of carbon lying far underground were squeezed by the overwhelming pressure above them, and baked by intense heat. The carbon took on different shapes and textures: when the pressure is as great as it is more than sixty miles down, plain carbon can be transmuted into crystalline gems like diamonds. For a long time this remained an entirely mysterious process; men coveted diamonds but did not know where they came from. In 1813 the great English scientist Humphry Davy demonstrated that diamonds were made of carbon. For a century scientists tried and fraudsmen claimed to make diamonds artificially. Only when this was at last achieved was the process by which natural diamonds are created fully understood.

These stones lay far too deep to be brought to the face of the earth by folding or cracking of the surface. Several million years ago, at some time after the diamonds had been created, igneous activity weakened the earth's mantle and the molten interior forced its way from the depths to the surface. As the fierce mixture of liquid rock and gases pushed upwards it carried with it the stones, whose formation was completed by this violent journey. These eruptions were not volcanoes: the liquid rock did not flow over the surface like lava but spurted into the air, fell back, and set hard. The resulting 'pipes' running up to ground level are irregularly cone-shaped like ice-cream cornets, widening as they approach the surface though often closed at the top.* Sometimes the exploding mixture did not form a neat cone but broke out in fissures.

When they were first formed these pipes stood high like small hillocks. Millions of years have weathered them away, carrying diamonds far and wide. Like gold, the stones were often water-borne and would settle as alluvial deposits in river beds. At the surface the diamond-bearing rock softened with time into a friable, yellowish ground, but below it kept its original condition, hard and blue. In 1871 imperial authority in London decided that the various names for the new mining camp in Griqualand West in the British colony at the Cape were either unpronounceable or undignified. The new town was officially named after the Colonial Secretary of the day, Lord Kimberley. Then the name of the mining town was adapted as the formal name of the diamond rock: blue ground became 'kimberlite', named indirectly after a Norfolk village far from any diamond fields.

This blue ground or kimberlite is the original matrix for all diamonds everywhere but it had not been encountered before the South African diamond rush. The diamonds of India and Brazil – the two main sources until the 1860s – had been found alluvially, washed out of their blue ground; one reason why the geological history of diamonds was still unknown at the time they were found in South Africa, when ignorance proved expensive. Since then kimberlite pipes have been found throughout the world – in Siberia, China, North America. In Africa they occur along the whole range of the eastern plateaux, from Sutherland, north-east of Cape Town, to Mwanza, on Lake Victoria in Tanzania. But of more than 150 pipes in Africa only two dozen are 'payable', the crucial miners' word meaning that minerals are not only present but are present in such quantities and with an ease of extraction that makes it profitable to mine them. About 500 miles north-east of the Cape of Good Hope and rather less from the Atlantic to the west two rivers meet. Inside the acute angle formed by this confluence where the Vaal flows into the Orange is a spot where within three miles of each other four great diamond pipes come to the surface which are not payable merely but immensely rich. There South African history was transformed.

* See illustration, p. 36.

Until that moment at the end of the 1860s it is hard to exaggerate
how remote, how poor, how unimportant, and how empty South
Africa was. Man was born in Africa but large societies came very late
to the south. After the last Ice Age Neanderthal man lived there and
left his remains not far from the Cape and just north of the Zambezi.
There is evidence of iron mining several thousand years ago near
what is now Mbabane in Swaziland.* Gold was mined also. The
Egyptians sent an expedition in search of gold to the Land of Punt
which was probably the mouth of the Zambezi; ancient gold artefacts
have been found at Mapungubwe in the northern Transvaal.

But for all such signs of economic activity, southern Africa had a
tiny human population until well into the Christian era. Africa as a
whole was far from underpopulated: in the Old Stone Age it held a
fifth of the human race, and by AD 200 had perhaps seventeen million
inhabitants. But not in the south. For ten thousand years only the
Bushmen roamed between Lake Victoria and the mouth of the
Orange, scattered and scanty nomadic tribes who lived very simply
off the land and asked only to be left alone. The southern sub-
continent was the empty end of the earth.

In other lands mining for minerals and gems, including gold and
diamonds, had become an important affair. Diamonds have always
had a special eminence. Though they lack the heavy colour of rubies
or emeralds, their peculiar fascination lies in the way light plays on
them. That was so even when the art of cutting gems so as to increase
their brilliance was in its infancy. Until recently their role was
purely decorative and nonfunctional. It was during the great age of
Kimberley and Johannesburg that a true 'use' was found for diamonds which until then were 'not a prime necessity of life, not at least
for anyone below the rank of countess or cotton-spinner's lady', as one
Kimberley digger put it.

Even as he wrote, his words had ceased to be quite true. At the
beginning of the nineteenth century diamonds were still a purely
aristocratic adornment. As the century progressed they were demo-
cratized. The diamond engagement ring is a footnote in the story of
bourgeois society. Wherever that society spread, in Europe – most of
all in England, Germany and France – and in North America,
diamond rings became increasingly commonplace among the mid-
dling orders. By the 1860s the potential demand for diamonds outran
production. Most diamonds came from Brazil, which produced a

---

* On a site where the Anglo-American Corporation now has a large opencast iron mine.

modest 180,000 carats* in 1867. The price was thus kept up and sales restricted. A new source could mean more plentiful and cheaper diamonds, and still wider sales.

Like diamonds, gold is highly decorative. From the time when it was first found and fashioned it was prized above all other metals. It is almost the softest of metals, easy to work into delicate shapes, and is the most ductile: a single grain of pure gold, less than one four-hundredth part of an ounce, can be drawn into a wire five hundred feet long. It is also curiously beautiful, with a colour and lustre that never decay. For all of these reasons gold became a symbol of supreme value. Croesus, the king of Lydia in the sixth century BC who left a name for huge wealth, left a more important legacy to classical antiquity and its heirs: gold coinage. Thenceforth for Greeks, Romans, Europeans, gold had an intrinsic value conferred on it which nothing could take away. It could not clothe men like wool, arm them like iron, warm them like coal, feed them like corn. But it was the most useful of all, as it could buy any of the others.

Gold coinage lapsed with the Dark Ages but the Venetians and Florentines reintroduced it in the thirteenth century. It played a useful part in the growth of trade during the Renaissance, an important part in the early mercantile age, an indispensable part in the birth of industrial capitalism. For two centuries Great Britain was the foremost trading and making nation. She introduced mature commerce, industrialism, capitalism to Europe and then to the rest of the world, and for those two hundred years she was on the gold standard. From 1717 her currency, the pound sterling, was convertible into gold and vice versa. An ounce of fine gold had had an official value of £4 4s 11½d at the Royal Mint. Outside England silver remained the principal monetary metal until the Napoleonic wars. During those wars Great Britain suspended the gold standard and the value of gold rose. Then the old standard was reimposed and the single gold standard, or monometallism, gained throughout the nineteenth century at the expense of silver or of bimetallism, the mixture of both metals as means of exchange.

By the second half of the century the gold standard was a potent symbol, and a problem. An increasing number of countries wished to adopt the gold standard, the more so after silver was demonetized in Europe in 1873. The gold standard was a sign of economic virility. In the 1870s France, Germany and the United States adopted the standard. Austria-Hungary and Russia followed later. But although

---

* A carat weighs 3⅛ grains: some 125 carats weigh an avoirdupois ounce, 4,410 a kilogram.

gold ensured stable currencies it obviously trammelled the fast
expanding economies of the industrial world if gold was in short
supply. And it was. World production increased sharply from an
annual average of 674,000 ounces in 1831–40, to 1,819,600 in
1841–50, to 6,350,180 in 1851–55 (the Californian gold-fields were
discovered in 1848, the Australian in 1851), but had not improved on
that figure by 1880, falling in fact to 5,729,300 ounces in 1876–80.
Shortage led to the dream of cornering the gold market, of gaining
control over the whole supply, in one country at least. The American
financier Jay Gould tried to corner gold in 1869 by bribing President
Grant's brother-in-law through whom he hoped to restrict government supplies of gold.

Everywhere the shortage was keenly felt. Following England, the
German and American industrial revolutions were getting under way.
And so even as the South African gold-fields were being opened up
one of those most closely concerned with them could say that 'there is
no doubt that sufficient gold is not available for all the countries
which have adopted the gold standard'.

Meantime southern Africa had changed: the handful of Bushmen who
lived there were joined in the early centuries of the Christian era by a
great migration of Bantu peoples from the inside corner of the
continent, the rain-forest hinterland of the Gulf of Guinea. This was
an epic movement of peoples comparable to the other *Völkerwanderung*
at much the same time, the westward migration into Europe. For
southern Africa especially, being peopled in large numbers for the first
time was a portentous development. The Bantu arrived on the
Limpopo and pushed on into the lush grazing lands of what is now
Natal, settling not in tiny groups like the Bushmen but in hundreds
of thousands, millions even.

One day South Africa was to become part of the empire of Europe
and Europeans would settle there. But it was unlike the other
seemingly empty spaces into which Europe expanded, North
America, Australia, Siberia. Seemingly, because all these places had
autochthonous inhabitants; but their numbers were too small and
their societies too primitive to resist aggression, conquest, settlement
and sometimes extermination by Europeans, or to play an important
part in the new colonial economies. So with the original South
Africans, the pygmy Bushmen and the Hottentots of the Cape. Once
the white man arrived, they dwindled away through disease and
through violence: 'The South African Boers,' a Boer once wrote, 'from
their first contact with these strange people, came to the conclusion
that extermination was the only policy to pursue.'

The Bantu did not obligingly dwindle away and they were not easily exterminated. They throve and multiplied and resisted conquest as long as they could until overcome by superior military technology. And once subdued and drawn into the new South African economy the role of these blacks was anything but peripheral. 'What an abundance of rain and grass was to New Zealand mutton, what a plenty of cheap grazing was to Australian wool, what the fertile prairie acres were to Canadian wheat, cheap native labour was to South African mining.' The blacks were to be South Africa's great raw material.

As the Bantu moved down the east coast, Europe began its contacts with Africa. The Portuguese were the first to reach the Cape: Camoens describes Table Mountain in the *Lusiads*. But they preferred the more hospitable shores and people of Mozambique round the coast, the more so after a Portuguese viceroy was killed by Hottentots at Table Bay in 1510. In the course of the sixteenth and seventeenth centuries other sailors stopped at the Cape and sailed on. No one wanted the land of southern Africa for itself: there was nothing there worth having. But the later seaborne rivals of the Portuguese, the Dutch, English and French East India Companies, all had exhaustingly long journeys to make by sea from northern Europe to the riches of the East. Fatality rates were very high on the voyage, with sailors living off a wretched diet, and revictualling stations were invaluable.

Thus the Dutch East India Company chose the Cape as a base. In 1652 a party of Dutch under Jan van Riebeck landed to establish a storehouse, granary, vegetable garden and hospital, and to buy what meat they could from the Hottentots, all for the sailors on their voyage to the Company's rich eastern empire based in Java. There was no colonial purpose behind the settlements. When they looked beyond Table Bay to the rugged mountains inland the first settlers did not think of spreading far. Nevertheless, unintended, the colony grew. The first Dutch were joined by German and French Huguenot families.

Then immigration ceased. Slavery began. These slaves came for the most part from Angola, Madagascar and the East Indies. They were used in households, in artisans' workshops, in the fields, as the settlement spread out into the valleys and hillsides of the Cape peninsula, then towards Stellenbosch, where corn was grown and wine was made. The Cape never knew full-scale plantation slavery, but the increasing use of slaves for all forms of menial work led to a stagnant, underdeveloped economy.

There was no squeamishness then about sexual contact between black and white. Liaisons between the Dutch and their slaves, black and brown, produced the ancestors of the 'Cape Coloureds' of today, who are made up of roughly one part European blood, one part African, one part Asian. For that matter, most true 'Afrikaners' have some coloured blood. In addition, as the colony moved further out, white settlers and Hottentots produced a people who lived on the edge of the Cape Colony and who in the nineteenth century were known bluntly as the Bastards.

Throughout the eighteenth century there was no further immigration to the Cape. All who had settled there had been obliged to become Dutch-speaking and Calvinist and so a small, homogeneous, indeed inbred community developed at the far end of Africa. The only notable product of the settlement was its wine, which soon became famous: Jane Austen mentions Constantia. Its happiest creation was 'Cape Dutch' architecture with its pretty gabled farmsteads. Cape Town was much praised by visitors at the end of the eighteenth century. It had grown into a handsome city, and remained one until the twentieth century. All the same, it was barely a true outpost of Europe.

Under the Company's autocratic rule Cape society was sterile and immobile. There was no commerce to speak of. Slaves were kept in their place by brutal punishment. On the fringes of the settlement there grew another economically backward but independent-minded group, the Boers, some of whom became 'trekboers', travelling farmers, or migrant pastoralists. With their ox-wagons and their herds, their families and their servants and sometimes their slaves, they drove further and further from the Cape. They went in search of grazing land and to escape the oppressive authority of the Company. They pushed north through the first ranges of hills to sparse uplands. They pushed east along the Little Karroo and the Great Karroo, south and north of the Swartberge, on to the well-grassed country of the Sundays river valley and then towards the Great Fish river.

Then in the 1780s Boer met Bantu, whom they knew as Kaffirs. A state of intermittent hostility between them began. By the nineteenth century these conflicts were dignified with the name of the Kaffir wars, but to begin with they were little more than reciprocal cattle raids. Company officials like Governor Plettenburg tried to halt the eastward expansion of the settlers, but it was just to get away from such as he that the Boers had trekked. More to the point was the large number of blacks inland: the Dutch colony at the Cape had barely any African population. North of the Orange river, east of the

Fish, in what are now Natal, the Free State, Lesotho and the Transvaal, were thriving and populous African kingdoms.

At the turn of the century, during the Revolutionary and Napoleonic wars and in consequence of them, Great Britain acquired the Cape. She hung on to it, and consolidated her rule in the early nineteenth century, not because it was in itself desirable as a colony but so as to deny to the French a strategically important base. By the standards of other European dominions – of India, of the West Indies, not least of the American colonies which England had lost a generation before – the Cape remained a bitterly poor place, still rudimentary in agriculture and commerce, insignificant in population. In 1806 there were some 26,000 Dutch and 30,000 slaves in the colony, as well as maybe 30,000 Hottentots, at a time when the population of the United States was touching four million.

British rule was no less autocratic than the Company's and it brought fresh vexation for the Boers. For one thing, the British took the part of the native peoples against the settlers. That was nothing new. It had been one of the grievances in the Declaration of Independence that the London government would not give the American colonists a free hand to deal with the 'merciless Indian savages'. The British enraged the Boers by their attempt at even-handed treatment of master and servant. In 1813 a commission investigated rumours of violent usage of coloured slaves and servants by their Boer masters. The subsequent use of coloured troops to attempt to arrest and even to shoot a white farmer provoked the frontier rebellion of Slagter's Nek in 1815. Five Boer rebels were condemned and hanged. They were not forgotten.

Traditional Dutch officials were suppressed and when at last slavery was abolished in the 1830s, the slave-owners were compensated inadequately and in clumsy fashion. The trekboers more than ever wanted to escape from British rule to the interior. Between 1835 and 1837 parties of the most irreconcilable Boers left the Cape Colony on the Great Trek. They were no more than ten thousand in all. Crossing the Orange river, as trekboers had been doing for at least ten years in search of grass during seasons of drought, they drove on across the plain between the Orange and Vaal, to the northern river or to the Drakensberg mountains to the east. One party of trekkers was led by Andries Hendrik Potgieter. This group included a boy just turned ten in 1835, Stephanus Johannes Paulus Kruger.

Another voortrekker, Piet Retief, published a manifesto which in one respect resembled the Declaration of Independence. It hoped that 'the English Government has nothing more to require of us, and will allow us to govern ourselves without its interference in future'. The English government did not see it as simply as that. For Lord Glenelg, the Colonial Secretary, the motives of the trekkers were 'obvious . . . they are the same motives as have, in all ages, compelled the strong to encroach on the weak, and the powerful and unprincipled to wrest by force or fraud, from the comparatively feeble or defenceless, wealth or prosperity or dominion'.

Strong and weak were maybe too simple terms to describe Boers and blacks. The trekkers had strayed into an area of turbulent conflict between the black peoples of southern Africa. The Mfecane – 'crushing' in Zulu – of the 1820s was a vast upheaval, a series of violent wars and migrations which dispersed and rearranged the Bantu tribes and made the Zulus under Shaka predominant. Now another tribe had arrived among them, a white tribe, weak if reckoned in numbers but strong with their possession of ox-wagon, horse and musket. This could only make the hinterland of Great Britain's South African colony more unstable than ever. And here was a problem.

For the best part of a century, from the first seizure of the Cape in the 1790s until the 1870s, British policy in southern Africa was entirely defensive. There was, as it seemed to the British, no reason for expansion in Africa. As late as the 1860s the only British possessions in the continent apart from the southern tip were the little entrepôts along the western coast. And why else? Anyone could see why the first British empire had been conquered. It did not need a Barbados planter or a Bengal nabob to recognize riches when they were to be found. Until the last decades of Victoria's reign most Englishmen would, however, have agreed with a later historian that 'the enormous areas of tropical Africa appear impressive on the map; but of most of them the plain truth is that they had remained so long ownerless because they were not worth owning'.

'Wider still and wider shall thy bounds be set' was far from the motto of successive London governments, who wished to limit the growth of the colony at the Cape. In 1820 a group of settlers, the riff-raff of the English countryside, had been planted in the eastern Cape, but this was not from expansionist zeal. The settlers came partly to dilute the Dutch population, partly to act as a buffer between the Colony and the Bantu beyond the Great Fish, Keiskamma and Kei rivers. When they appealed for military support in the frontier wars in which they inevitably became involved the settlers got a dusty answer. They complained of harassment by the blacks; Sir James Stephens of the Colonial Office replied 'if men will settle in the neighbourhood of marauding Tribes they cannot, I think, claim of their Government that at the National expense they should be rescued from the natural penalty of their improvidence any more than the vine dressers at the foot of Vesuvius can expect indemnity against the effects of an eruption'.

As with the 1820 settlers, so with the Boers; or more so. The British authorities waited as the trekkers fought their way north and east, establishing fledgling 'republics' as they trekked. Some parties were wiped out by the blacks but there were victories on the other side. The most famous was at Blood river 16 December 1838, 'Dingaan's Day' when Pretorius avenged the previous killing of Retief, routed Dingaan's Zulu army, and established the Republic of Natalia. This vexed the British because it threatened to give the Boers an outlet to the sea. To avert that Great Britain annexed Natal in 1845.

The trekkers still there had to submit to British tradition, including the notion that the colony was officially colour-blind, a contrast with the Boer republic set up north of the Vaal in 1846: its articles proclaimed 'no equality in church and state' between the races and excluded 'bastards [mulattos or 'Coloureds'] . . . down to the tenth degree' from the Volksraad, the legislative assembly of the republic. For another quarter century South Africa settled into an uneasy political equilibrium. The British grudgingly accepted the independence of the feeble inland Dutch republics, as long as they stayed feeble and stayed inland. In theory Great Britain still claimed that the trekkers had not renounced their status as subjects of the Queen merely by quitting her realm. In practice the two 'republics' were left alone. One called itself the South African Republic or Transvaal, since it was to the north of the Vaal river. Its autonomy was recognized by the Sand River Convention in 1852. The practical independence of the Orange Free State between the Vaal and Orange was granted two years later. These states, as they called themselves, were large territories, heavily populated by black Africans, sheep and cattle, very thinly populated by the Boers who notionally ruled them.

The complexity of the South African situation was recognized in 1846 when the office of High Commissioner was created, usually to be combined with the Governorship of the Cape Colony. His job was to oversee the broad British interest throughout South Africa, dealing with the colonies themselves, with warring black tribes, with obdurate Boers. London still had no notion of developing the colonies. In 1841, for example, 23,950 people left the United Kingdom for Canada, 14,552 for Australia and New Zealand, and all of 130 for the Cape. By 1864 the entire population of the Cape Colony was still less than half a million. Little more than 180,000 of them were white, less than the population of Chicago at the time. The entire white population of the Transvaal was scarcely more than 20,000, smaller than then remote American towns like Memphis or Milwaukee. Those Transvaalers were almost all Dutch or Afrikaners and spoke the 'Taal', the decayed and simplified form of Hollander Dutch which was to become the literary language Afrikaans. And despite the 1820 settlement and decades of British rule the white population of the Cape Colony as well was still overwhelmingly Dutch.

A small Jewish community existed in South Africa in the early nineteenth century. Jews had come to South Africa in the Company's days but not until 1841 was there a *minyan*, a quorum for divine service, in Cape Town. A few Jewish families established themselves in trade there, and at Port Elizabeth to the East, and even journeyed into the interior. The Mosenthal brothers, Joseph and Adolf, had come to South Africa in 1839. They bought a firm at Graaf Reinet, moving it subsequently to Cape Town and then to Port Elizabeth. They were big dealers in ostrich feathers*as well as wool and leather. Inland, the Lilienfeld brothers, Martin and Leopold, established themselves as the chief traders at Hopetown on the Orange.

They traded among a rural community: the population of South Africa mostly lived in the country. Not merely rural, its economy was pastoral. Wine was made and corn was grown in the Cape, and sugar and cotton had begun to be grown on plantations in Natal by the 1860s. To work these plantations indentured labourers were brought from India, adding another ingredient to the heady racial mixture of the country. But most farming was a matter of grazing sheep and cattle and the chief export was wool, the only serious sign in fact of economic activity in South Africa to the outside world.

The Boers were not sophisticated agriculturalists. In colonies and republics they held large farms, up to 6,000 acres, and sometimes two, an upland and a lowland farm, moving flocks seasonally from one to the other, as with the Spanish *mesta*. Exports of wool increased from 144,000 lb in 1834 to almost 5.5 million lb in 1851, and then dramatically to 25 million lb in 1862 as the Civil War disrupted American supplies. But peace in the United States brought slump to South Africa. The country was ill-equipped for economic recession, with primitive financial institutions, rickety local banks liable to collapse in a storm or even in a stray financial ill wind. Even before the slump began land was dirt cheap in the interior. In 1860 a large

---

* This trade was important to the South African economy: Julius Mosenthal of the next generation wrote a book on the subject.

farm at Vooruitzigt near the Vaal which was to become world-famous was bought by the brothers Johannes and Diedrich de Beer for £500.

If there had been any sign of mineral wealth in South Africa the British might have taken more interest in the country, but there was not. In Namaqualand on the Atlantic coast north of Cape Town copper had been mined in a desultory way on a small scale for more than a hundred years. Prospectors hoped to find more. They had long been looking for gold in South Africa, but with no success. In 1853 Pieter Jacob Marais searched for gold in the Transvaal, following another prospector, J.H. Davis. Marais had the permission of the Volksraad of the South African Republic. Andrew Wyley, a geologist in the employ of the Cape administration, prospected with the encouragement of the Orange Free State in 1856.

But the burghers of the republics were in two minds about gold. On the one hand their countries – which barely deserved the name – were perpetually indigent, state bankruptcy often looming in the none-too-far distance. They needed money for the modest administrative needs of the republics and to fight against the black tribes, who vastly outnumbered the Boers inside the two republican territories, let alone in South Africa as a whole, and who had to be subdued at frequent intervals by commandos. On the other hand, it was all too easy to see that if gold was found it would destroy the Boer way of life.

That way of life was without parallel anywhere else, even in the American West or the Australian outback. This handful of Dutch farmers were spread as thinly as possible over the land they had conquered. Boer love of solitude, not to say Boer misanthropy, was a byword:

> His neighbours' smoke shall vex his eyes, their voices break his
> rest.
> He shall go forth till south is north, sullen and dispossessed,

as Kipling put it. Far from his neighbours' smoke the Boer led a simple, backward life among his family with often no book in the house but the Bible. His house lacked any modern comfort. His manners were primitive. The dirtiness and squalor of Boer life was a constant theme with English travellers. Few other people would have seen the attraction of this way of life; to the Boers it was an idyll they had chosen, for which they had come very far. Gold would mean progress, the modern age, and must surely destroy the idyll.

That threat was far off in the 1850s. Davis found no gold, nor did Wyley, although his report to President Boshoff of the Orange Free State said that there might in fact be gold in the country, in the

northern rather than the southern part, but lying deep down: as it
happened a prescient conclusion. Two years later the Prussian
explorer Karl Mauch told of his first travels north of the Limpopo. He
even had some success. Gold was found at Tati, and it inflamed the
belief that there were untold riches in 'Zambesia' (i.e. Southern
Rhodesia, now Zimbabwe), the plateau between the Limpopo and
Zambezi. That belief was to be a persistent chimera for a generation.

If a large gold-field had been found in the 1860s, it would have
changed the way the Boer republics saw themselves, and it would have
changed the way the British government saw South Africa. The old
anti-expansionist sentiment in London was beginning to break down,
but very slowly. Bits and bobs were added to British South Africa –
the desert land up to the Orange river in 1847, Kaffraria east of the
Great Fish river in 1866 – but a good many Englishmen still agreed
with Lord Grey: the new territories were 'not merely worthless, but
pernicious – the source not of increased strength but of weakness –
enlarging the range of our responsibilities, while yielding no additional resources for properly sustaining them'. That might indeed
have been the verdict on South Africa as a whole. The British tried one
expedient after another to lessen their imperial responsibilities there.
A legislature was granted to the Cape in 1853, to be followed by
responsible government, in an attempt to be rid of the burden.

In the late 1860s the idea of federation for South Africa began to be
floated. It was hoped that this would solve the political difficulties of
relations with the Boer republics and African kingdoms. Federation
was advocated *faute de mieux*. It might have solved the problem. In the
end it was forestalled by the depression which indeed postponed all
attempts to bring South Africa further into the nineteenth century.
There were many bankruptcies. The Cape parliament set up a
Retrenchment Committee to cut expenditure. The Dutch republics
were in a feeble state, and the Orange Free State the worse of the two
after fighting another war with its Basuto neighbours.

No further economic development was likely in the foreseeable
future. Two short lines from Cape Town were the only railways in the
sub-continent. In 1866 a telegraph link was established from Cape
Town to Grahamstown but no further, and a direct submarine cable
link to London was not to come for another thirteen years. Depression
ended the modest scheme of road-building which had been under way
since the 1840s. This lack of modern transport was felt with special
sharpness in a country less well endowed with natural means of
communication than almost any other. The land around the Cape
and to the east is hilly but passable. Inland lie several ranges of craggy

and inhospitable mountains. River valleys form barriers to cross-country travel without offering any compensating advantage: there is not a single navigable river south of the Limpopo. The Orange is by far the longest river but in the months between rainy seasons it is little more than a dry course, and even after the rains it cannot carry boats.

Small wonder that men despaired of South Africa. It was not a country; scarcely even in Metternich's phrase a geographical description. It was a number of miscellaneous territories, a ramshackle collection of polities, a congeries of peoples, united by little more than poverty. It was a land without millions and certainly without millionaires. At the beginning of 1867 no one dreamed of making his future in that faraway and empty land.

That was about to change utterly. The transformation of the country and of so many lives began in 1867 by the banks of the Orange. In this dusty upland country on the fringe of the Cape Colony there were a few settlements. One was Colesberg, eighteen miles south of the Orange. A further 170 miles north-west was Griquatown, capital, if it could be called that, of Griqualand, home of the mulatto Griqua people. In between these two was Hopetown, founded in 1853 and named after the secretary of the Cape Colony, Major William Hope. It was a Dutch town in a Dutch countryside where almost all the white inhabitants spoke the Taal.

One such was Sievert Christian Wild. With his stepson Schalk Jacobus van Niekerk he worked a farm called De Kalk, north-west of Hopetown. In December 1866 van Niekerk had decided to sell his portion of De Kalk. A neighbour called Daniel Jacobs wanted to buy and van Niekerk visited Jacobs's farm to talk business. There he noticed the children playing 'klip-klip' or fivestones with pebbles. Looking at one of the stones he said, 'Dars a mooi klippe voor en borst spelt' ('There's a pretty stone for a woman's brooch'). The children's mother told him that it had been found by Erasmus, one of her sons; van Niekerk, who said that he had a modest collection of 'mooi klippe', semi-precious stones, was welcome to keep it.

A few months later van Niekerk showed the stone to a friend who took it to Hopetown. The idea that it might be a diamond was mentioned and laughed at, but it was taken to Colesberg from where it was sent in an unsealed and unregistered envelope to the nearest geologist, Dr William Guybon Atherstone, more than two hundred miles away in Grahamstown.

He reported that it was a diamond of $21\frac{1}{4}$ carats.

After corresponding with the Colonial Secretary, Richard Southey, Atherstone sent the stone to Cape Town. On 19 April it was sent to

London where it was valued at £500. A replica of the diamond, now known as the Eureka, was sent to Paris to be displayed at the Cape Colonial stand at the Universal Exhibition.

For a couple of years South Africans were unsure whether their land held a treasure trove. Many did not believe it. As was to happen less than twenty years later on the Rand, the diamond finds were greeted with scepticism so vigorous as almost to suggest unconscious hostility. In 1867 Sir Roderick Murchison of the Museum of Practical Geology in London said he would stake his reputation that not so much as a matrix of diamonds existed in South Africa. The next year Professor James R. Gregory of London University visited areas where diamonds were said to be found, but 'saw no indication that would suggest the finding of diamonds or diamond bearing deposits in any of these localities. The geological character of that part of the country renders it impossible . . . that any could have been discovered there.'

The learned gentlemen were about to be proved wrong. In March 1869 a Griqua shepherd brought a stone to Schalk van Niekerk, who bought it for five hundred sheep and ten oxen. He took it straight away to Hopetown. The stone was a fine white diamond of 83½ carats. In Hopetown Lilienfeld Brothers bought it for £11,200. It was taken to Cape Town where Southey laid it on the table of the Cape Assembly and grandiloquently declared: 'Gentlemen, this is the rock upon which the future success of South Africa will be built.' Then it was sent to London for cutting and, now known as the Star of Africa, was sold to Lord Dudley for £30,000*. A sleepy, forgotten land was about to awake. The diamond rush was on.

---

* In 1974 the Star was auctioned in Geneva for £225,000 – less than half in real terms (such is inflation) of what Dudley paid for it.

2

In the Early Days

They called it a rush, but the word was scarcely right. Apart from the Californian Forty-niners who had crossed America overland rather than by ship round Cape Horn, no prospectors had ever had such a slow and arduous journey as those who first made their way to the banks of the Vaal. From the Cape to the diamond-fields is the best part of six hundred miles and the landscape of the South African interior makes for hard travelling. After the first ranges of mountains comes the Great Karroo. The climate of the Cape is temperate and drought is not a problem. Inland it is different. The Karroo is eerily beautiful, but parched.

That beauty was captured by Olive Schreiner in the first famous South African novel, *The Story of an African Farm*. She watched the 'dry, sandy earth, with its coating of stunted "karroo" bushes a few inches high, the low hills that skirted the plain, the milk-bushes with their long finger-like leaves'. The diamond rush took place only a few years after

> The year of the great drought, the year of 1862. From end to end of the land the earth cried for water. Man and beast turned their eyes to the pitiless sky . . . month after month, the sun looked down from the cloudless sky, till the karroo-bushes were leafless sticks, broken into the earth . . . and only the milk-bushes, like old hags, pointed their shrivelled fingers heavenwards, praying for the rain that never came.

Even in better years the land was still dry, 'and the earth itself was naked and bare'. After the Karroo and the little settlement of Victoria West, some hundred and fifty miles remain before Hopetown and the Orange. Port Elizabeth in the eastern Cape is nearer Griqualand as the crow flies, but it is still 250 miles to Pniel up the Great Fish valley, to Cradock, and across the plain to the river.

The first hopeful diggers were Boers. They lived near at hand, in the Cape or the two republics. They knew how to trek across this land, in search of grazing then, of diamonds now. They had the equipment, wagons and 'spans' – teams – of oxen, guns for shooting game, picks and shovels. A handful of men at the diggings became hundreds, then thousands. In the year of the river diggings' boom, more than 10,000 came. The Boer majority diminished and became a minority. There was an attempt to keep some diggings for Boers only but the other diggers overturned this by a show of force. As the news spread more came to the diggings from further away, England and Ireland, Germany and Russia, and disappointed but still optimistic prospectors from the American and Australian gold-fields. Before long five ships were fitting out in Boston to take would-be diamond diggers to the Cape. Not only hopes but some capital also was needed to reach the fields. The cheapest steamer fare from London to Cape Town cost £31 10s; the wagon journey from the Cape to the fields was £12.

The first diamonds had been found by the Orange near Hopetown. Then other finds were made on the banks of the Vaal sixty miles to the north. Late in 1869 stones were found at Klipdrift on the north or right bank of the Vaal; not long afterwards prospectors from Natal found diamonds on the opposite bank at Pniel. Diggers were soon at work for scores of miles along both sides of the river. They established claims in the koppies (little hills) along the banks, moving large boulders to get at the gravel in which the diamonds were found. This gravel had to be taken to the river in cartloads and washed. It was usually loaded into a cradle holding two or three sieves, which were rocked as buckets of water were poured over. Then the mixture was emptied on to a sorting-table, where the digger ran through it with a wooden scraper. Diamonds showed up brilliantly and were seldom missed on the sorting-table; the larger gems, a digger reported, were often found in the sieve or even in the act of digging. When a large stone emerged there was shouting and running and celebration. But although the diggers did not know it, this was not the way to find large quantities of stones.

On the way from Hopetown to Pniel the diggers' trek took them past farms where travellers often stopped: hospitality was a Boer principle. Bultfontein was one such farm half way on the uplands above the river, owned by Cornelius du Plooy, Dortsfontein another, owned by Adriaan van Wyk but still known as Dutoitspan*after a previous owner, Abraham du Toit. Diamonds may have been found and sold at Dutoitspan as early as 1868; so J.B. Robinson later claimed. They were certainly found at Bultfontein in September 1869. A government official visited the farm and wrote to Southey on

---

* A 'pan' is a pond or small reservoir for keeping water on the arid uplands.

4 November that diamonds were abundant at a 'farm called "Toispan"'. By 14 November these stories had reached Leopold Lilienfeld who dashed to Bultfontein in the hope of buying the farm. He found that du Plooy had just sold it to a speculator from the Orange Free State called Thomas Lynch. Lilienfeld saw a way round this. He pointed out to the pious du Plooy that he had made the sale on a Sunday when no business should be done. The first sale was cancelled and forthwith, on a weekday, Lilienfeld bought the farm. 'The Jews have got ahead of us again', a frustrated prospector complained. More diggers moved to Dutoitspan. At first van Wyk would only let his fellow Boers prospect, but a diggers' revolt overcame his resistance.

The diggers of 1869 thought that inland diggings were second-best or a temporary billet. No one knew where diamonds really came from but all diamond-prospecting lore said that the stones were found by watercourses. The men who first dug at Dutoitspan did not know that there were diamonds literally embedded in the walls of the farm, that they were sitting on top of incomparable, countless riches. And so they soon moved on to join the other hopefuls along the Vaal.

A whole line of settlements grew up by the tree-lined bank of the river. Some men lived in covered wagons, some in bright white canvas tents, some in shanties built of reeds, some slept on the open ground – when they could. From April to August the weather was cold – the river diggings were at 4,000 feet – and from September to March it often rained. Few of these early diggers made much money; most of the stones found here were worth less than £1. But they lived a gregarious and merry life. They were not comfortable – it was said that out of the six thousand first arrivals twenty had a mattress – but there was a comradely atmosphere. Crime was rare. Visitors remarked on how little violence the diggings saw compared with the mining camps of Australia and California. Murder and violent robbery were scarcely known and neither was lynch law.

Order was kept after a fashion by the two local police forces. On the north bank of the Vaal the Frontier Armed and Mounted Police of the Transvaal were smartly dressed in dark brown corduroy with peaked leather helmets. From their camp on a little koppie outside Klipdrift they came to chastise the occasional malefactors, with none-too-gentle punishments. Petty thieves were tied up to a wagon wheel and given four dozen lashes of the cat. Elsewhere criminals were put in the stocks. On the other bank the Orange Free State Police were a rabble, drunken, dissipated, seedy-looking reprobates, in garments of every shade. They were a poor advertisement for the state which with some geographical plausibility claimed jurisdiction over the part of Griqualand West lying inside the angle formed by the Vaal and Orange. In any case, whose writ ran at the diggings was a question yet to be decided.

When the diggings began, Pniel was little more than a village, with most of its buildings of wood and iron and a few of brick. Its centre of attraction was Mrs Jardine's hotel, which had a permanently busy bar, and a store selling all the necessities of a digger's life. In one room more than a dozen men slept on mattresses pushed against each other; two tiny 'married compartments' were divided by a piece of green baize. Drink could always be had at a price but food was a problem. The best source of protein at the diggings was game. All around and about the course of the river were hares, partridges and plovers, and the profusion of gazelles and small bucks that are South Africa's glory: gemsbok, blesbok, steinbok, springbok, kudu. But there was a critical shortage of good vegetable diet and this laid the diggers open to the curse of the river diggings, 'camp fever', whose causes were poor diet and poor sanitation. There were no sanitary arrangements at all to speak of and in the age when the great killers like typhoid and cholera had not been controlled that was a grievous omission. At frequent intervals fever swept through the camp, killing men and laying more low.

These were the years celebrated in Boyle's *To the Cape for Diamonds* and Angove's *In the Early Days*. Few who survived became famous: most of the great names of South African mining had not yet arrived. Two had.

When the first diamonds were found Joseph Benjamin Robinson was near by in the Orange Free State living as a wool merchant and cattle-breeder. He was then in his late twenties, having been born in 1840 at Cradock, high up the Great Fish valley in the eastern Cape. J.B.R. claimed to be descended from 1820 settlers, which was probably true although his enemies liked to deny it. By 1868 he was installed on the Vaal at Hebron and was a commanding figure before the rush began in earnest. He was to remain such a figure, respected, feared and hated, in South Africa and England for another sixty years. He had many advantages when he reached the diggings: his proximity, his farmland and livestock as a form of capital, his good relations with the Free State Boers. No one ever accused J.B. Robinson of an excess of sympathy for his fellow men, but if there were any he warmed to it was the burghers of the two republics, like him independent, tough, harsh in dealings with white men, harsher with blacks.

He was impossible to miss, a blue-eyed man of six feet who rarely laughed, or drank, or talked much, very quick-witted, intimidating to many men and attractive to many women. All his life J.B.R. inspired stronger feelings than any of the other mining magnates. Radically different verdicts on him were recorded. The extremes range from his official hagiographer Leo Weinthal to his most bitter denigrator Louis Cohen. For the one J.B.R. was 'a robust, well-built man, alert, pugnacious, full of energy, with florid features, strong jaw and keen blue eyes, which pierced you like a diamond drill . . . [there were] little secrets about "J.B.R.'s" generosity that would have have surprised his critics'.

To Cohen his appearance was less impressive:

Sour visaged and unsympathetic, he looked as yellow as a bad apple. [His wool-dealing trade was] sometimes called Boer Verneuking*. (If you don't believe me, ask Leopold Albu, who swears that when he is knighted he will call himself Lord Verneuker with Robinson's permission.) There are some interesting tales told of J.B.R.'s business acumen while dealing with the Boers. If I had space to narrate them all you would not think that grand people are always great. Robinson was never popular with anybody in Kimberley; he had no personality, no magnetism, but resembled a mortal who had a tombstone on his soul. . . . Notwithstanding that he was cold as a fish, there was no denying his admiration for the fair sex in his salad days. But surely . . . if he did run after the petticoats that's no proof of love.

Weinthal had good reason to write his encomium – he was a paid hack – and Cohen had better reason to dislike Robinson. But it is the hostile verdict that wins. Throughout his life almost no one liked J.B.R. although he captivated women and won notoriety as a 'seducer of men's wives and daughters'. He was impossible to work with, and no joint enterprise between him and other entrepreneurs ever lasted long. He was enormously successful, but success brought him no joy.

He was made a Justice of the Peace by his friends in the Free State; and upheld the law as vigorously as he did his own interests. He worked hard at his claims on the farm Adamantia. Mostly the diggers worked their own claims with comrades. A few local Griqua were employed. Robinson was the first to see the value of more and cheaper labour. Through his contacts in the Free State he brought in black labourers to work on his claims: it was a momentous development.

* *Verneuk* is 'swindle' in Afrikaans.

J.B.R. had shown how mining, an individual and small-scale occupation, could become a labour-intensive industry. That was a beginning; soon blacks flooded towards Kimberley from all the neighbouring territories, near and far. They came to earn cash with which they could buy guns, or cattle for a dowry.

There were already other entrepreneurs in the making among the rough diggers. Charles Dunnel Rudd arrived at the Vaal in 1870. Born in 1844 at Hanworth Hall in Northamptonshire, and educated at Harrow and Trinity College, Cambridge, Rudd seemed a rare bird among either early diggers or later diamond magnates, an upper-class Englishman. He was a good athlete, a cricketer at school, a champion racquets player at Cambridge* and a long-distance runner until his health broke down. In 1865 young Rudd was sent to Cape Town and then on to Natal to enjoy the supposedly health-giving South African climate and to shoot big game. In 1868 he married Frances Chiappini in Cape Town. And so, like J.B.R., Rudd was near at hand when the river diggings opened up. But his first venture was not successful. He returned to Cape Town with camp fever and no money. After raising a stake he tried again.

Neither man made real money on the Vaal river banks under the jurisdiction of the Transvaal. The president of the South African Republic, Pretorius, hoped to establish his authority more firmly over Klipdrift. He granted three men a monopoly of diamonds on the north bank for twenty-one years. It was not the last time the Transvaal was to issue concessions or monopoly, and not the last time it was thus to infuriate the mining community.

The north-bank diggers revolted and declared an independent Diggers' Republic. One of their number, Stafford Parker, a large, bewhiskered former seaman, was elected president. For several months President Parker ran his republic with surprising efficiency, or at least firmness. Diamond thieves were flogged, cheats were ducked in the river, and there were the stocks for drunkards and for the prostitutes who were now, inevitably, congregating at the diggings. The Diggers' Republic was not scrupulous about public finance and some of the monies it collected ended up in its officials' pockets, but then the same could be said of the Transvaal.

The Republic finally capitulated to Pretorius and the Transvaal. Whereas the British authorities had consciously held back from this edge of the colony, they now began to move forward; it was scarcely a coincidence that this movement took place as the diamonds were

---

* In the inter-university singles final he beat R.T. Reid of Oxford, later a Liberal — and 'pro-Boer' — a statesman who became Lord Chancellor Loreburn.

found. Until then Griqualand West* had enjoyed an ambiguous political status, with uncertain boundaries. It was a debatable land. The Griquas had held some 'inalienable lands' but had seen these eroded by moneyed pressure from outside. They knew also the hostility of the neighbouring Boer republics. Early on, when he had demanded independence for the Transvaal, Pretorius had asked the British with obvious resentment for 'the same privilege which you are giving to the coloured population, namely self-government'. The status of Griqualand West might have been kept in abeyance but for the finding of diamonds by the river.

A political settlement soon became more urgent. Most diggers had resolutely stuck to their preconceived notions about the origins of diamonds. If stones were found upland away from rivers, they must have been moved there by some freak of nature. It was slowly realized that the reality was the other way round. Towards the end of 1870 more diamonds were found by chance at Dutoitspan. This time their number must mean that the diamonds were not freaks. Even if the history of the diamonds was unknown, their presence was unmistakable.

The diggers left the river to scramble the score or so miles up country, pitched a new camp and started digging at the head of the pipe. At first van Wyk charged a royalty of a quarter of the value of diamonds dug on his farm. Then he granted claims thirty foot square which became the standard diamond claim, ten times the size of your grave, the diggers said. A royalty of 7s 6d a month per claim was charged. Before long claims were changing hands at anything up to £50. Most of the diggers had no money to speak of but a few had larger resources. Van Wyk was offered £2,600 for his farm. It was a fortune beyond the Boer's dreams. Not beginning to understand what the mine would be worth one day, he accepted. And so the end of 1870 saw the mines being frenetically dug at the neighbouring pipes of Dutoitspan and Bultfontein, a few hundred yards apart. To the north of the diggings a tented encampment was soon dotted with shanties, the beginnings of a mining town.

With the striking of the dry diggings a speedy settlement of the political question became essential. There were legal arguments and there were moral arguments but both were complex. The rights and wrongs of the Griqualand West question were as abstruse as those of Schleswig-Holstein. The Transvaal claimed the right-bank river

---

* This curious name became necessary after one Griqua chief led some of his people several hundred miles away to settle in 'Griqualand East', lying north of Kaffraria or Transkei. Griqualand East was annexed to Cape Colony in 1879.

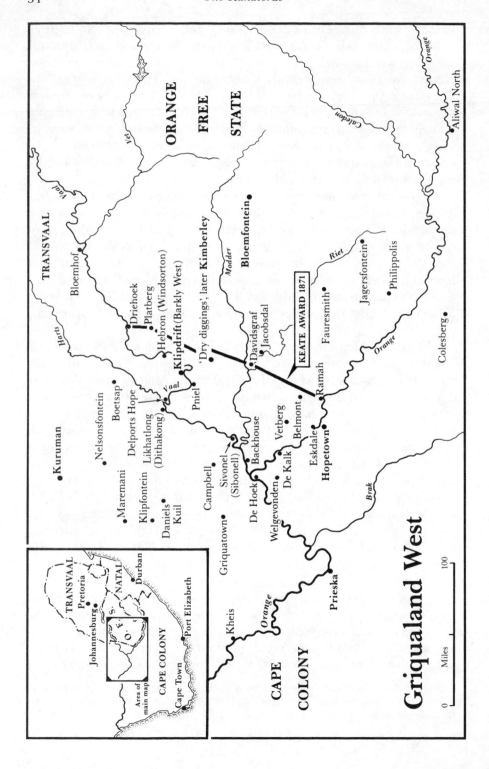

Griqualand West

diggings: all of four parties, the Transvaal, the Orange Free State, Waterboer's Griquas, and a black chief, Maharu, claimed the left-bank diggings around Pniel. The authority of the two Dutch republics scarcely ran in this no man's land, but then it scarcely ran in many parts of what was indisputably republican territory. In the absence of any effective political organization it could be said that there were good arguments for granting the dry diggings to the Free State and the river diggings to the South African Republic. It could certainly be argued that the wealth of the area should be preserved for the communities which lived there. But great wealth is its own argument and overbears moral and legal niceties. The Free State asked for independent arbitration by an outside power. The British refused.

The presidents of the two republics met Waterboer at Nooitgedacht in August 1870. They might have reached an agreement but a spoke was put in the wheels by the shadowy and sinister figure of David Arnot. He was a Cape lawyer, born of a Scottish father and a Coloured mother, who flits through the history of Griqualand West like a firefly. In December 1870 the new High Commissioner and Governor of the Cape, Sir Henry Barkly, set up arbitration proceedings under R.W. Keate, Lieutenant-Governor of Natal. The Free State refused to take part. They might have guessed at a foregone conclusion. They could not have known that Arnot had gone to the trouble of granting several farms to one of the four arbitrators, John Campbell, and had persuaded Maharu to take Waterboer's side.

While the arbitration was under way, its outcome became even more urgent. Two miles north-west of the first two mines was Vooruitzigt, another farm well known to diggers as a billet, and there lay two further pipes waiting to be discovered. In May 1871 diamonds were found at De Beers, as it was called from its previous owners. At once the diggers rushed there. The Free State authorities made some attempt to control the rush, but soon gave up and, with 10s. a month charged for claim licences, the market established some kind of order. Finally, on 17 July, the last treasure trove was found two thousand yards east of De Beers. The new mine was discovered by diggers from Colesberg who called it Colesberg Kopje. It was soon simply known as New Rush. None of the previous mines had been rushed so quickly and dug so feverishly as New Rush. Within less than six months it had been dug sixty feet deep and was producing nearly £5,000 worth of diamonds a month.

When they first found these koppies, broad shallow hillocks, the diggers still thought that the diamonds had been deposited on the surface by some freak, strewn through the sand as they had been at the river diggings. It seemed a miracle. At the river, diggers sunk pits in heavy gravel scattered with boulders. Now they found diamonds sprinkled through a light surface soil of decomposed yellow ground. The stones were so thinly covered with earth that after every heavy rain little crystals were washed free from sand and lay shining on the ground. These easy pickings continued for the best part of sixty feet down.

Then the loose, sandy soil ran out. Hard blue rock was struck. To begin with diggers assumed that blue ground meant the end of diamonds, and after skimming the surface they would move on to another claim. Then some more inquisitive than the others tried to cut into this hard ground. It turned out not to be rock-hard but friable. It split with the pick and crumbled with the spade. And it was rich in diamonds. In fact the further down the ground was dug the higher the proportion of large stones it yielded.

Each little claim was a separate undertaking the size of a broadened squash court, with a regulated space left between claims for access. Some cut sloping adits, up and down which a wheelbarrow could be pushed. Others cut steps going down, smaller and smaller, to reach the bottom and centre of the claim by a kind of staircase. But as the sloping sides of the stairs themselves consisted of diamondiferous ground it was obviously more lucrative to cut straight down the side in a perpendicular shaft. The only way to clear the floor was by filling a bucket with broken ground which was then lifted to the surface by a

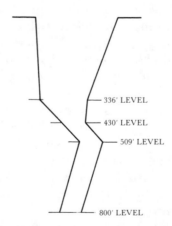

336' LEVEL

430' LEVEL

509' LEVEL

800' LEVEL

Section through a diamond pipe

mate or a black labourer. At the surface the cut rock was battered with shovels to break it up and then passed through a sieve. Then it was sifted a second time through a cradle or rocking sieve, and sorted.

Before many months had passed each of the mines presented the same sight, a trellis of roadways and betwixt them a crazy patchwork of claims dug at varying depths. Conditions of work were grim. The diamond fields are bone dry for most of the year, cold in winter and horribly hot in high summer, January and February, when the temperature can often go well above 100°F in the shade.

Hot winds blew the red dust from the surrounding veld in clouds over the workers, and these dust blasts were mixed with the powdered white limestone and pulverized cement of the ridge, shaken through their sieves and blown in the faces of the miners, inflaming their eyes, clogging their noses, and even coating their skin through their clothes. So fine was this powder and so sharply blown that it penetrated even hunting-watch cases and few watches could be kept running after a month on the diggings.

As the diggers continued their uncomfortable, but profitable, work, in October 1871 Keate ruled in favour of the Griquas and Waterboer asked Barkly to take over the territory in the name of the British crown. This was not the end of Arnot's work. The territory which had been awarded was bordered by the angle of the Orange and Vaal rivers and on the east the third side of the triangle was a line through Ramah on the Orange to Platberg by the Vaal. It became clear that this border did not include all the diamondiferous area. Now Arnot persuaded the British that the boundary ought to lie further east. And so the border was redrawn to include every diamond it could within Griqualand West, which became a Crown Colony in October 1871. It was absorbed into Cape Colony in 1880.

This little passage of colonial history is complicated and arcane – but instructive. The non-white peoples had been played off against one another, and against the Boers. The Griquas soon found that any victory they had won was illusory. The erosion of their own land-holdings meant that they would soon be absorbed into a new economy as propertyless labourers. The British government had entered upon a new 'forward' or aggressive policy in South Africa by picking up Griqualand West. Gladstone's government in London had embodied the traditional Liberal principles of peace, economy and reform, including a distaste for overseas expansion. It was after Lord Kimberley became Colonial Secretary in July 1870 that the idea of federation as an answer to the South African conundrum took a new and strong hold on the political imagination. All of the territories of South Africa could come together in a peaceful federal union; the Dutch republics would be seduced rather than ravished. The acquisition of Griqualand West was a step in this process and with the discovery of diamonds a very convenient one.

The ruthlessness with which the imperial authorities could act was a portent for the future. At the same time it did little to make the republics more amenable to seduction. A less proud and suspicious people than the Boers would have looked askance at the dealings of Arnot and Keate; still more so when President Brand's attempt to have the award submitted to outside arbitration was frustrated by Arnot. Arnot retired to Eskdale where he had been granted thirty-six farms, as well as a million acres on the diamondiferous west bank of the Vaal.

The diggers, not for the first time, thought that they had had their way. Most of them had been disposed towards British sovereignty; that was why Parker's Republic ended. Some favoured the Free State's claim to the diamond-fields because of the Boer government's unsentimental way with blacks. There were non-white claim-holders in the early days, mostly Cape Coloureds and Griqua. The white diggers all disliked this and maintained that every non-white was an actual or potential participant in the already flourishing IDB* market. Besides that, 'It would be almost impossible for white men to compete with natives as diggers; there were differences between their living expenses . . . the difference between the general wants, necessities, character and position of the two races utterly forbid it.' This was the nub. Blacks could live and work more cheaply than whites. For two hundred years white-ruled South Africa had been more or less a serf society. Now for the first time the black man was seen not only as the white man's servant but as his economic rival. A fateful corner had been turned.

This competition was felt more sharply because of the diggers' mode of life and very high living expenses. By the end of the year a permanent settlement was forming on the fields, or rather two. One on the northern side of the Dutoitspan and Bultfontein mines was to become Beaconsfield; another, overtaking it in size, a couple of miles to the north at Colesberg Kopje or New Rush, was to become Kimberley, and the mine itself was to change is name from New Rush to the Kimberley mine. Wagons and tents gave way to iron shacks, and those in turn to more solid buildings. Further away were the

---

* Initials constantly used both for the market – illicit diamond buying – and for one who engaged in it – illicit diamond buyer.

locations for black workers, growing at an even faster pace as men flocked to the fields from the Basuto homelands and further away in the Transvaal or even from the Xhosa-speaking territories east of the Cape. One witness said that by the end of 1871 there were already 10,000 blacks working in the Kimberley mine alone, and nearly 5,000 whites. It was reckoned that there were in all some 50,000 men on the fields. Each of the four mines was by now an awe-inspiring sight, 'the whole resembling a hive of busy human bees, bustling and elbowing, creeping and climbing, shovelling and sieving'.

The townships may not have been violent but life there was not dainty. At Dutoitspan men thronged liquor bars from Benning's and Martin's hotel, the chief meeting-place, down to the lowest dives. Drinking was an expensive pastime. 'Drinks', which meant Cape brandy or worse, cost sixpence, but a bottle of decent beer cost 2s 6d, when it would have been 3d in London. It was said that even then the bar scarcely showed a profit. Meals were half a crown, board and lodging 12s 6d. But the price of drink did little to reduce consumption. There were few genuine doctors on the fields then, maybe only two, though many quacks. One qualified doctor, R.W. Matthews, reckoned that more than two thirds of the men he treated had their condition exacerbated by alcohol. His gruesome duties included dealing with madmen – there were many on the fields whose minds were unhinged by tension and failure and there were to be many suicides – who, under whatever degree of restraint was necessary, were despatched to an asylum in Grahamstown.

Not all was drunkenness, failure and despair at the dry diggings. Rudd was rebuilding his position after an initial lack of success, a spell at Cape Town as a diamond dealer, and a disaster when a parcel of diamonds addressed to Vienna was stolen in the post. He was an imposing figure, 'a tall erect man, slender in build having fine dark eyes, thick fair hair, and a well-trimmed dark beard'.

He was not as conspicuous as Robinson. J.B.R.'s own characteristic story that he had helped discover the dry diggings may or may not be true, but he was certainly one of the first at Colesberg Kopje. He opened a new business as a diamond merchant 'prepared to give the highest prices for all Descriptions of Diamonds and other Stones', in partnership with a young colleague, Marcus Maurice, who acted as his courier to London with parcels of diamonds in a red flannel body belt under his shirt. It was uncomfortable but at least it prevented a repetition of Rudd's misfortune. Robinson was reckoned by most diggers to be honest, a man who worked long hours and scarcely ever drank. But he could be as violent sober as other men drunk. His sharpest early quarrel was with Ernest Moses, dentist, diamond dealer and scurrilous newspaper columnist (a combination which could only have been found at Kimberley). Moses's writings enraged Robinson and one day Moses was passing Robinson's office when J.B.R. rushed out with a horsewhip and thrashed him in the street. Later the two apologized to each other, but the incident did not endear Robinson to the community.

Until now Robinson had enjoyed a special eminence on the fields. He was the first really able, industrious and determined entrepreneur to start making his future there. He was far from being the last. By the end of 1871 it was certain that the dry diggings were not a temporary freak. Griqualand West as a whole had yielded 153,410 carats in 1870, 403,349 in 1871. Around the mines, camp was giving way to town, the biggest in southern Africa outside Cape Town and despite its mushroom character the only industrial town in the sub-continent. Like filings to a magnet, thousands of black Africans were drawn there looking for work. So were more and more hopeful diggers, colourfully described by one of the chroniclers of early Kimberley:

> Each post-cart and bullock-wagon brought its load of sordid and impecunious humanity. Rabbis, rebels, rogues and roués from Russia and the Riviera . . . Unfrocked clergymen, with the air and the soul of sinners, who had never known the strain of work; broken stalwart soldiers . . . caring for nothing but billiards and brandy . . . and divorcees with a variegated past printed on their features, who had plucked the periwinkle blossom not wisely but too well.

All these incomers hoped to make their fortunes; most would not; just a few would.

3
Three Men

Throughout England, across Europe, in North America, in Australia, millions heard for the first time of South Africa; thousands decided to go there to seek their fortunes. The trickle was turning into a flood. For the first time South Africa was a popular destination.

The prospectors journeyed in times of dramatic change. The United States was recovering from the Civil War and about to begin its industrial explosion. Europe was in turmoil after the Franco-Prussian war, with France making an awkward start to the Third Republic, and two new national states coming into being: Italy united at last with the taking of Rome by the Piedmontese, and the new German Reich created by Bismarck about to take off industrially. Its union seemed to release hidden energies which were put to work both inside the country and outside, not least in South Africa to which some of the new Kaiser's ablest subjects migrated. In England too it was a time of excitement, in political life with Gladstone's first ministry from 1868 to 1874, and in cultural and literary life: 1871 was the year of *Middlemarch* and the opening of the Albert Hall.

One young fortune-seeker was David Harris, a Jewish boy from Canonbury in north London, not yet twenty when he sailed from Southampton in 1871. He had borrowed £150 from his mother 'to proceed to the South African diamond-fields, where, I said, I felt sure I would make my future. I had the optimism of youth; the confidence of improving my position – one of the traits of my race.' Sixty years later David Harris wrote a list or necrology of men he had known on the fields: 'Cecil John Rhodes, C.D. Rudd, F. Stow, J.B. Robinson, B.I. Barnato, Woolf Joel, Anton Dunkelsbuhler, J. Wernher, L. Breitmeyer, F. Baring-Gould, Max Michaelis and Sigismund Neumann. Of these there are only two alive today – Sir Max Michaelis and myself. Alfred Beit and the two brothers Albu arrived in the early "eighties" ', he added incorrectly. These were the men who were to supplant the independent diggers, to transform diamond mining, ultimately to change the course of South African history. Two of them were born in South Africa – Frederick Stow or Philipson-Stow, as well as J.B.R. – and five in England: Rhodes, Rudd, Barnato, Joel, Baring-Gould, and Harris himself; the other eight in Germany.

Three of these men were destined for special fame. They were almost exact contemporaries, born within fifteen months of each other, and severally arrived in South Africa within the space of five years. They grew up on the diamond-fields, first as rivals, then as colleagues. They moved on to the Rand gold-fields and, having acquired the same dominance there as at Kimberley and made further colossal fortunes, they died within the space of nine years, between the ages of forty-five and fifty-three. They were Cecil John Rhodes, Barney Barnato and Alfred Beit.

The first of them to reach the fields was Cecil Rhodes. He was born in 1853 in the small town of Bishop's Stortford in Hertfordshire, thirty miles north of London, where his father was vicar for more than a quarter of a century. The family had been yeomen farmers for generations. Early in the eighteenth century they moved from Staffordshire to St Pancras, now part of central London but then in open fields.* They prospered, buying more grazing land and more businesses. By the early nineteenth century they had ascended from the yeomanry through the professional upper-middle classes to the edge of the gentry.

The Reverend Francis William Rhodes, Cecil's father, was educated at Harrow and Trinity College, Cambridge, and inherited a decent competence. He educated his sons liberally: the first two boys, Herbert and Frank, went to Winchester and Eton respectively. Cecil might have followed either, but because of his poor health he stayed at home at Bishop's Stortford and was educated at the local grammar school, where he was industrious but undistinguished, and he left at sixteen in 1869.

The intention was that he would complete his education with his father before university. Mr Rhodes's hope, indeed, had been that all his sons would go into the Church; in fact none did. But shortly Cecil fell ill with what was diagnosed at the time as pulmonary consumption or tuberculosis, and like Rudd he was sent to South Africa because of his health. In fact this malady, which was to dog him all his life, was probably what doctors now call atrial septal defect and laymen call a hole in the heart. The illness was certainly serious but it

---

* In 1890 as an act of piety Rhodes erected a family tomb in the churchyard of the old church of St Pancras (not the Greek Revival church on Upper Woburn Place but the building of many centuries' construction with some Anglo-Saxon traces which lies behind the railway terminus).

is remarkable how little it affected Rhodes's energies over the next thirty years. If anything, its effect seems to have been to give him a sense of urgency. Never knowing when mortality would catch up with him was one of the forces which made Rhodes a driven man.

The eldest son of the family was Herbert, who had gone to Natal under a scheme of assisted emigration. He had been granted 200 acres of farmland where he was growing, or trying to grow, cotton. Cecil sailed to join him and reached Durban on 1 September 1870. In Natal he learnt that Herbert had caught diamond fever and was dividing his time between cotton fields and diamond-fields, without much success on either. Cecil stayed on the farm until the harvest in July 1871. In October he set out to follow his brother.

He covered the four hundred miles in an ox-cart with a few tools, some provisions and several books. His destination was Colesberg Kopje: 'an immense plain, with right in its centre a mass of white tents and iron stones, and on one side of it, all mixed up with the camp, mounds of lime like anthills'. He continued this description of wonder in a letter home:

> a small, round hill about 180 yards broad and 220 feet long; all round it a mass of white tents. . . . It is like an immense number of ant heaps covered with black ants. . . . They have been able to find no bottom yet and keep on finding steadily at 70ft. You will understand how enormously rich it is, when I say that a good claim would certainly average a diamond to every load of stuff – a load being about fifty buckets.

Herbert had acquired three claims. But though an early bird he was not a persevering one, and before long he returned to Natal leaving his eighteen-year-old brother in charge. Other diggers soon noticed this tall, fair boy, blue-eyed and with somewhat aquiline features, wearing cricket flannels.

Rhodes was a solitary, uneasy youth. Then and later he was a lonely man, 'fond of a glass or two' – his drinking was to be talked of – but who could 'when jolly with the bottle talk like a Mirabeau. For the fair sex he cared nothing.' He was often taken for a man older than his age, and struck those who did not know him as a queer fish, moody and silent until he suddenly spoke a 'thought' several times in a staccato voice. All his life he was more a striking than an attractive figure, with piercing eyes, high-pitched voice, and nervous, feminine laugh. He made few friends but those he made were devoted unless and until a breach came. Of all the mysteries of his career and character this trait is the most mysterious. In the age of imperialism

Rhodes was to become a public idol and also the object of execration. It is possible to understand both, while his charm quite fails to communicate from beyond the grave. Yet there it was, a personality which inspired intense loyalty and even love.

He found three particular friends and colleagues on the fields in the early days, one of whom was also to figure largely in South African history. John Xavier Merriman was an English bishop's son aged thirty in 1871 and already a member of the Cape parliament. For the years 1870–75 Merriman was a digger at Kimberley whose hopes were never fulfilled. He became in turn Rhodes's friend, his political colleague and then his bitter foe. He had none of Rhodes's gambling flair or love of commercial adventure – he found Kimberley 'repulsive' and was hopeless with money – but had what Rhodes never had, principle and public honesty.

Rhodes's second associate on the fields was John Blades Currey, an Englishman who had migrated to South Africa in 1850. But the most important partner Rhodes found was Charles Rudd. The two dug and sieved and sold together and they had fair luck. 'I found a $17\frac{5}{8}$ carat on Saturday, it was slightly off (in colour), and I hope to get £100 for it. Does it not seem an absurd price?' He soon reckoned that he was averaging 30 carats a week. Of course, a few large stones were worth much more than many small ones amounting to the same weight; and already the price of diamonds was fluctuating rapidly. Before long Herbert returned, bringing with him another brother, Frank, who was killing time before he went into the army. Neither Rhodes nor Rudd much enjoyed the digger's life. Rhodes one day broke the little finger of his right hand while humping buckets and could never afterwards give a firm grip in a handshake. For his part, Rudd's health made it impossible for him to join in the drinking which was an important part of business as well as social life at the diggings. That was left to Rhodes.

From digging the two soon branched out into other enterprises, one of which was pumping water. As the mining of diamonds became a more sophisticated business, more ancillary services were needed. The Big Hole, as New Rush or Kimberley had come to be called, was going deeper in a fantastic series of steps, sometimes compared to a Stilton which had been hacked out with a spoon. It needed continual pumping to keep it free from water.* Rhodes brought the first large pump to Kimberley, and along with wire ropes and machinery for hauling from the depths of the mine, this service formed

---

* When surface mining finally ended at Kimberley the Big Hole was left to nature and became what it is today, a great man-made lake.

a large part of their business throughout the 1870s. They also added to their interests refrigeration, then in its infancy, and an obvious boon in sweltering Kimberley.

Their ice-making machine was brought up from the Cape by mule wagon. February and March of 1872 were record wet months and an early winter set in. The two were nearly ruined. But the following summer was especially hot: 'the machine paid for itself in three months, and we sold it at the end of the summer for more than it cost'. In that hot summer diggers found Rhodes turning the handle of an ice-cream bucket while Rudd sold it from a packing case at the corner of the Diamond Market. They charged sixpence a wineglass, with an extra sixpence for a slab of cake. But they had not forgotten diamonds. Rudd's share of the profits from ice-making came to £1,500. With this, his first capital, he bought into the Baxter's Gully claim on De Beers mine.

Throughout that first decade their business was much in the hands of Rudd. Before he had left for South Africa it had been Rhodes's intention to go up to Oxford. He now had the resources to embark upon an extraordinary and unique academic career. After another breakdown in health and a trip to the Transvaal he returned to England in 1873 with ample money to pay for his education and went up to Oxford, where he was accepted by Oriel College, to its subsequent benefit. For the next eight years he led two lives, spending long periods in Kimberley, where he accumulated more money, expanded his business, and gradually built an empire of claims at the Big Hole; and keeping terms at Oxford whenever he could. His travelling between mining town and university city was enormously inconvenient and expensive. At Oxford he lived in his college; at Kimberley he and Herbert lived in lodgings called the 'West End', presided over by Major Drury of the Cape Mounted Rifles.

As an undergraduate Rhodes worked intermittently. He made more friends, among them Rochfort Maguire and Charles Metcalf. But he was never gregarious, and particularly disliked female company. When obliged to attend dances at Kimberley he would have only a few dances, and then with the plainest girls he could find. And yet by the end of his time at Oxford he undoubtedly enjoyed something of a social success, member of the Bullingdon and Master of the Drag. Sometimes, when talking to his acquaintances, he would amuse them by absent-mindedly putting his hand in his pocket and pulling out a fistful of diamonds. Maybe Rhodes had learnt early, the first of the Randlords to do so, an important truth: that although England is a land of strong class consciousness, of many snobberies and of innately felt social distinctions, there are few barriers which money cannot surmount.

On his way to keep his first term at Oxford, Rhodes's ship crossed with another, outward bound. Among its passengers was a young Londoner, born Barnett Isaacs in 1852. His grandfather was a rabbi and his father an Aldgate shopkeeper, whose two sons and three daughters were brought up in rooms over his shop. Barnett was educated at the Jew's Free School along with David Harris, one of his large cousinage. His sister Kate was married to Joel Joel, landlord of the King of Prussia public house. They had three children, Barnett's nephews, Isaac, Woolf and Solly. The Isaacs boys worked at the pub: the elder, Harry, as barman; Barnett as odd-jobman – he also helped his father in the shop and hustled on the streets. Both brothers loved the music-hall, and Harry progressed from backstage to the footlights as an amateur conjurer. As he found work at the Cambridge and other halls, he took with him as stooge and straight man his brother Barnett, always known as Barney. The story goes that while Harry was taking a curtain call the stage manager called from the wings, 'Barney too'. The catch phrase stuck, and turned into a name. Barnett Isaacs became Barney Barnato. Soon the two of them were billed as the Barnato Brothers.

Their cousin David had left for the diamond-fields and then came home with much money (won, as it happened, by breaking the bank at roulette rather than by digging) and with tales of limitless opportunity. Harry followed him to South Africa in 1872. On 16 September the Mutual Hall in Cape Town billed 'Signor Barnato, the Greatest Wizard Known'. Like his brother, he was also a modestly successful prizefighter, fighting at two shillings a bout. Soon he was playing in New Rush with some success. Harry Barnato had certain natural advantages: though 'strikingly vulgar', he was a good-looking man. Before long Barnato dabbled in diamonds and became a budding koppie-walloper, the name for a small dealer who bought from the diggers.

Jokes about his two callings were unavoidable. In November there was a large diamond theft. Someone wrote to the *Diamond News*: 'If anybody wants to steal a parcel of diamonds, he must be as quick of hand as Signor Barnato!' Harry replied jocosely to those who feared that 'in my capacity as a diamond buyer . . . I might combine the attributes of the "Wizard" . . . I keep the callings distinct, the only approach to "magic" in the former business being the astoundingly high price I am always prepared to give for good stones.' Spoken in jest, but this was not the first time that the name of Barnato was to arouse suspicion.

The following year Barney joined his brother at Kimberley. He brought with him a little cash and forty boxes of cheap cigars supplied by his uncle Joel. With this capital he started in business. Both brothers were energetic, engaging, quick-witted – useful qualities for koppie-walloping. They knew nothing about the diamond business at first, or no more than that diamonds were valuable and greatly increased their value on the way from the sorting-table to the jeweller's window. The brothers were not poor for long. In 1874, the year after Barney's arrival, they formed Barnato Brothers, 'dealers in diamonds and brokers in mining property'. By 1876 Barney had made £3,000 and had bought a block of four claims in the Kimberley mine. For six years the brothers accumulated their 'war-chest', the capital for larger future operations.

For the rest of his life Barney inspired hatred as well as envy and awe. He himself later said that in those early years 'every form of slander and insinuation was heaped upon me. They compassed me round and laid snares for me. Men of the Diamond Fields, you can never know the bitterness they caused me.'

There is always a fascination about how any self-made millionaire made his first small fortune. Very often those who remember his early years do so with hostility and envy. This was peculiarly true of Barney Barnato. Certainly he was 'hot' as a financial operator even by the standards of Kimberley at the time. It is impossible to say how far he deserved all the accusations that were made against him in his first years in Kimberley, most of all the charge that he was up to his neck in IDB. Maybe Barnato made too much money too quickly, particularly for one with no training in or flair for diamonds as such. And yet he was more than just a street-smart hustler or a sharp operator. His native wit went further than that. He was one of the first to grasp two hugely important facts: that the Kimberlite blue ground went deep and was rich in diamonds all the way; and that only a combination or cartelization could control the output of diamonds and thus the price, and prevent the periodic slumps which disrupted Kimberley throughout the 1870s, driving diggers to drink, flight and suicide.

The story of Barney Barnato shows that no subtle intellect was required to make a fortune, only cunning and shrewdness. The last of these three men to reach the diamond-fields was very different. Maybe 'genius' should strictly be reserved for the fine arts and scholarship, but if it can be extended at all to the art of finance then Alfred Beit was a genius. He was born in 1853 some six months after Barnato and six months before Rhodes (who, as it happened, shared a birthday: 5 July). He was the second of the six children of Siegfried Beit and Laura Hahn. His father was a merchant importing silk from Lyons to Germany. The Beits were Sephardic Jews who in the course of the second Diaspora had emigrated from Portugal. In the seventeenth century they had settled in the Hanseatic port of Hamburg where Alfred was born. He later described himself with some exaggeration as 'one of the poor Beits of Hamburg' whose father found it difficult to pay for his schooling and who had to leave before he had finished the *Realschule*. In fact, Beit showed no aptitude as a scholar. His family had high hopes for his brothers; for Alfred a place was to be found in the business. Other brothers went into business in Germany with considerable success. Otto, born when Alfred was twelve, followed his elder brother to Kimberley and fortune.

These brothers were the product of the energetic and rising world of German Jewry. For all Alfred's lack of scholastic achievement he was a tribute to his teachers. Most of the Germans who went to South Africa had been well grounded in business method and no doubt began at school to learn the English which became their principal language in manhood. Their correspondence is astonishing testimony to the proficiency in it acquired by most of them. We can imagine the atmosphere of the Beit family home: cosy, dutiful and industrious, a Jewish 'Buddenbrooks'.

Another and more substantial family of German–Jewish merchants were the Lipperts. The head of the family was David Lippert, whose firm had important trading interests in South Africa before the diamond-fields were opened. His wife, by whom he had three sons, Edward, Ludwig and Wilhelm, was Susanne Hahn, sister of Laura. So Alfred Beit's first job was working for the Lippert firm's Amsterdam branch, to learn the diamond trade at its great centre. Then Beit was sent as Lippert representative to South Africa.

From the beginning Beit showed an uncanny aptitude for sorting and valuing diamonds, which was the foundation of his career. More than merely a knack of knowing what diamonds were worth, it was a flair for examining stones, and an extraordinary memory. The story was often told later at Kimberley of how a shady dealer tried to sell an uncut diamond to Beit, who recognized it as a stone which had passed through his hands more than five years before.

He reached Kimberley in 1875. Here the ugly duckling shed its down. A small, shy, unprepossessing young man was transformed into a formidable entrepreneur.

The effect of his new environment was to locate within him an entirely different business spirit, and to stimulate into action entirely new sources of brain-power and ability. Within two years after his arrival this environment had converted the boy of mediocre talents, the lethargic, undistinguished young apprentice, into a man of impetuous unremitting energy and enterprise.

The description is fawning, but all the same not far from the truth. But the transformation did not affect his shyness. Beit always remained a social cripple, caring as little for women as Rhodes. Even in masculine society he was undemonstrative and, at first, unimpressive. That was one reason why in his early years at Kimberley he was little heard of.

Heard of or not, he was intently building up his position. As well as his business representing Lippert, Beit became representative of the Paris house of Jules Porges, destined to become a powerful force at Kimberley, albeit a shadowy one. He became partner to Porges's first man on the fields, the Julius Wernher with whom Harris had sailed. Like the other embryonic magnates, Beit was operating on two planes, dealing in diamonds and at the same time buying claims in the mines.

No one ever threw the accusation of sharp practice against Alfred Beit, at least not until much later when the Randlords had become bogeymen. By his own account, he built up business by a combination of shrewdness and fairness. When he reached Kimberley he found that even five years after the diggings had been opened very few people knew anything about diamonds.

They bought and sold at haphazard, and a great many of them really believed that the Cape diamonds were of a very inferior quality. Of course, I saw at once that some of the Cape stones were as good as any in the world, and I saw, too, that the buyers protected themselves against their own ignorance by offering generally one-tenth part of what each stone was worth in Europe. It was plain that if one had a little money there was a fortune to be made.

By now Siegfried Beit had recognized his son's ability and lent him £2,000. Beit used the money to buy not diamonds but 'stands' as they were called, sites for business and residential buildings, for which there was a fierce demand. He let a dozen offices, really corrugated iron shanties, for a vast rent, £1,800 a month in all. Twelve years later he sold the ground on which the shanties were built for £260,000.

In the 1870s the three men knew of rather than knew each other. Like most people in Kimberley, Rhodes did not take Barnato very

seriously at first and shared the general view that he was at worst a
scoundrel and at best a buffoon. There was of course a clash of culture
as well as of personality. With all his personal oddness, his half-
hysterical mannerisms, his brutality in speech, his political obsessions, Rhodes counted as an Englishman and a gentleman. Barnato
was indelibly exotic and plebeian. Kimberley was developing its own
social sense, and when the barrier fell between rough and
respectable, there was no doubt on which side Barnato stood.

Like others before and after in similar circumstances he emphasized
rather than played down his roughness. He had come to Kimberley as
a huckster and he stayed one. He played the buffoon, the cheerful
tummler. He was a performer, his life led both metaphorically and
literally before the footlights. For years he played in amateur theatri-
cals. One favourite part was Sir Henry Irving's in *The Bells*, another
was Bob Brierly in *The Ticket-of-Leave Man*. For several years he
followed another of his boyish enthusiasms, the boxing ring. Barney
Barnato went through life bald-headed, as in a prize fight. He was
coarse and common, never losing his Aldgate accent and never
acquiring any interests outside money and sport. And he didn't care.
His philosophy was simple: 'If you are going to fight, always get in the
first blow. If a man is going to hit you, hit him first, and then say, "If
you try that, I'll hit you again".'

Rhodes met Beit in 1879 when they were introduced by William P.
Taylor, a minor speculator who had started dealing in diamonds at
the age of fourteen. Some time later Rhodes passed Beit's office late
one evening, and found him still in: 'Hullo!' said Rhodes, 'Do you
never take a rest?' 'Not often', said Beit. 'Well, what is your game?'
asked Rhodes. 'I am going to control the whole diamond output
before I am much older.' 'That's funny,' said Rhodes, 'I have made
up my mind to do the same; we had better join hands.' And so they
did.

Between Rhodes and Barnato, then, there was a distant and wary
respect. Between Rhodes and Beit there was a much closer tie,
admiration on one side, something close to adoration on the other:
Beit fell for Rhodes in every sense. Each saw in the other, developed to
a high degree, the qualities he himself lacked. Rhodes the intellectual
posturer venerated Beit's mental faculties. Beit for his part was
helplessly shy, *chétif* in appearance and gauche in manner. Never in
his life could he be at ease with others, but would sit fidgeting and
spreading waves of awkwardness. He must have recognized his own
great intellectual superiority over Rhodes; what he envied and loved
in the other was his commanding and leonine personality.

Between Beit and Barnato there was unconquerable hostility. Here was another complication. Rhodes was a gentile. Beit and Barnato were Jews, as were most of those who came to dominate mining at Kimberley and then on the Rand. The fact would have been commented upon at any time. It was especially conspicuous when the Jews were approaching their social zenith and their financial heyday. Before long the Randlords would become world famous and, to their critics, a byword for oppression, chicanery and rapacity. That so many of them were Jewish was something their enemies rarely forgot.

The story of the European Jews in the three or four generations between emancipation and catastrophe is extraordinary. England saw only a small part of the episode: in 1870 Isaacses, Joels, Harrises and Phillipses belonged to a community of fewer than 50,000 Jews in England. The British Jews had not been ill-used by Continental standards. Their position was equivocal, and yet the *outré* figure of Disraeli could become Prime Minister. Another Jew, Sir George Jessel (to whom Barney Barnato claimed to be related), became Master of the Rolls in 1873.

The Isaacses, Harrises and Joels grew up poor but not poverty-stricken, in a society which was escaping from the ghetto. Lionel Phillips, who became a partner of Beit and Wernher, came from a rung higher on the social ladder, son of a modestly well-off merchant. None of them had much education and what they had was old-fashioned: they were 'taught more Hebrew than English'. But they were free to travel and seized the opportunity. Their lives were comparatively unhampered.

Compared that is with the Jews of central Europe. The great mass of Jews still lived in eastern Europe, in Galicia, Hungary, Romania, but most of all in Poland and the Russian Pale of Settlement. Here they were still unemancipated spiritually as well as physically, living in their own communities, the ghetto or the *shtetl*, dealing with gentile society for commercial purposes but severely segregated from it, practising their own religion and maintaining their own cultural traditions, often wearing distinctive dress, speaking their own language, Yiddish, often persecuted but sustained by a strong sense of community: a community which changed little until it was destroyed. That was where the bulk of those who are now South African – and British and American – Jews sprang from. But with the significant exceptions of Isaac Lewis and Sammy Marks, none of the first generation of mining magnates came from such a background.

By contrast more than half a dozen – Beit, Adolf Goerz, Edward Lippert, Sigismund Neumann, George Albu, Max Michaelis, to say nothing of various brothers and cousins – were German Jews. They were all born within the borders of what became the German Empire under Kaiser William in 1871. To those can be added two shadowy figures born under the Habsburgs, Jules (first Yehudi, then Julius) Porges from Prague, and A.H. Nellmapius from Budapest.*

The situation of the Jews in Europe west of the Elbe was very different from that of their brethren in the east. The nineteenth century saw the process of emancipation well under way. It had begun before the French Revolution, in the 1780s when Joseph II of Austria freed the Jews and 'called into existence the most loyal of Austrians'. The Revolution accelerated the process. The ghetto was opened and not closed again, even in the period of reaction which followed Waterloo. In this new Europe Jews were soon ascending giddy heights, not only in commercial, intellectual and artistic life – an ascent which reached its pinnacle in Vienna at the end of the century – but in public life also. 'The philosophers have only interpreted the world in various ways; the point is to change it.' In fact, as Marx wrote the words, fellow Jews were starting to change the world or at least to run it. Disraeli was only one example.

In all this there was an irony and a threat. The French Revolution was ambiguous. The forces unleashed in 1789 were turned against all the vices of the *ancien régime*, 'old custom, legal crime, and bloody faith', including the subjection of the Jews. And yet revolutionary liberalism went hand in hand with nationalism. At Valmy the French triumphed as much through a new national spirit as through zeal for the Revolution. The Jews were the first to benefit from the overthrow of old custom, but one day they were to find that the new nationalism was a worse enemy than prescriptive injustice.

That became steadily clear in Germany throughout the nineteenth century. In 1819 there was a spasm of anti-Semitism, the 'Hep-Hep' movement, a taste of what was to come. Things were little better by 1862 when 'to our educated German Jews the feeling of hatred towards the Jews displayed by the Germans has always remained an unresolved puzzle', as Moses Hess said in a proto-Zionist pamphlet. The great crash of 1873 which began on the Vienna Bourse took place just as the young magnates-to-be were moving towards the diamond-fields. A good few of those who had grown rich in the boom before the slump were Jews, like the 'railway king' Strousberg, and the collapse

---

* Two others should not be added, however: Julius Wernher and Hermann Eckstein, sons respectively of a railway engineer and minor public functionary in Hesse and of a Lutheran pastor in Württemberg. Many, not only anti-Semitic enemies of the Randlords, assumed that anyone with a German name was a Jew; these two were gentiles. .

led to a new wave of anti-Jewish feeling. Within six years the great Prussian historian Heinrich von Treitschke wrote a famous article declaring that 'the Jews are our national misfortune'.

The German Jews who went to Kimberley knew this well enough. They were not persecuted as they might have been several generations earlier and as they would have been two generations later, but a good many of them knew what it meant, as did Leopold Albu at his school in Berlin, to suffer 'on account of the racial feeling called "Judenhetz"'. They could not guess what the future held in their own country – nor the twist in the story by which they would stimulate another kind of Jew-hatred in England.

What happened in South Africa was a small part of the effervescence of Jewish genius. Earlier on it was impossible to see how brilliant the flowering would be (although an intelligent anti-Semite like Richard Wagner guessed at it). It was an astonishing phenomenon, comparable to other collective explosions of genius – Periclean Athens, Rome at the end of the Republic, Quattrocento Tuscany, Elizabethan England, Georgian Scotland – but stranger than any of those. And the Jews whose genius expressed itself in the art of finance were the objects of a particular hostility, not only Strousberg but Bismarck's banker, Bleichröder, Sassoons, Guggenheims, Rothschilds. In France Jew-hating was to culminate in the Dreyfus affair. In the United States too there were ripples of hostility: Kimberley's first decade saw a famous scandal when in 1877 Judge Henry Hilton, the new owner of the Grand Union Hotel in Saratoga, told the immensely rich New York banker Joseph Seligman that as a Jew he was no longer welcome at the hotel.

And so in South Africa. There also the Jews were accompanied or even preceded by the usual generalizations: the country was 'a great hunting ground for that tribe. They fatten on the heavy, credulous Boers.' Quite soon at Kimberley the stereotype of the koppie-walloper – or the IDB – was a 'Jew from Whitechapel', or Poland. They are found in the early memoirs, of Angove, Boyle and Peyton and in fictional descriptions of Kimberley also, W.E.T.'s *The Adventures of Solomon Davis* or George Griffith's *The Memoirs of an Inspector*. There was certainly an influx of Jews from London, and from eastern Europe, before and even more after the pogroms which began in 1881.* Soon there was a more settled Jewish community at the fields. It opened a synagogue whose first *chasan* was given to playing the

---

* 'Lithuania' would have been more accurate than 'Poland'. By contrast with American or British Jews, most South African Jews today are descended from 'Litvak' emigrants who came from a quite small area of Lithuania.

violin at his congregation. He was a German called Albu, uncle of the brothers George and Leopold.

Despite the popular belief, most Jews in Kimberley were not light-fingered koppie-wallopers and, in contrast to another fable, only a few became millionaires: there were far more losers than winners on the fields. One of the losers had a revenge of sorts when he recaptured the early days of Kimberley, and later of Johannesburg, in his funny, acidulated and abusive memoirs. Louis Cohen was born in 1854, within a couple of years of the three great men. He grew up in his native city of Liverpool. Cohen reached Kimberley in 1872. Before long he had made the acquaintance of the Barnato brothers.

Soon Cohen and Barnato were in partnership. They began dealing early every morning, at 7 a.m. at the Scarlet Bar, and they reckoned to make £20 to £30 in a morning. Cohen implies that they dealt honestly but also that standards of honesty were not high. A character in an early diamond-fields novel put it simply: 'Everyone in Kimberley, take my tip, is more or less connected with IDB.' The charge was so loose that there was no defence against it. How prevalent IDB was is impossible to say. Many believed that it was disastrously rife, and the vehemence with which some miners denounced this 'custom of the country' may have been in part Freudian projection, or pot-and-kettle:

> Many of the claimholders, I repeat, commenced their diamond field life as professors of the same art, and when, after robbing the Dutchmen of their ground and their diamonds they became mine-holders, then they discovered, to their intense horror and indignation, the heinousness of a crime which they once had practised themselves so successfully.

Illicit dealing had a more serious consequence than blackening a few illustrious names. It provided a weapon with which the diggers attacked blacks. Many black labourers on the four mines succumbed to the temptation to pilfer stones. When they were caught they received rough justice, a severe flogging with the sjambok. No white digger was punished for treating his servants too roughly. Punishment was reserved for those suspected of buying stones. In December 1871 at New Rush suspects had their tents burned down by a posse which further enjoyed the 'pleasant diversion of chasing negroes round about, aimlessly punching their heads when caught'. The following month saw the first recorded death of a black labourer. A digger, 'indignant and disgusted at the laziness' of his black, had been kicking him when a diamond dropped out of the Zulu's loincloth. The black was tied by the neck to a pole by the digger who went for help and on his return found the labourer dead at the bottom of the claim.

The diggers had more ambitious ideas than kicking the blacks. They wanted to ban them from claimholding, to have a minimum punishment of fifty lashes for any black found with a diamond which he could not account for, and to introduce a system of 'passports' for black workers. This was a favourite scheme of Alfred Aylward, an unusual and colourful man even on the fields, a large black-bearded Irishman, an agitator, a journalist who not only edited the *Diamond News* but was correspondent for the *Daily Telegraph* of London. He claimed, probably untruthfully, to have been imprisoned as a Fenian; Sir Garnet Wolseley of all unlikely people called him an amusing ruffian; he was a man to reckon with.

The British authorities were unwilling to accede to the diggers' proposals for disciplining blacks. Many officials sincerely believed in a tradition of colour-blind paternalism. When regulations for curfew, searching and flogging were drafted a Colonial Office official minuted them 'Regulations monstrous'. But Aylward and the diggers had their way. First, forty-six black diggers at Bultfontein had their licences suspended. Then Proclamation No.14 of August 1872 gave the agitators all they wanted: searching and flogging for any black labourer found on the camp without a pass. The *Diamond Field*, rival to Aylward and the *Diamond News*, lamented, 'concession is become the order of the day on the Fields. The diggers have only to ask and they receive.'

It was not quite true that the authorities were in awe of the diggers. Two men saw that the August Proclamation stored up trouble for the future. One was Richard Southey, a veteran administrator, almost sixty-five when he arrived as new Lieutenant-Governor of Griqualand West at the beginning of 1873. J.B. Robinson lent his carriage and horses to bring Southey and his wife to the New Rush camp. Southey's government secretary was John Blades Currey, a civil servant, the man who had six years earlier tried to arouse European interest in Cape diamonds, Merriman's friend and briefly Rhodes's colleague. Both men began on terms of mutual distrust with the diggers.

For his part Southey saw what the diggers wanted and deplored it. They wanted to exclude all persons of colour from the exercise of franchise as well as the right to own claims 'and to grant the privilege to white persons purely because they are white'. But as 'until recently nearly all the land in the Province belonged to persons of colour . . . there is great injustice in attempting to deprive them'. This was a private communication to his chief Barkly. Had the diggers seen it they would have despised Southey all the more. As it was the stage was set for sharper confrontations between imperial officials and diggers.

In any case New Rush, or Kimberley as it became in July 1873, was depressed and edgy after a slump the previous May and a more severe one about to break. Life went on for the diggers who remained, drinking, squabbling, brawling. At the centre of many a fight were Barney Barnato and his sidekick 'Lou' Cohen. There were many evenings like one at the Gridiron, a restaurant owned by a former hussar called Richardson. There, sitting with a pretty woman, was Captain Hinton, a petty local official, and his younger brother. The two young diamond dealers tried to talk to 'the Brighton Beauty'. 'Hot words passed with the Hintons, and I departed with a lurid and sanguinary description of the Hebrew race ringing in my ears.'

The koppie-wallopers walloped on for all that and for the next four or five years 'Beit was very little heard of, Rhodes was digging in De Beers, Robinson had a diamond office in the main street, where, dour and sour, he sat waiting for customers; Lewis and Marks were busy on the Kopje, Woolfie Joel was doing broking, Jack Joel was doing anything and Barney and I were doing everybody.' These magnates in the making were before long to emerge and dominate Kimberley.

4

No Spot More Odious

On 13 October 1873 Cecil Rhodes had matriculated as an undergraduate of Oxford. That Michaelmas was the first term he kept at Oriel. He returned to Kimberley early in 1874 under sentence of death: a doctor had examined him and given him six months to live. He survived for the best part of thirty years, and the next two of them were spent entirely on the fields where he remained until Easter 1876. They were critical years. Kimberley itself as much as young Rhodes was reckoned to be mortally stricken in 1874.

The prices of diamonds oscillated wildly and the mines were peculiarly liable to extremes of boom and slump. The first slump of 1872 had been followed by fresh finds but they only served to depress the price still further. With each collapse more diggers were weeded out. This process was even welcomed. Reporting in May 1873 that the market 'I am happy to say' had again turned down, Merriman added the hope that 'the losses incurred by a number of local speculators will make them more cautious'.

Then the crash of 1873 sent its ripples out from Vienna as far away as the Cape, and the diggers' fortunes sank very low. Banks in Cape Town desperately called in loans and the population on the fields plummeted to as few as 6,000 white men. Even without the effects of European recession the fields were caught in a vicious circle and so remained throughout the 1870s. The value of diamonds mined had increased steadily, even dramatically: from £24,813 in 1869 to £153,410 in 1870, £403,349 in 1871, £1,618,076 in 1872 (the first full year's mining in the New Rush or Kimberley), £1,649,451 in 1873. Then, after a plateau, the sum dropped to £1,313,334.

These were baleful figures, especially since they conveyed a fall in value. Production in volume rose steadily; it was the price per carat which kept dropping. The price would sometimes return to a healthily payable 30s a carat, but just as often it slumped below 20s which was ruinous, most of all for the small digger with high unit costs of production and with the continuous heavy expenses of life in Kimberley.

The larger the stone the higher of course the price per carat: a twenty-carat stone was worth much more than twenty of one carat. But even then the real profit eluded the individual digger. In August 1873 a 53-carat stone had been found and sold for £300; it was almost immediately resold for £1,000. The digger's unreflective instinct was to dig for more stones, but this only kept the price depressed.

For those who remained, the fields were a lively enough place. The tensions of boom and slump meant a community of men living on their nerves who wanted to spend their winnings from the mines on simple pleasures. The three chief recreations of the place were drinking, gambling and sex. The first two went together despite a very unpopular attempt by Southey to suppress gambling throughout Griqualand West. In canteens and hotels play ran high: in the Red Light in Main Street, the company whose billiards room 'would make you open your eyes and close your pockets', or the Blue Posts, or the Hard Times or the Perfect Cure, or the Irish resort, the Cavan kept by Paddy Murtough. Then there were 'gambling halls' proper, like Ashwells or the one in Stockwell Street. One casino owner was said to have made £40,000 in a few months. Some diggers made fortunes at the tables. More lost them. Few were immune from gambling fever. Barnato was a notoriously high player.

At first there had been few women at the diggings, very few white women indeed. Before long they followed, and there were plenty of black whores to cater for the diggers. There were white women also. The San Francisco doggerel ran:

> The miners came in '49, the whores in '51,
> And so betwixt the two of them they bred the native son.

So it was in Kimberley which soon filled up with 'Often undesirable ladies, who, all the same, were most desirable.' One nicely-dressed, bright-eyed young woman from Cape Town was well known 'and did not conceal the fact that she desired to be better known'. One evening at Graybittel's saloon she drank with the boys and then asked to be put up for sale, 'not for life of course'. Standing on a champagne case she was knocked down, after a keen auction, to John Swaebe for £25 and three cases of champagne. He took her to his frame tent across the road. 'But after half an hour had elapsed, "the boys" got round the tent and carried it bodily away, thus exposing to view the amorous pair.'

There was more here than tomfoolery. The woman was an octoroon, or 'Coloured'. In the early days on the diamond-fields dealings between white diggers and non-white women were easy. As elsewhere in Africa and Asia, it was in part the coming of more or less respectable white women which ended that. Coloured 'mates' were put away. This did not matter to the celibate Rhodes or Beit, but it did to Barnato. His mate was a girl called Fanny Bees, a barmaid and actress. She was described as an 'Afrikander of the St Helena type', a euphemism for Coloured. Her black hair and olive complexion had caught other eyes than Barney's. Sometime in the mid 1870s she became his mistress and, more than fifteen years later, his wife.

Although sexual relations across the colour line are an important episode in the story of South Africa, they were in reality less important than the struggle for power at the mines. Throughout the early years at Kimberley the miners had squabbled and fought among themselves and to make a common front against the authorities seemed well-nigh impossible. But one thing united these rowdy and fissiparous diggers: 'The only point at which the malcontents join issue is niggers being allowed to hold claims.' With IDB as a pretext, rioting and tent burning continued intermittently throughout 1874 and 1875. Currey had warned that the surrender of 1872 was stirring up more trouble and it seemed as if his misgivings were about to come true.

In March 1875 the diggers, led by Alfred Aylward, issued a Proclamation, itself a reply to a Proclamation by Southey against 'taking illegal oaths or assembling in arms'. A black flag was hoisted near Kimberley mine and Southey signalled to Barkly for more troops. On 12 April, William Cowie, a canteen keeper, was tried at the resident magistrate's court for supplying arms to Aylward. His conviction precipitated a great meeting of armed diggers. As ever, black claim-holders were the real grievance. For a moment it looked as if revolution would break out. But the incipient revolt was defused by the coolness of the magistrate. The diggers dispersed, and Aylward bolted to the Transvaal.

The immediate victors after the Black Flag revolt appeared to be the diggers. Even those like the *Diamond Field* who were foes of Aylward agreed on the central question: the attempt to raise blacks to an equality with their masters must mean 'financial ruin for the whites, moral ruin for the natives'. Despite the half-hearted mutterings of British officials there was to be no more talk of equality. The wishes of the Diggers' Protection Association were gratified by the report of an official inquiry into the financial position of the mines. The authorities accepted the diggers' submissions that too many free blacks at the mining camp were the root of crime. Black IDBs and black whores were rounded up and expelled. Southey's liberal experiment was ended and blacks were forbidden to hold claims or to wash debris. There was now no role at all on the fields for blacks except as labourers. The pass laws became an established feature, and so, from 1875, did the 'compound', the barracks where black labourers were confined. Passes and compounds were all part of a process of bringing the black African tribesman into an industrial economy or, as an official report put it, 'of introducing manual labour under skilled supervision among a class who otherwise would have lived in contented idleness'.

Government responsibility for the mines and the mining community was rapidly diminished. That pleased the diggers. They did not see that government withdrawal left a vacuum which could only be filled by company mining. Even before the Black Flag rebellion the perceptive Merriman had seen this: 'Everything points to the reign of the capitalist.' More and more individual claims were being combined.* Another who saw which way the wind was blowing was Robinson. In December 1876 he advertised for such claimholders as might be inclined to join him in amalgamating their ground for the purpose of working it upon co-operative principles. The idea was sound but he was not the man for it. Never was anyone less emotionally apt for any task of co-operation than the cantankerous and selfish Robinson. His personality had been on open display a few months before. He had bought a diamond from Anton Dunkelsbuhler, a respected diamond merchant. It transpired that the diamond was 'hot' and was likely to feature in an IDB case, although Dunkelsbuhler had bought and sold it in good faith. Robinson demanded security against its confiscation; would not accept a letter of credit; applied for a writ to prevent Dunkelsbuhler leaving Kimberley on a business trip; was almost set upon by an enraged mob; and was once more execrated in the press.

For the moment Robinson stepped out of the public limelight and turned his attention to his private life. In October 1877 he married Elizabeth Rebecca Ferguson, daughter of one of the pioneer diggers. The wedding was in one sense a surprise: Robinson was a thirty-seven-year-old satyr, who had never given any indication that he might be seeking domestic quiet. In another way it may not have been surprising. The story went that Robinson had been obliged to marry Miss Ferguson. So Louis Cohen believed, and he stored up the titbit for more than thirty years to use against his enemy.

---

* Until 1872 there was a limit of two claims which could be jointly owned. Then the limit was raised to ten, and in 1876 the limit was abolished.

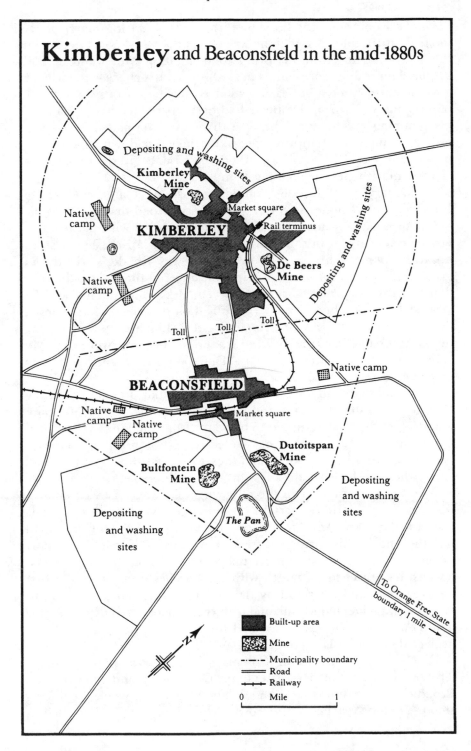

Kimberley and Beaconsfield in the mid-1880s

Depositing and washing sites

Kimberley Mine

Native camp

KIMBERLEY

Market square

Rail terminus

Depositing and washing sites

De Beers Mine

Native camp

Toll Toll Toll

BEACONSFIELD

Native camp

Native camp

Native camp

Market square

Dutoitspan Mine

Bultfontein Mine

Depositing and washing sites

The Pan

Depositing and washing sites

To Orange Free State boundary 1 mile

Built-up area

Mine

Municipality boundary

Road

Railway

0   Mile   1

Unlike Robinson, Rhodes showed no sign of getting married. He too had his troubles. By 1876 he was one of the best-known figures on the fields. Starting from his proto-amalgamation with Rudd he had acquired more and more claims in De Beers. But a check came when a government commission heard evidence that the operative in charge of pumping water from De Beers had been offered a bribe to damage his pumping machinery. The operative was summoned and when ordered to name the tempting speculator, he wrote down on a piece of paper 'Mr Cecil Rhodes'. In Rhodes's absence Rudd denied the charge, and Rhodes did so himself on his return.

In January 1876 Rhodes and his accuser met at the Court of Enquiry where Rhodes said that he had forwarded the matter to the public prosecutor who would charge the operative with perjury. He was indeed charged, but the matter mysteriously lapsed. This was an unsatisfactory conclusion. The prosecutor was Rhodes's friend and executor, Sydney Shippard. A suspicion lingered that he would have done all that was necessary to clear Rhodes's name if it could be cleared and that he – and by implication Rhodes – had chosen discretion instead of valour. Rhodes went to England to keep all the terms at Oxford for the next year and to lick his wounds: 'My character was so battered at the Diamond Fields that I like to preserve the few remnants.'

Successful diamond men by now made a regular practice of visiting London to arrange business and in September 1876 Barney Barnato and his Joel nephews returned to the Cape from England. They sailed on the same ship as a young man called Robert Standish Sievier. Few had heard of him then, nor would they for years. But the day would come when the Joels would know his name all too well.

By contrast a traveller in the following year was very well known, possibly South Africa's, certainly Griqualand West's, most famous visitor to date. Anthony Trollope was sixty-two in 1877 and approaching the end of his strange career, part Post Office official, part novelist. He had published more than fifty books but his success as a novelist had reached its zenith with the Barchester novels published between 1855 and 1867 and by the mid 1870s his star had begun to fall. He supplemented his income with travel books. His departure for South Africa was itself a mark that interest at home in the hitherto neglected Colony was waxing.

His fame preceded him. Colonials read books; shortly before Trollope's arrival a leader-writer in the *Cape Times* could allude to Mrs Proudie's snobbishness, and know that the allusion would be understood. Trollope reached Cape Town, which he gloomily described as 'a poor, niggery, yellow-faced, half-bred sort of place, with an ugly Dutch flavour about it', and after visiting Durban he travelled from Pietermaritzburg to the Transvaal. His companion on this journey was 'a gentleman of about a third of my own age, who had been sent out by a great agricultural-implement-making firm with the object of spreading the use of ploughs and reaping machines'. This was almost certainly George Farrar. He was eighteen, having just come from East Anglia, where he had grown up as a representative of a family engineering firm, Howard, Farrar & Co. He stayed on. Although he played no part on the diamond-fields he was to be an early pioneer on the Witwatersrand, and to become not the least rich and powerful of the gold-mining magnates.

The two men reached the diamond-fields in October. On the way there they had stayed at Boer farmhouses. Like other English travellers of the time, Trollope was struck by the primitive backwardness of the Boers, and also by their touching simplicity and their generous hospitality. Whatever he may have expected from Kimberley, he disliked what he found, 'a most detestable place'. His antipathy was only stimulated by his host, Lanyon, who did nothing to conceal from his distinguished visitor the contempt which he felt for the place himself. In any case, Kimberley embodied what Trollope most hated: easy riches, gambling, social decay – everything which he had condemned in *The Way We Live Now*. Certainly there were several prospective Augustus Melmottes on the fields.

At first sight Kimberley did not even have meretricious glamour. There was a 'little village' at each mine, 'very melancholy to look at, consisting of hovels or drinking bars, and the small shops of the diamond dealers. Everything is made of corrugated iron, and is very mean to the eye.' It was nearing high summer, with no rain for months, and the temperature at 96 in the shade, 161 in the sun. 'I seemed to breathe dust rather than air.' When Trollope saw Dutoitspan it comprised 1,441 mining claims held by 214 claimholders. White men were paid between £3 and £6 a week. About 1,700 blacks were employed working in the mines and on processing the rock, at wages of 10s a week and food. As Trollope noted, the cash wage was more than many rural labourers in England earned. Bultfontein had 1,026 claims in the hands of 153 claimholders and employed about 1,300 blacks.

He did not visit De Beers but went straight on to Kimberley. Thanks to his visit we have a vivid description of the Big Hole in its great period. It was surrounded by a rim which looked at first to be made of slag or 'debris', but was in fact natural. Ascending this 'you look down into a large hole. This is the Kimberley mine. You immediately feel that it is the largest and most complete hole ever made by human agency.' The area of diamond claims occupied nine acres, but digging had forced back the area actually dug open to an extent of twelve acres.

> Then suddenly, beneath your feet lies the entirety of the Kimberley mine, so open, so manifest, and so uncovered that if your eyes were good enough you might examine the separate operations of each of the three or four thousand human beings who work down there.

The side of this bowl was smooth except where there were incrustations similar to those at the bottom,

> among which ants are working with the usual energy of the ant-tribe. . . . Near the surface and for some way down, the sides are light brown . . . the light brown has been in all respects the same as blue, the colour of the soil to a certain depth having been affected by a mixture of iron. Below this everything is blue. . . . But there are other colours on the wall which give a peculiar picturesqueness to the mines. The top edge as you look at it with your back to the setting sun is red with the gravel of the upper reef, while below, in places, the beating of rain and running of water has produced peculiar hues, all of which are a delight to the eye.

Around the edge stood large high-raised boxes into which the broken ground was deposited. It was brought up by aerial tramways: 'Wires are stretched taut from the wooden boxes slanting down to the claims at the bottom – never less than four wires for each box, two for the ascending and two for the descending bucket. As one bucket runs down empty on one set of wires, another comes up full on the other set.' The buckets were large iron cylinders sitting upon wheels running on the wires. First of all the tramways had been worked by black labourers, then by horses, then by steam engines. These tram wires descended from everywhere around the rim of the mine to the centre at the bottom. The buckets went to and fro with 'a gentle trembling sound which mixes itself pleasantly with the murmur from the voices below. And the mines seem to be the strings of some wonderful harp – aerial or perhaps infernal – from which the beholder expects that a louder twang will soon be heard.' Down below 3,500 blacks were at work, digging the soil which was loosened by blasting every evening after the workers had left the mine at six o'clock.

What Trollope was describing was one of the first great opencast mines. It was smaller than the modern type of opencast mine; the

Rösing uranium mine in Namibia operates on a far vaster scale, where the measure is in thousands of yards across rather than hundreds, and where the ants are beetles: enormous dumptrucks, tiny in the distance. But the mining technique cannot change in principle. Shelves or steps are still knocked out by blasting.

The Big Hole then had one feature unlike any modern mine:

the sub-division into claims and portions. Could a person see the sight without having heard any word of explanation it would be impossible, I think, to conceive the meaning of all those straight-cut dikes, of these mud walls all at right angles to each other, of those separate square pits, and again of those square upstanding blocks, looking like houses without doors or windows. You can see that nothing on earth was ever less level than the bottom of the bowl, and that the black ants in traversing it, as they are always doing, go up and down almost at every step, jumping here on to a narrow wall, and skipping there across a deep dividing channel as though some diabolically ingenious architect had contrived a house with 500 rooms, in not one of which should there be a pair of stairs or a door or a window.

As well as the tramway, one firm, Messrs Baring-Gould, Atkins & Co., had already gone to the expense of sinking a perpendicular shaft with a tunnel below to enter the mine, and Trollope went down by the shaft. The mine had originally been divided into 801 claims but half of these had never held any diamonds. There were at the time of Trollope's visit 408 existing claims, variously valued for the purposes of taxation. Some at the west end were worth only £100 each, but on the south side were twelve claims valued at £5,500 each.

A contemporary map of the mine divided into claimholdings shows what even Trollope cannot desribe in words – the fantastically complicated pattern of ownership which had developed, something to make the proprietary rights in a Côte d'Or vineyard seem simple by comparison. Some claims were minutely subdivided, with individuals holding as small as one-sixteenth of a claim. In all the mine was divided into 514 portions. The inconvenience of mining by claims was obvious and Trollope showed unconscious foresight when he said that while at Dutoitspan and Bultfontein the workers were scattered, at Kimberley 'everything is so gathered together and collected that it is not at first easy to understand that the hole should contain the operations of a large number of separate speculators. It is so completely one that you are driven at first to think that it must be the property of one firm.'

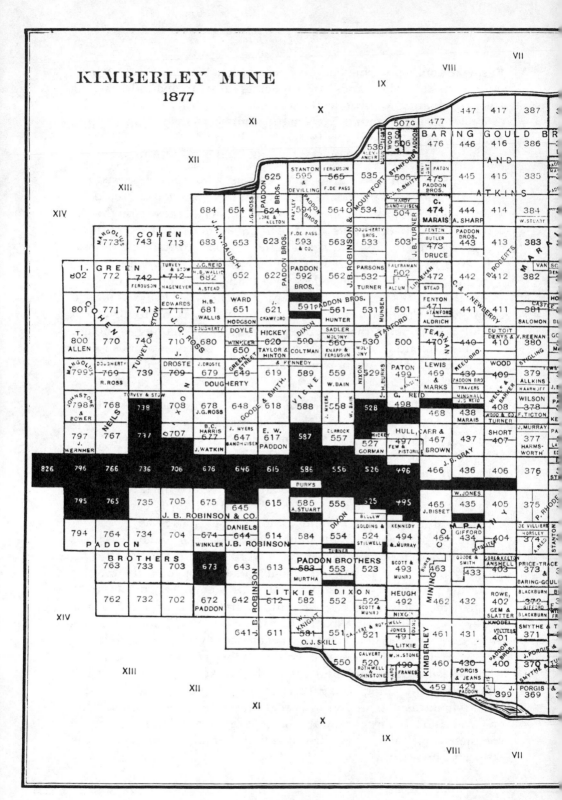

KIMBERLEY MINE
1877

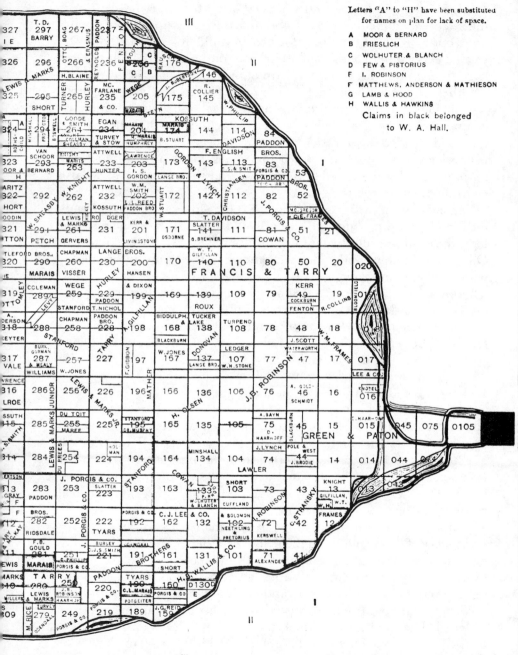

There were two pressures pushing the diamond-fields towards amalgamation, technical and financial. The technical pressure was the increasing difficulty – soon to become the impossibility – of mining huge open pits as if they were owned in hundreds of smallholdings. Even if every other condition had been favourable it would have been hard to maintain this primitive form of ownership.

Other conditions were not favourable. The second consideration, less obvious but more important, was financial. Unchecked mining by competitive diggers maximized output which meant that the price of diamonds tended downwards. The only answer the diggers knew was to dig for yet more stones. But if the mines could be combined, the output could be controlled and reduced and the price per carat also controlled – upwards.

Amalgamation was some way off but Kimberley was already inexorably dividing into those who would win and those who would lose, those who would be famous and those who would be forgotten. On the one hand were the men well known in their time at Kimberley who now only flit through the pages of memoirs and old newspapers. There was 'a Jew called Ikey Sonnenberg [who] was for many years about the best-known character in South Africa', as an exotic visitor, the Irish peer Lord Rossmore, said. Sonnenberg never made a fortune, but remained one of the characters of Kimberley and then Johannesburg, where twenty years later Louis Cohen saw him wearing 'the actual ancient white topper I saw him sport in '73, and sucking, I could swear, the identical antique pipe'.

Where there had been only two or three doctors, there were by the late 1870s many more, much in demand even if their general level of medical skill was not high. One of their number, Hans Sauer, himself a graduate of Edinburgh, sneered that all but one of the other Kimberley doctors 'were provided with diplomas obtained by passing the medical examinations of the second- and third-rate medical colleges scattered about England, Scotland or even America'. R.W. Matthews was one of their number, the man who so deplored the prevalence of drunkenness on the fields. Another of Sauer's foes was to have a more interesting career. Dr Rutherfoord Harris had come to Kimberley in his mid twenties and had soon established his reputation as a very clever and able man but an ambitious intriguer who could not be trusted much further than he could be watched. His abilities were to bring him into the orbit of Rhodes, one of whose henchmen he became, and his love of intrigue was to bring him unwelcome fame twenty years after he first came to South Africa.

Of all these medical men, 'the Doctor'* meant only one: Leander Starr Jameson, who joined Dr James Prince in partnership at Kimberley in 1879. He immediately established himself as a popular, competent, no-nonsense physician. Men liked him, and women liked him even more if Louis Cohen is to be believed. A man who complained that his wife was childless was advised to consult Jameson 'with the result that, hey presto! before the year was out, and on the first of April too, he became the proud pater of bouncing twins. The Doctor was, indeed, a life giver.'

And on the other side were the men who were building up their position on the fields not merely for transient fame but for vast wealth and power. After his brief stint at the *Independent*, Lionel Phillips had moved into the real business of the fields. J.B. Robinson had been impressed by the young man's flair. Phillips was in his early twenties, a popular figure with the serious mining entrepreneurs who met at the Craven Club and was on the way to making his first fortune: a fortune which he was to lose and win over and over again.

Anton Dunkelsbuhler had begun at the fields as an agent for Mosenthals. Before long 'Dunkels' (always his informal name on the fields, later formally adopted by him) was in business on his own account. He was helped by a peculiarly sharp business brain. It was said that he slept with a ready reckoner by his bed; certainly no one in Kimberley could more quickly work out the value of so many diamonds of so many carats. He belonged with so many others to a large, loose German-Jewish cousinry. His nephew Gustav Imroth was to come to Kimberley in 1880, when he was eighteen. Imroth's mother was a Hirschhorn; Fritz Hirschhorn, in the late 1870s still a boy in Frankfurt, was to be an important diamond magnate, though dwarfed in turn by his first cousins, the Oppenheimers.

Two brothers were fast learning the diamond trade. Leopold Albu had gone first from Berlin to South Africa. In 1876 he was joined by his younger brother George. The following year found them set up as diamond merchants at Kimberley. They expanded into claimholding, buying into the two Beaconsfield mines, Dutoitspan and Bultfontein, but the name Albu did not join the first league of mining magnates at the diamond-fields. Nor did the names of Lewis and Marks: they were losers at Kimberley. Isaac Lewis and Samuel Marks were cousins, and they were Jews, but unlike the Beits, Imroths, Lipperts, Oppenheimers, they came from far east of Germany, from the Lithuanian *shtetl* of Szegind-Neustadt.

---

* That was his South African soubriquet; 'Dr Jim' was a fancy of the British press.

In 1868 Sammy Marks went to South Africa and with help from the small local Jewish community started up as a travelling pedlar. He was joined by Lewis, his brother-in-law, and they set out for the diamond-fields. At first they were in business as general traders but soon moved into diamonds, as merchants and then claimholders. They built up some capital and used it to advance money.

Until the mid-1870s little European capital had been invested in Kimberley. Slowly the European diamond trade began to take a closer interest in the mines. Jules Porges was the leading diamond merchant in France, maybe in Europe. He had been born the son of a jeweller in Prague in 1838. Moving to Paris he very soon became a dominant figure in the diamond trade. When the Griqualand West fields were opened up he needed representatives in South Africa. The first was Julius Wernher, who had travelled out in 1871 on the same ship as David Harris. He was another German, born into 'an old and reputable Protestant family' in Darmstadt. He had served with the Prussian Dragoons during the Franco-Prussian war and he looked the part: a massive, heavily bearded man of great physical strength. Much by chance Wernher had been recommended to Porges who was looking for an assistant to go to South Africa. Wernher arrived in January 1871 when he was not yet twenty-one.

Within little more than a year he could write home that he had become indispensable to the firm which would one day grow into the greatest business in South Africa. Wernher rapidly built up the business and in 1876 persuaded Porges to visit the fields and to make large purchases in the Kimberley mine. These claims were consolidated with those of Lewis and Marks as the Compagnie Française des Mines de Diamant du Cap de Bonne Esperance, the French Company. It was capitalized at 14 million francs and by 1878 was paying 16 per cent dividends. The partnership between Porges and Wernher on one side, Lewis and Marks on the other, was uneasy and by 1881 the Lithuanian cousins had been squeezed out. Their defeat at Kimberley was not quite a blessing in disguise, but the cloud had a silver – or rather golden – lining. They were to be among the earliest successful magnates in the Transvaal.

Gradually Kimberley shed the features of a camp. Tents were replaced by corrugated iron huts and then by solid houses; but there were still few families. Groups of friends lodged together like undergraduates. Until 1880 Rhodes lived in digs with a group of young bachelors sometimes called the 'twelve apostles'. Then he moved to share a primitive cottage with Neville Pickering, the young secretary with whom he enjoyed an intense friendship. They lived opposite the Kimberley

Club which was opened in August 1881 to replace the informal canteens and smaller clubs and which soon became the centre of social life for many of the mining community.

In the same way in 1879 Julius Wernher shared lodgings with several men including one he described as a cheery, optimistic fellow of extraordinary goodness of heart and of very great business ability. By now they had become business partners. Beit had much prospered. His neat little figure was recognized everywhere in Kimberley, riding a large horse, and on one disastrous occasion a bicycle. As well as successfully dealing in stones Beit had bought more and more into De Beers, where the friendship with Rhodes begun in 1879 had come to fruition.

The distinction between diamond dealer and claim owner was increasingly blurred. So was the distinction between individual entrepreneur, partnership and limited company. This was the age when private ownership was being superseded by the joint-stock company system. South African mine owners were in the forefront of this development, using company formation as a device for raising capital. At the same time several of the very largest concerns in Kimberley and then on the Rand remained private firms until well into the twentieth century: among these the one which linked the names of Wernher and Beit was to be the greatest.

In 1880, Wernher went to London where he tied up the loose ends of their huge and ever-expanding business by organizing a diamond syndicate. In 1884 Porges and Wernher formed the London firm of J. Porges & Co. to deal in diamonds and diamond shares. For the next few critical years Beit was its sole representative in Kimberley, as well as playing his part in the French Company, and also in De Beers where he soon joined the board. The London Diamond Syndicate established under Wernher's guiding influence soon succeeded in stabilizing the price of diamonds.

Stabilization was one step. The next was to manipulate prices absolutely, that is to say upwards. To that end another link in the great chain of monopolization was needed. The Syndicate controlled final distribution; amalgamation of the mines was necessary to control output. The great diamond empires of Rhodes on one side and Beit and his partners on the other were within sight of amalgamation if they acted in concert.

The other team in an epic contest for control of the mines had taken shape. Barney Barnato progressed from koppie-walloper to ownership. In 1876 he bought six central claims in the Kimberley mine for the enormous sum of £30,000. Within two years years they were producing £1,800 a week, making Barnato one of the richest men, perhaps the richest man, on the fields.

He still liked to appear on the stage: in *The Orange Girl* in June 1876, two years later in the title role of *Othello* when in Kimberley of all places the line 'Haply, for I am black' produced pandemonious uproar. At the end of the evening Barnato would go to drink at the London Hotel owned by ''Arry', his brother, which afforded 'board and entertainment to men and beasts (beasts especially) at moderate prices and on strictly cash principles', 'the favourite abiding place of all the pleasant emigrants from Petticoat Lane and the fair but unfortunate land of Poland'.

In 1878 Barnato entered politics. He stood for the Kimberley municipal council and won a seat in an election influenced, the *Diamond News* said, by 'bribery in its most direct and objectionable form'. Barney Barnato was still part joke, but in other ways an increasingly formidable figure.

But J.B. Robinson seemed even more formidable and his political ambitions were more serious than Barnato's. In the 1879 elections he joined the municipal council. When the new council met, Barnato proposed Robinson for election as mayor by his fellow councillors. The election was irregular, was contested, and went to the High Court. There Robinson won and became mayor. In itself this was a portent. Among the proto-magnates there were bitter rivalries, but when it was them against the rest – the diggers, the largest population of the fields – they closed ranks. And so when after the absorption of Griqualand West four seats were awarded to it in the Cape Town legislative assembly, Rhodes, rather than fight Robinson publicly, took one of the seats for Barkly West, leaving Kimberley to J.B.R.

The Kimberley over which Robinson presided as mayor was now a solid municipality. A visitor like the American actor Stephen Massett touring South Africa and contributing a column to a newspaper back home, was struck by some resemblance to mining towns in California, but also by its orderliness. No visitor ever found Kimberley lovable, or even comfortable. It remained in many ways a sordid town. The rich could insulate themselves from dirt and discomfort. That was less easy for the poor whites who were now forming a distinct class. These were the original diggers, the losers. By 1880 it was reckoned that three quarters of the claimholders of a few years before were working as salaried overseers. For them life on the fields was less merry.

Life was least pleasant of all for the mass of black workers. On and off the compounds their conditions were grim. In 1879 the annual death-rate among whites was 40 per 1,000. Life was twice as dangerous for blacks at 79 per 1,000. That reflected the hazards of life on the mines where life was certainly cheap, and also the squalor of life generally at Kimberley and Beaconsfield. One of Barnato's first duties when he became councillor was to join the sanitary board. This was no sinecure. The board was indeed the most important arm of municipal government in a town with no sewers, where carts collected the nightsoil, where nevertheless cesspools were littered all over town, a hazard to the health of poor white and black alike. Another sanitary inspector had complained that the heaps of debris round the mines 'afford ample screens for Natives and I regret to say Europeans – Dutch – to render the place as pestilential as possible'.

It was this contrast between the sordid background of life on the fields and the glittering foreground occupied by the emerging millionaires which struck Trollope. He did not forget his visit. One of the characters in his last completed novel, *An Old Man's Love*, written on his return to England, is John Gordon. He has made his way to the diamond-fields. 'If there be a place on God's earth in which a man can thoroughly make or mar himself within a space of time, it is the town of Kimberley. I know no spot more odious in every way. . . . The white man himself is insolent, ill-dressed and ugly.' And yet, 'If a man be sharp and clever, and be able to guard what he gets, he will make a future there in two years more readily than elsewhere. John Gordon had gone out to Kimberley and had returned the owner of many shares in many mines.'

In real life several men had become large mine shareholders. But they were only a small fraction of the diggers who had flooded to the fields ten years before. For most of them the 1870s had been a time of struggle, excitement, hollow victory and ultimate disappointment. Their victory had been over the blacks at the mines. Seeds had been planted in the soil of South African history, which would grow and blossom later. The first real industry in the sub-continent had seen the first industrial colour bar, the first appearance of a white working class, and for the first time the systematic disciplining of a large black labour force by pass laws, searching and compounding. This the diggers had helped to bring about – for what?

A man who had known Kimberley in the 1870s cast a cynical eye back. The diggers had dreamed a dream.

The diamond-fields were to be the working man's paradise, where he should not be insulted by the presence of the capitalist or be jockeyed out of his property by the astute promoter, but should be free to follow his new vocation working – vicariously and by medium of the nigger – in a land where it should seem always afternoon, a land of equality where one man should be as soon as another and rather better, a land where drink should be reasonably plentiful, where the attention of the Police should be confined to niggers, and where tall hats and starched collars should be unknown.

That dream had blown up like a soap bubble, and burst like one. The reign of the capitalists was at hand.

5
The Ridge of Gold

For years there had been dreams of gold in the Transvaal, fascinating dreams for fortune-seekers, perplexing dreams for the Boers, enticing dreams for the British. The British had acquired the Cape not quite in a fit of absence of mind, as the familiar phrase goes, but without any purpose beyond establishing a base on the way to India. Until well into the last quarter of the nineteenth century the chief British concern in the sub-continent remained the naval base at Simonstown, across the Cape peninsula from Cape Town. The colony which was its hinterland expanded away from the sea, and the British were faced with a succession of negative decisions. They had no desire at all to conquer a greater South Africa, with all the problems that would bring. They had followed the trekkers inland so as to make sure that no Boer republic which could overshadow the British colony was born, and to make sure also that the Boers had no outlet to the sea – were trapped, impotent. It was for that reason that the British extended their sway along the Natal coast towards Zululand and to Portuguese Mozambique beyond: they wanted to keep other powers from this coast and from a potentially dangerous alliance with the Boers.

Thus the beginning of the diamond age had found South Africa a puzzling jumble of British colonies, Boer republics and independent African kingdoms. To unravel this puzzle became an object of British policy. Not out of any instinct of 'imperialism': that scarcely affected the calculations in Africa even in the late 1870s, nor even when the diamonds were found. The scramble for Africa had not yet begun, and to men of the age the question was not so much why it had not, as why it ever should.

The answer that regularly suggested itself to the South African conundrum was federation, a bringing together of all the parts of the jigsaw puzzle, the construction of a loose political unit which must crucially include the two Afrikaner republics. As long ago as 1858 Sir George Grey, Governor of the Cape, had started to feel his way towards federation. And in 1867 his successor Sir Philip Wodehouse presciently said that if the Dutch republics were to be absorbed it would be more easily done soon while they were weak than in the future when they might be much stronger.

In 1874 Gladstone and Kimberley were succeeded as Prime Minister and Colonial Secretary by Disraeli and Carnarvon. Carnarvon sent the historian J.A. Froude on two ostensibly private visits to the Cape, visits which had the meddlesome and ineffectual quality found when academics dabble in politics. He warmed to the Boers, and he too commended federation, urging it on the South Africans he met.

The trouble was that federation was an answer to an insoluble problem. The possibility of bringing together the parts of South Africa by consent was an act of faith. Even those who favoured federation recognized this some of the time. Federation was working in the two great settlement colonies of Canada and Australia, but to extrapolate from those examples was misleading. Even the French Canadians were more co-operative than the Cape Dutch, let alone the burghers of the two republics. No other colony, no other country, had the peculiar human composition of a small, self-confident settle community living among a much larger autochthonous population. The trouble with the South African jigsaw was that even when they were forced together the pieces did not fit.

The most awkward piece was the Transvaal. For decades it had stood aloof from the rest of South Africa, not to say from the rest of the world. The Transvaalers displayed an acute touchiness not only towards the British authorities, the 'English flag' from which they had trekked to escape, but towards any foreigners. In 1851 Thomas Baines travelled from Grahamstown to explore in the north. As a draughtsman and painter he has left the clearest visual images of the South African interior in the mid nineteenth century. But it was not only pretty landscapes he was looking for, or so the burghers thought, and they were probably right. At Mooi River Dorp, what is now Potchefstroom, Baines and his companion were turned back. He was followed by Marais, Davis, Wyley and Mauch, but for all their adventures they came away empty-handed.

The searchers who followed found what they were looking for: Edward Button, George Parsons, and James Sutherland who had some experience of gold-prospecting in California and Australia. Button found gold at Spitzkop in 1870 and at the end of the year the Volksraad passed a resolution which empowered the government to reward those who found precious minerals or stones. Then in the new year Button made a further find at Potgietersrus. The Natalia was the first gold mine in the Transvaal and 1871 saw the first gold rush. Both finds were on a small scale, but they were a beginning. Button became gold commissioner of the Republic. In February 1873 Thomas McLachlan and two partners made a larger find in the Lydenburg district.

These early finds of gold were near the eastern escarpment, where the great Transvaal plateau breaks and drops away to the low veld and the coast, a country of wild beauty. Diggers now poured into Delagoa Bay (later Lourenço Marques, now Maputo), the southern-most part of the Portuguese territory of Mozambique. Delagoa was the nearest port to the landlocked South African Republic, whose government saw it as the Transvaal's natural outlet to the sea. But until the railway was built, nearest was not very near. The 150 miles from the new gold-field to the port were not so far as the crow flew but they crossed a mountain range and then one of the roughest and most disease-ridden bushlands of Africa, a veld populated by savage animals, malarial mosquitoes and tsetse flies. Some diggers never made it.

Those who did pitched a mining camp. The gold-field had a more pronounced American flavour than Kimberley because so many of its inhabitants had travelled on from California. An American called MacDonald was prominent among them. Some months after the field was proclaimed it was visited by Thomas François Burgers who had become President of the Republic the year before. He was struck by the number of Scottish names among the miners and said that the camp should be called 'MacMac'. The name stuck. He granted the five hundred or so diggers the right to send two representatives to the Volksraad.

This was a fateful concession, running against the traditional principles of the Transvaal, but Burgers was desperate. The Republic's finances were now shakier than ever, its 'blueback' paper currency hopelessly devalued. He had arranged a loan of £60,000 from the Cape Commercial Bank and now saw a way of backing a new, solid currency. The following year, 1874, the first coins were minted from Transvaal gold. Two nuggets, the Emma of $16\frac{1}{2}$ ounces and the Adeline of 23 ounces, were among the total of 256 ounces of gold sent to be minted in London as 'Burgers pounds' with the republican coat of arms on the obverse and the President's head on the reverse.

The Transvaal government was now embarked on a delicate game. If gold-mining and gold-miners could be controlled, the riches they dug would give the Republic real independence. Crucial to this goal was a railway line to the coast at Delagoa, an escape from the encroaching British Empire. For this very reason the plan was opposed by the British government. The reforms which Burgers had attempted to introduce offended conservative Boers. African risings were a perennial challenge, not so much militarily as financially: the peripheral black chiefdoms never seriously threatened the South African Republic, but rebellions like Sekhukhune's in May 1876 were crippling to the impoverished state. The truth was that the Transvaal was not really a 'state' at all, scarcely more so than the neighbouring African kingdoms or for that matter the present-day countries of independent Africa. A few score thousand burghers ruled this vast territory only in the sense that no one else did. The burghers would fight and even die for their Republic, but not pay for it. From time to time African areas were 'pacified' by a violent Boer commando, or *chevauchée*, but they were not settled. Nor were the diggers easily absorbed into the Republic.

The consequences of the gold finds were not hard to see. Sir Garnet Wolseley reported to London in 1878 that 'The Transvaal is rich in minerals; gold has already been found in quantities, and there can be little doubt that larger and still more valuable gold-fields will sooner or later be discovered. Any such discovery would soon bring a large British population here.' The burghers rightly saw the diggers as a threat, and for their part the diggers were hostile to the Pretoria government, forming a 'defence committee' and keeping up vituperative criticism through their papers, the *Gold Fields Mercury* and the *Transvaal Argus*.

Before long the centre of activity on the Lydenburg fields shifted. In the valley of the Blyde River some way away from MacMac a digger struck gold. One of the diggers who followed there was Cecil Rhodes's brother Herbert, the Wykehamist cricketer, who had preceded Cecil to the diamond-fields and dug with him. Always restless, Herbert had gone back to the cotton farm in Natal, leaving his younger brother behind to make good. Herbert Rhodes came to Lydenburg with a gang of cronies from Pietermaritzburg in Natal. These fortune-seekers inappropriately called themselves the 'Pilgrims' and thus named the latest mining camp Pilgrim's Rest.

Herbert did not find his fortune there, any more than at Kimberley. Too easily bored, he lacked the application to stay in one place. Having made a faint mark at two mining towns he drifted on into Portuguese territory where he was arrested for trying to sell a gun. Then he went further north. Staying near Lake Nyasa he met with a bizarrely unpleasant end. While opening a keg of rum in his tent the keg burst and the spirit caught fire, wrapping him in flames. By the time a doctor arrived Herbert Rhodes had died of his burns.

A far more important arrival at the Lydenburg diggings was Alois Nellmapius, a young man of twenty-six when he reached South Africa in 1873. He was born in Budapest in 1847 of a Jewish mother and father who was baptized but who may also have been Jewish. Nellmapius made his way to Africa by way of Holland where he briefly attended a Jesuit college. At some stage also he studied as a civil engineer. Nellmapius's training gave him a great advantage: almost uniquely among the diggers he knew how to use dynamite. Early mines had been cut by hand and in appallingly laborious fashion: a five-mile drive on a mine in the Harz mountains had taken scores of years to cut. Then in Switzerland in the seventeenth century controlled gunpowder explosions had been used for the first time to make mountain roads. Gunpowder was inconvenient because of the quantities needed.* Nitro-glycerine had the power, but its instability made it impossibly dangerous.

In 1866 Alfred Nobel at last succeeded in making a stable mixture of nitro-glycerine and clay which he called dynamite. It was to transform mining and civil engineering, and warfare as well. Among the first people to learn this were black tribes in and around the Transvaal whose hopes of revolt were literally blasted away by the new invention. Blasting was used on the Kimberley mines but for comparatively simple open-cast mining, blowing loose slabs of friable blue ground. At Pilgrim's Rest Nellmapius skilfully blasted rocks at his claims. Alluvial sluicing, another digger said, had been a lottery. Nellmapius was the first to sluice ground in large quantities and in a systematic manner, employing twenty white men and two hundred blacks in regular day and night shifts. This was the beginning of a career. From then on Nellmapius's genius was not to be limited to mining. He saw that if the mines succeeded at all they would transform the archaic, pastoral republic and would lead to – and need – a large network of services.

Although there would be a railway one day, what was needed immediately was a road. Early in 1875 Nellmapius persuaded Burgers and the Volksraad to approve a scheme for a daily transport service between diggings and port. He was granted a farm every fifteen miles along the route to rest men and animals, a land grant of 3,000 morgen† in all. The Lourenço Marques and South African Republic

---

*   In 1855 when Pot Rock in New York Harbour was blown up it took ninety tons of powder.
†   Traditional Rhenish measure of land adopted by the Boers = 2.11 acres.

Transport Company was registered, a track was cut and on 12
February 1876 the first goods were delivered to the diggings from
Delagoa Bay. At first ox-drawn wagons were used, but oxen were
susceptible to the tsetse fly. Before long Nellmapius had replaced
ox-wagons with black carriers, 'but one of the spirited enterprises of
which he was the originator'.

There were to be many more. Nellmapius's cordial relations with
the Republic government bore all kinds of strange fruit. In 1878 he
had persuaded Pretoria to allow the opening of the first synagogue at
Pilgrim's Rest. By 1881 his friendship with the Republican government meant that he could originate another enterprise which was not
only spirited but spirituous, the sole right to distil liquor in the
Republic in return for an annual royalty of £1,000. In the same year
he was granted a monopoly for the manufacture of gunpowder. But
this was already an obsolescent market. What the miners needed in
large and ever-increasing quantities was dynamite: whoever secured
that monopoly would have a licence to print money.

Before Nellmapius's next coup the Transvaal was turned upside
down. Sir Bartle Frere was sent to the Cape as Governor and High
Commissioner to prepare for federation, while in late 1876 the
Governor of the neighbouring colony of Natal, Sir Theophilus
Shepstone, was instructed to look into the affairs of the Transvaal
with a view to annexation. Evidence of internal disarray and collapse
was not far to seek, and in April 1877 the Transvaal was annexed in
the name of Queen Victoria and the British Empire.

If the British thought they could absorb the Republic so easily they
were soon proved wrong. Annexation was a short-lived and unhappy
episode. Burgers's political career abruptly ended and a new mood of
Boer militancy and national obduracy began, embodied in the person
of Paul Kruger. He visited London to argue the burghers' case against
annexation. He also, significantly, visited Holland and Berlin. The
British had sent Owen Lanyon as acting administrator to the
Transvaal, where he achieved the difficult feat of making himself more
unpopular than he had been at the diamond-fields.*

Then in 1879 the British picked a quarrel with the independent
kingdom of Cetshwayo and invaded Zululand. The campaign had
profound and unintended consequences for the Transvaal. First the
British were defeated. A British regiment was destroyed by Zulus at
Isandhlwana. With it died for the Boers, among others, the legend of

---

* Among both diggers and burghers his swarthy complexion and the fact that he had spent
some time in Jamaica led to rumours that he was less than really white; he was in fact born in
Belfast of English and Irish parents.

British invincibility. Six months later the Zulus, in turn, were finally crushed, removing the threat to the flank of the Transvaal which had always tempered the Boers' ambitions against the British. The two battles could not have been better designed to stimulate republican ambition.

The following year the Liberals returned to power in England. Kruger assumed that Gladstone would annul annexation. He did not. Instead Lanyon stepped up his efforts to make the burghers pay their taxes, a task which had defeated the Republican government. At the end of 1880, under the triumvirate of Kruger, Joubert and Pretorius, the Transvaalers rose to throw off British rule. The British were defeated at Majuba and in despair the British government retroceded independence to the Transvaal. In the year after this 'First War of Independence' the burghers elected Paul Kruger President. And so this strange, obstinate and cunning – *slim* is the Afrikaans word – man became ruler of his country for its last two decades of independent existence, years during which he saw it transformed, subverted and overthrown.

The stories of diamonds in Griqualand West and gold in the Transvaal have several uncanny parallels. In both places there was, for a long time, a half-conscious instinct that the fields would be found. In both there was a wait between first discoveries and the finding of the field. In both there was a subsequent episode of setback – blue ground, pyritic ore – following by breakthrough. In both, once the finds had been made, there was an ever-increasing tendency both for financial and technological reasons to drive out the small individual digger. But then there are differences of time-scale. The movement towards monopoly took far less time in Johannesburg than it had in Kimberley, partly from the very fact of Kimberley's having been first. By contrast, the interval between the first finds and the great finds was much longer in the Transvaal than at Kimberley: four years between the diamonds found on the bank of the Vaal and the mining of the kimberlite pipes, and fifteen years between the first east Transvaal finds and the striking of the Main Reef on the Witwatersrand.

The Rand was there, waiting to be found, but men seemed to hold back from finding it. That seems fanciful. But the history of failures to find gold is an extraordinary one: the story of Karl Mauch walking across the Main Reef; the story of A. W. Armfield, appointed Inspector of Gold-Fields under the brief British administration of the Transvaal who in 1878 prospected right along the Witwatersrand and reported that there was only a slight indication of gold; the story of

Gardner Williams of De Beers, one of the greatest mining engineers of his age, declaring in 1886 that the Rand reefs were worthless.

In the early 1880s the next find was still a long way away from the Rand. As early as 1874 Thomas McLachlan, the first at Lydenburg, told Burgers that the Kaap valley was a promising gold-field but the great discovery had to wait another eight years. The diggers moved from Duivel's Kantoor, then to the Kaap valley, and finally in the south of the district was the strike at Barberton. Previous gold finds had flattered to deceive. This was the real thing at last. It came in the same year as the London Convention. The diplomatic recognition of the South African Republic's – albeit conditional – independence coincided with the first loud tickings of the time bomb within the Republic. When the Sheba Reef was hit, the Transvaal looked to have found a gold-field ranking with California and Australia. There was surface gold in nuggets but there was also a quartz vein of exceptionally rich quality; this became known as Bray's Golden Quarry after the digger Edwin Bray. The Quarry returned 5,000 ounces of gold from 13,000 tons of ore. Before long the Sheba Company was paying shareholders a 170 per cent dividend, and £1 shares had gone up to £105.

Until then the miners in the east Transvaal had been true diggers, individual fortune-seekers staking their claims and working them just as had been done on the diamond-fields in the early 1870s. Now to Barberton as before to Griqualand West came a new immigration of entrepreneurs. One was Edouard Lippert, son of a Jewish businessman from Mecklenburg who had interests in South Africa and who was married to Alfred Beit's mother's sister. Young Edouard (or Edward as he later called himself) had hoped to become a musician, but decided to follow his cousin and exact contemporary Alfred to South Africa along with his brothers Ludwig and Wilhelm. Edward walked from Delagoa to the Barberton Fields.

The two men shared a house until they had a row over a business deal. Lippert moved from Barberton to Kimberley where he joined Beit in the Porges firm. Then he returned to the Transvaal, to Pretoria and a career that was quite different from that of most of the other mining magnates.

Following Lippert to Barberton came Abe Bailey. He was South-African-born, from Cradock in the East Cape, of Yorkshire descent on one side, Scottish on the other. He went home and worked in England briefly and then came back to South Africa in his late teens to run his father's shop. Bailey (christened Abraham but never known as anything but Abe) was twenty-one when he too felt the lure of gold and set off for Barberton. There he became manager of the Transvaal Gold Exploration Company.

There were others who had been briefly, or just obscurely, on the diamond-fields before coming on to Barberton. Still another German Jew, Sigismund (or sometimes Sigmund) Neumann, had gone to Kimberley in the mid 1870s and worked as a diamond merchant with the firm of V.A. & E.M. Littkie. He moved on to the Rand where he became a powerful gold-mine financier.

So too did Hermann Eckstein. Rather older than Beit, he was born near Stuttgart in south-west Germany in 1847, the son of a Lutheran pastor. He made his way to Kimberley and became a mine manager, running the Phoenix Diamond Mining Company in Dutoitspan. At the fields he became a close friend of Beit's: they spent their evenings together in what was known as the Old German Mess. And so when Wernher and Beit wanted to send a representative to Barberton they chose Eckstein. He managed their company, De Kaap Goldfields, but it was not a success.

Another who went straight to the new gold-fields was destined to become a diamond rather than a gold magnate. Thomas Major Cullinan was born at Elandsport in the Cape, son of a settler from County Clare. He had served in the Ninth Frontier War when still only in his mid teens, and was little more than twenty-one when he came to Barberton in 1884. There and in Johannesburg, he was as yet little known but he was another whose name would become in the true sense world famous.

Friends and enemies used to call James Percy FitzPatrick an Irishman. So he was by descent in the same sense as Cullinan. His father was a Tipperary man, one of the first generation of Irish Catholics to take advantage of emancipation and make a career in the professions or in the service of the British Empire. Percy was born at King William's Town in the East Cape in 1862. He went to school in England, to Downside, and then in Grahamstown, before joining a bank in Cape Town, listening to stories of diamonds and gold in the north. In 1884 he trekked to Barberton. His first job was not in the mines themselves but as a transport rider on the new road. He stored up the memories of those days and would tell stories to his children. More than twenty years later he turned bedtime stories into a book. Several of the Randlords wrote books; none ever wrote a book as famous or successful as FitzPatrick's *Jock of the Bushveld*, the story of the veld, of hunting kudu and rietbuck, of snakes and baboons, of a man and his dog. One reader, Theodore Roosevelt, spoke for many when he called it the best of all dog books.*

The 'last trip' from Delagoa described in *Jock* was a disaster for FitzPatrick who lost almost all his oxen from disease. He reached Barberton, where he was taken on to work for two partners, Cohen and Graumann, at £15 a month. They owned among other interests an hotel, which FitzPatrick managed, and a weekly paper, the *Barberton Herald*, on which he worked. This association sealed FitzPatrick's future. The partners' most important business activity was stockjobbing. In fact dealing in shares was almost the principal activity in Barberton, one of the ways in which it connected Kimberley with Johannesburg. Shares fluctuated wildly, from a few shillings up to ten pounds and then down again. Although the occurrence of good pockets was quite genuine, there was simply not enough gold to justify the existence of a company with a big capital.

Many such companies were floated, all the same. Barberton saw not only the first serious gold mining in South Africa but, just as significant, the first great gold share boom. Capitalist methods, in the strict sense, had grown up at Kimberley slowly along with the diamond mines. When European capital came to the diamond-fields there was a formal structure of company promotion and share-dealing which culminated in the great amalgamation. But in any case, the capitalist gambling instinct was gratified at Kimberley by simply dealing in the stones themselves, as their prices rose and fell: commodity broking. There could be no commodity broking in gold, a product whose price was fixed. And so the opportunity for investment, speculation, arbitrage, had to come through shares in mining companies.

By late 1885 more than a hundred different shares were quoted on the Barberton stock exchange, with the market dealing in a total of more than four million shares. The exchange made up in exuberance for what it lacked in financial sophistication. Sometimes it stayed open until midnight. The commissions it landed now came from London and Paris. For the first time South African shares caught the public imagination in Europe. Shares shot up, higher and higher. The Sheba mine itself reached £100. At least this was a real mine producing real gold, which was more than could be said for many of the other Barberton shares. It was a great boom, and a great bubble.

And a great burst when it came. A handful of mines went on

---

* The acerbic young William Plomer remarked that Jock loved the veld, had soulful eyes and bit niggers – 'a true South African'.

producing gold and their shares paid good dividends. Most did neither. Thousands of investors in England, France and the Cape were left with nothing but richly decorated, worthless share certificates. The Barberton adventure did nothing to build public confidence in South African mines and one result was that when the really great gold find was made there was a reluctance to back it. Subsequently the promotion of mining companies was more careful and more skilful. Or it may be that the promoters, some of them, merely grew more adroit and learned ways of removing the investing public's money steadily and painlessly.

The great scare had an unintended effect. There was now indigenous South African capital in Kimberley itself. Several of the leading diamond magnates investigated the new gold-field. With the advantages of proximity and of expert advice they were much more adroit than European backers. Barnato lost a lot of money at Pilgrim's Rest and was chary of new gold-fields. Rhodes was busy with his schemes in the north. Both were limbering up for the amalgamation struggle in Kimberley. Robinson had no money to invest.

Beit's investment in Barberton was especially cunning. Around 1880 he had hired J.B. Taylor as a diamond broker, the younger of two sons of an early digger who had played some part on the old mining board at De Beers. The two brothers, James Benjamin and William Peter, both made careers on the diamond-field. J.B. followed his brother to London when they tried to catch on to the first enthusiasm for diamond shares, publishing a financial paper, *The South African*, with Beit's backing. Returning to Kimberley in 1882 to find that the bottom had fallen out of the market, J.B. moved on to the East Transvaal, where he made friends with FitzPatrick and learnt something of mining engineering from Gardner Williams who was then at Pilgrim's Rest managing the Transvaal Gold Company.

Early on in the De Kaap rush Beit inspected the gold diggings accompanied by Jules Porges, who was paying another visit to South Africa, and Sigismund Neumann. 'Beit's energy when inspecting shafts and drives was really astonishing. From early dawn until dark he would ride and walk over the rougher country without showing signs of fatigue.' That was one of Beit's advantages: though a neurotic, he was no valetudinarian. He had chosen the right man in Taylor who closely investigated the mushrooming companies and the mines which they supposedly controlled. There were, Taylor realized, one or two solid properties – the Sheba, the Oriental – but little else. He impressed this on Beit who dropped Barberton like a hot brick. He lost money, of course, but being first out he lost far less heavily than others.

The East Transvaal boom collapsed and Barberton was soon little more than a ghost town. A few worthwhile properties continued. One was the Sheba, which was now securely in the hands of Sammy Marks. He had become a considerable personage in the South African Republic, a friend of Kruger's with fingers in a good many pies. Among his properties were a farm on the north bank of the Vaal, near what was to become Vereeniging. J.B. Taylor was invited to shoot there with his brother and three others. He went by coach from Barberton to Pretoria and then headed south by ox-wagon, a route which ran straight across the long ridge of the Witwatersrand.

Near the farm Langlaagte they were caught by a thunderstorm and outspanned. After the storm, Taylor and the others walked along the outcrop of a reef running east and west. The nature of the reefs struck him as being very unusual and he thought they might be worth sampling for gold. He knocked off three or four chunks with a view to crushing and panning them on his return to Pretoria. But when they eventually returned to Pretoria he found that his black 'boy' had been using the rock sample to support his pots and kettles over the open fire on which he cooked and had left them behind.

A few prospectors were working on the Reef but did not understand the conglomerate. They were looking for alluvial gold – many of the little streams which gave the Ridge of White Water its name were panned – or for rich, visible lodes as at Sheba.

Then in late 1885 gold was found on the farm Kromdraai on the west Rand and a public digging was proclaimed. Further east several prospectors were working on the central heights of the ridge. On 8 June 1885 the Pretoria newspaper *De Volkstem* reviewed the prolonged and unsuccessful search for gold and went on, 'Today the dark mantle of doubt has been removed and there is now a feeling of security in the certainty of the bright change that must take place here soon.'

Two brothers, Harry and Fred Struben, were digging on the farm Sterkfontein where some loose banket pebbles led him to think there were gravel beds or drifts in the neighbourhood that might be auriferous. Though no more than a self-taught geologist, Fred Struben guessed that the whole country had once been submerged and that the conglomerate beds might possibly carry gold as they did elsewhere. In 1885 Harry Struben organized a prospective syndicate in Natal to take over the various properties where the brothers had acquired mineral rights. The syndicate sent a representative north but he – another pessimist – returned in three months 'saying there was more gold to be found in the streets of Pretoria than on the whole Rand.

Now the Strubens ordered a five-stamp mill from England. They had grasped the fact that gold was not going to be found in nuggets or in alluvial deposits which could be panned, but embedded as faint grains in hard rock, which would have to be milled or crushed before the gold was extracted. But the first large-scale crushings were a disappointment. They were a long way from 'payable gold'.

Not far away was the farm, Langlaagte where Taylor had been caught in a thunderstorm, which was divided into four portions. On Portion D of this farm, owned by Petronella Francina, the widow Oosthuizen, at some time during April, the outcrop of the Main Reef was discovered. Quite by whom or just when is not known for sure, and matters not. At the time whoever located the Reef was far from boasting of it in public; instead there was a discreet but vigorous jostling for position. Mining rights for the different portions of Langlaagte were quickly granted.

The Gold Law of the Transvaal laid down that all rights of mining for and disposing of all precious metals and precious stones belonged to the State. Privately owned land could be proclaimed and thrown open. The Government did not have the right to throw open land without the permission of its owner. But the financial inducements were such that no owner of a farm – a backward, upland grazing farm – was going to refuse: he received half the revenue from licences paid by claimholders. The discoverer of a gold find was entitled to select six claims on the property.

By comparison with the Kimberley claim of 31 foot square, the Transvaal gold claim was 150 by 400 Cape feet, the shorter side along the strike or line of the reef. The Gold Law gave the owner the right to mark off one-tenth of his land as a *mynpacht*, a mining lease which was his own. He could mine it himself, or lease it, or sell it. A tenth is only a tenth. But as the Reef runs, a tenth of a large farm was seemingly all that was needed to exploit its gold. The *mynpacht* could be and was staked along the line of the outcrop. In the event this was far from the real truth about the nature of the Rand. It would not be long before miners realized that the Reef ran up from the bowels of the earth at an angle, or from the outcrop down and away. Stand on the Reef, facing away from the angle at which it broke surface, walk away for so many hundred feet: if a shaft was then struck straight down it would eventually meet the Reef once more.

Even so, the *mynpacht* system favoured the landowner. In the first place that meant the Boer, the farm owner; but really it meant the entrepreneur with capital. The mining policy of the Transvaal government since it regained independence had consciously favoured large capital, a policy reinforced by unintended effects.

Again in a curious resemblance to Kimberley, there was to be a broad pattern of causes – legal, social, technological and financial – which would spell doom to any chance of the individual digger holding his own. The process of takeover and amalgamation took place more quickly on the Rand than at Kimberley, partly because gold is less amenable than diamonds to individual mining. That was true of any gold-field, *a fortiori* on the Rand. A Californian miner might get rich quickly with large nuggets, or even several rich pans from the river. On the Reef huge quantities of ore had to be dug and processed to get gold. And the rule of large capitalists was also hastened because the process of capital formation had already taken place. The Witwatersrand gold rush was dominated not by a crowd of men who wished to make their fortunes but by a few who had already made them: the Rand imported its capitalists ready-made from Kimberley.

First among them was J.B. Robinson. By his own account he was sitting one day in 1886 in the Kimberley Club when he received a telegram from an associate called Evans telling him of the gold finds and urging him to come and investigate them. At that moment Robinson was a defeated man. He had not only been the chief loser in the De Beers amalgamation now under way, largely through his own obstinacy and vindictiveness, but more than that, he was horribly in debt, the Cape of Good Hope Bank holding £100,000 in nominal share values in two of his companies as collateral against loans. Financial collapse was not very far away. It was a bad moment even if there was no greater problem than retrieving his situation in Kimberley; seemingly a hopeless one for a new adventure in gold.

He was rescued and staked for the Rand by Alfred Beit who had little more reason to trust Robinson or wish him well than anyone else in Kimberley. Robinson went to see Beit and explained his troubles. With another man it might be described as throwing himself on Beit's mercy, but even in adversity Robinson was prickly and aggressive. Beit looked at Robinson's figures and saw that his shareholding ought eventually to see him through. He agreed to take over Robinson's position from the bank, and offered the collateral shares by telegram to Porges, who refused them; and so Beit took them on his private account. Beit further advanced him £20,000 for investigating and investing in the Transvaal. He never received any public thanks from

Robinson, who ever after bragged of being the first man on the Rand.
When he heard of J.B.R.'s bombastic self-advertisement Beit would
laugh.

He could afford to. In his complicated personality, mixed with
neuroticism, with extreme financial acumen, with acute shyness and
social ineptitude, there was certainly great kindness. Beit's generosity
to people in distress was no doubt exaggerated by his hagiographers
but it was real enough. He was always a soft touch for anyone with a
hard luck story. A little girl came to his office in 1880 with a note
which Beit read while, as so often, biting his handkerchief. Her
mother had written to say that she had been widowed and needed the
money to start a small shop. The girl went away with a cheque for
£250.

J.B.R. was not such a deserving case and Beit was acting in the
tradition of enlightened self-interest. Although Robinson was acutely
short of cash he had real assets in diamond-mining companies, and in
the struggle that was now under way in Kimberley these assets were
invaluable for Beit. As for the second loan, by initiating what became
the Robinson Syndicate Beit ensured that, having become the key
financial figure in Kimberley, and having avoided burning his fingers
too badly at Barberton, he would be in on the ground floor on the
Witwatersrand. As Sam Goldwyn might have said, a vague promise
from J.B.R. was not worth the paper it was written on. Merriman had
had just such an oral promise that he would be kept in touch on all
developments on the Rand with a view to investment. Robinson
managed to forget about Merriman. Beit on the other hand had an
undertaking which there was no getting out of.

Financially fortified, Robinson left for the north. He was not the
only traveller. Hans Sauer had received a letter from a medical
colleague who had recently moved from Kimberley to Pretoria, telling
him about the gold-fields and sending a pound or two of ore from the
Rand. Sauer decided to see for himself. He took the coach for
Pretoria ; in it he found Robinson. They both pretended to be going to
Pretoria but met again on the hotel stoop at Potchefstroom the
following morning, after the coach had left, and mutually confessed
their true destination. They were two stubborn men – both of them
were enthusiastic wielders of the horsewhip against their detractors – but they got on easily enough.

It was Robinson who struck first when they arrived. He took
lodgings with Petronella Oosthuizen, the widow, at Langlaagte, and
started to look around him. Within a week he was buying: Portion B
of the farm for £800, Portion D from the widow and her kinsman for a

rental of £450 a year together with an option to buy outright for
£12,000. He had the resources thanks to Beit, and he spoke the Taal,
Afrikaans, fluently. It was said at the time that he had gone further in
using his charm and that he had seduced the widow Oosthuizen
before buying her farm. It was the kind of rumour which always
gathered about J.B.R.

He bought heavily along the reef and proclamation of the field on
20 September found Robinson sitting prettiest of all. To the east of
Langlaagte he had a *mynpacht* on the farm Turffontein, which was to
be the central and eponymous mine of the Robinson Gold Mining
Company, and to the west more property: a total of seven farms and
more than 40,000 acres. Any cordiality between Robinson and Sauer
was short-lived. Writing more than fifty years later Sauer said that the
purchase from the widow 'cannot exactly be called generous'. But
Sauer was, if not embittered, certainly vexed by his own setbacks. On
his return to Kimberley he went to see Rhodes, who, not early in the
morning, was still in bed. After inspecting the samples, Rhodes and
Rudd advanced Sauer money to return to the Transvaal. Catching the
coach outside town for discretion's sake, Sauer found his two new
confederates themselves aboard.

Their visit to the Rand was not happy. Rudd used a method of his
own for treating ore, which produced little gold. Then he persuaded
himself that the main reef leader was a delusion, or even a hoax,
'salted' as so many claims had been salted with diamonds by
fraudsmen in the early days at Kimberley.

Sauer had arranged an option on some rich claims, but Rudd could
not be persuaded to take them up. They were bought instead for £750
by Beit who offered half at cost to Rhodes. Rhodes refused. The claims
were incorporated into the Robinson Gold Mining Company whose
shares were to rise from £1 to £80, and whose capital was to be
increased from £50,000 to £2.75 million. Sauer's acerbity about 'the
almost unbelievable stupidity of Rudd' is understandable. But it was
not Rudd's stupidity, or caution, alone. Rhodes was haunted by the
fear that the whole thing might turn out to be a 'frost'; he found gold
reefs disconcerting and hard to value after Kimberley where he knew
exactly how much a section of blue ground was worth; he was
unhappy on the Rand, which was still wild country, and on one
occasion he who had a morbid horror of snakes was nearly bitten by
one; and in the middle of his tentative exploration he rushed back to
be at the death bed of his beloved young friend Neville Pickering.

Despite his misgivings, Rhodes bought half of the farm Roodeport
and some other properties, some of which were worthless, but he had

the basis for a gold-mining business. In any case Rhodes had practical as well as sentimental reasons for returning to Kimberley. His battle with Barnato was now at its full fury. Paradoxically, Rhodes, whose wealth was enormously greater and more solid than Robinson's at this point, was strapped for capital with the struggle to buy rocketing diamond shares and had less time to spare than Robinson who had been the loser on the diamond-fields. Rhodes's caution was backed by expert advice. Gardner Williams had stopped at the Rand in 1886 on his way from Kimberley to Barberton where he was to report on a new gold find for Lord Rothschild. He was shown over the outcrop by Sauer who asked him what he thought as Williams was finally leaving. 'He looked at me for the moment and slowly said; "Doctor Sauer, if I rode over those reefs in America I would not get off my horse to look at them. In my opinion they are not worth hell room."' It was an opinion shared by other expert mining engineers; and in the light of past experience not without reason. The outcrops were certainly gold-bearing, but weak in quality. Most people supposed they could only be mined for quite a short distance downwards. One Australian mining engineer thought it geologically impossible that the reefs could run deeper than two hundred feet. If so, then there was gold but not very much, and it would not be long before it was exhausted.

Within four years these forebodings seemed to be fulfilled. A bonanza, an unparalleled eldorado on the Rand had come, and, so it seemed, gone. But once more appearances were deceptive. Momentous changes were now in motion. Men would soon flock to the Reef as they had to Kimberley. Wolseley had foreseen this. At the end of his despatch he added, 'The time must eventually arrive when the Boer will be in a small minority' in South Africa. For that reason he had advocated taking over the Transvaal. That episode had been unhappy; Wolseley could not have foreseen that it would be tried again and again, by imperial authority working hand in glove with a group of financiers who even as the Rand was discovered were preparing their base at Kimberley by means of monopoly.

6
Bubble and Amalgamation

The early partnership between Barnato and Louis Cohen had ended acrimoniously although the two were later reconciled. When Cohen's play *The Land of Diamonds* was put on in Kimberley Barnato took 'the Jewish part, Mike Jacobs, and right well he acted', the author said. Then Cohen left Kimberley empty-handed to go north to the Transvaal, visiting the diggings at Pilgrim's Rest, and then working as manager of a theatrical troupe. He even served for a spell with the British forces on their way to fight the Basuto but his military career ended ingloriously when he broke a rib falling from a Cape cart, and he returned to Kimberley in 1881. In that turbulent year President Garfield of the United States was assassinated, and so was Tsar Alexander II, his death precipitating the great wave of pogroms which drove so many Jews to flee from the Russian Empire, some of them to the Cape. South Africa saw the upheaval of Boer victory and of retrocession.

And Kimberley experienced a year such as it had never known before. On his return Cohen found the place greatly changed. The old-time digger, the farmer-digger, the gentleman-digger, had almost disappeared.

> In their place had sprung up a mushroom breed of financiers who were destined in the near future to put their hands deep in the pockets of the British public, and form the trade of a brood of costers and aliens whose business methods later on made South African company promoting a vehicle for wholesale plunder and chicanery. In 1881, that Bubble year . . .

There was poetic licence in Cohen's description and he suffered from the distorting effect by which absence makes changes seem more dramatic than they are. The individual digger had in fact been all but an extinct species for some time. But 1881 certainly was a bubble year, and a new breed of financiers had certainly sprung up. If the old-time digger enjoyed a St Martin's summer in the 1880s it was on the east Transvaal gold diggings. While Barberton had its heyday, another story was unfolding in parallel on the diamond-fields. By the end of the decade, gold-diggers from the East Transvaal and diamond magnates from Kimberley would converge on the Witwatersrand. Meantime the great issue of who controlled the diamond-mines had to be resolved.

In 1880 Barney and Harry Barnato had spent six months in London where they had floated Barnato Brothers. They returned early in December. In March they floated four new mining companies, and were soon enjoying a remarkable success. In July they published a quarterly report for the first time, claiming to have mined 10,328¼ carats in three months, worth £17,478 6s 3d. A dividend of 9% for the quarter was paid. Even Barney Barnato's many critics were silenced. The fact that Barnatos were paying a dividend at all distinguished them from many or most of the newly-floated companies. The costers had learnt a new trick: that company promotion and share pushing might be almost completely unrelated to the mining which was supposed to be their ultimate object. For some time in 1881 it was reported that mining operations were all but suspended and the sole topic was the share market. At the beginning of the year there were twelve public companies on the fields. Within six months there were seventy-one, in which more than £8 million had been 'invested'. Even when this great bubble of company promotion was not actually fraudulent it was conducted on unsound lines. When the inevitable crash came its effects were dramatic. The shares of a market leader like De Beers fell by more than half, from £45 to £20. The broader effect was to wipe out still more small men and to clear the field further for the giants.

Of Barnato's rivals, Rhodes had other preoccupations in 1881. That year he entered the legislative assembly in Cape Town as the Member for Barkly West. His constituents were almost all Dutch-speaking graziers whom Rhodes ingratiatingly told, 'My ancestors were keepers of cows'; it is one of the marks of his character, a sign of the loyalty he could inspire, that these Boers still returned him up to the end of his life, even after the Jameson Raid. In the parliamentary chamber he was a conspicuous and curious figure, ignoring the rules of procedure, calling others by name (rather than 'the honourable Member for . . .'), wearing tweeds contrary to convention, and speaking erratically in his squeaky voice.

His political ambitions were still unformed, but he already showed ambition beyond the interests of the mining industry, which was what the honourable Member for Kimberley, J.B. Robinson, who entered parliament at the same time as Rhodes, confined himself to. J.B.R. supported the government on the issues of the day, disarming the Basuto and providing a regular supply of black labour. The mines, he said, were good for the blacks. 'One of the first lessons to be instilled into them will be respect for the laws of *meum* and *tuum*. The Diamond Fields have, in my opinion,' – and surely he was right – 'done much to accomplish this.' When Charles Rudd, the 'Member for De Beers', joined the Assembly soon afterwards he, too, spoke even more openly on the need for control of black labour.

In October 1881 Rhodes returned to Oxford for his final term, the only undergraduate then or ever with an income of £50,000 a year who was also an elected legislator. With Oxford behind him Rhodes had one ball fewer in the air to worry about but he soon picked up others. His career was expanding inexorably.

For J.B. Robinson by contrast the tide was beginning to ebb. In September, ever litigious, he brought a libel action against the *Standard and Diggers' News* and was awarded derisory damages of fifty shillings. His fellow citizens' enthusiasm for J.B.R. had worn thin. He was still very strongly placed and even now could have had a successful part to play in the coming struggle for power if it had not been for his pathological obstinacy and inability to work with others.

By 1883 there were no more than forty-six companies and fifty-six private holdings on the four mines,* still a good number but a long way from the hundreds of claims which had existed not many years before. A comparison between maps of the Kimberley mine in 1877 and six years later tells the story. There are still plenty of different companies but no longer the Burgundian confusion of 1877. The Central held the dominant position which its name suggests. It had reached that eminence much less through the activity of its head, Francis Baring-Gould, than that of far the ablest and most dynamic member of its board who was, needless to say, Alfred Beit. Inability to work with others was never Alfred Beit's problem. With all his neuroticism he was the most subtle and persuasive of operators. Even so, as he sat in the Old German Mess with Wernher, or drinking champagne and stout with Rhodes at the Blue Post, few realised how powerful he was.

Diverse ownership led to ever more acute problems as shafts were sunk for the first time to mine underground. The shafts and underground drives of one company could be blocked by an antagonistic company. As Barnato was to put it, once the amalgamation had been achieved:

*   Kimberley: 11 companies, 8 private; De Beers: 7, 3; Dutoitspan: 20, 21; Bultfontein: 8, 24.

To work on the underground system you must have a mine intact. For instance, I do not think the underground workings of De Beers Mining Company, before it became De Beers Consolidated, were a success. The Victoria Company caved in, why? And the Oriental as well, and why? Because there was one company working against the other, and if one was on a level of 500 feet, the other perhaps was on a level of 450 or 400 feet.

Another reason for combination was mine management in the broadest sense. The economic control of black labour was not as important at Kimberley as it would be on the Rand and although wages were often cut the diamond-mines were always to pay more than the gold mines. Nevertheless, the recruitment and proper disciplining of blacks were important to Kimberley. The miners were more concerned than ever about the sapping of their profits by IDB and looked to the government and the legislative assembly in Cape Town for assistance. The diamond trade was already surrounded by a host of regulations and licensing systems. Since 1874 diamonds could be bought legally only by licensed dealers and brokers. This did little to stop IDB. Some reckoned that it was costing the mines £3 million a year although this figure – almost twice as much as licit production – seems far-fetched.

The opportunity for theft was as great or greater outside as inside the mines. Blue ground was brought up to the depositing and washing sites where it was left to decompose in the sunlight. Then each day a portion of the now friable rock was taken by labourers and broken up by other labourers who had easier pickings at this stage than had those at the bottom of the mine. 'The Kaffirs are the great thieves', Trollope said, or was told, and it seemed that no punishment deterred black labourers, neither long spells in gaol nor ferocious floggings even though the rewards were small: the labourer who had first secreted the stone would only receive a tiny part of its worth from IDB.

By 1880 a special court had been set up to deal with IDB. All diamonds leaving the fields had to be sent in registered parcels. Then the Diamond Trade Act of 1882 entirely altered the balance of the law by placing the burden of proof on anyone found with a diamond he could not account for. As Lord Randolph Churchill remarked when he visited South Africa a few years later, the ordinary presumption of law in favour of the accused person disappeared, and an accused person had to prove his innocence in the clearest manner, instead of the accuser having to prove his guilt. Abandonment of the principles of English law went further. A 'trapping' system was introduced, the use, that is, of *agents provocateurs*. This drastic step was disliked, but it was defended by both Rhodes and Merriman. Up to the end of 1885, 19,272 carats of illicit diamonds were recovered. Merriman still complained that, 'Words cannot depict the flourishing state of the Illicit Diamond Buying trade and the consequent misery of the honest shareholder'.

If no further action could be taken against dishonest buyers, the illicit trade could be attacked again from the other end, by controlling the mine labourers. In 1882 B.V. Shaw of the Kimberley Detective Department made a close study of IDB. He noted that at every stage of the proceedings, until the soil was finally deposited in the washing machine, 'the facility for theft is very great'. He pondered an answer. Despite attempts to discipline and compound blacks the conditions in the mines in the 1870s – single diggers or partnerships working blocks of claims, blacks going casually to and fro – had not made systematic searching easy and the right had been exercised mostly to search suspicious individuals in exemplary fashion. As the mines expanded in scale, supervision of workers became more thorough. The searching of all workers was first recommended in 1880, and was recommended more strongly by Shaw, who concluded, 'I regard compounding of natives in barracks as the only absolute remedy.' The implementation of his recommendations would mean the revenge of the Black Flag rebels.

That was not long coming. Regular searching was authorized in 1883 and was gradually enforced. So also was Shaw's other suggestion, the compounding of blacks. As long as they were free to go from the mines to their own abodes there was still an opportunity for purloining gems by careful secretion or by the simplest method, swallowing. The only 'absolute remedy' was to keep workers in compounds where they could be constantly supervised and removed from all temptation. As a last refinement, at the end of their spell of work they could be detained and purged like snails so that when they finally left they carried nothing of value intestinally.

The first compound which had been opened by the Central Company in April 1875 had proved a success. Within a few years the final searching was perfected. At the end of their stint the workers were shut in the compound where 'they remain in a perfectly nude condition, save for a pair of fingerless leathern gloves, which are padlocked to their hands, for some ten days'. Missionaries were allowed to visit them 'but are requested not to worry them unduly. They are enjoined to impress upon the native mind two simple

Christian precepts – the virtue of obedience and the dignity of labour. Thus the natives receive religion in moderate doses, and in like manner their allowance of Kaffir beer is regulated on strictly reasonable lines.' The same writer added that, 'Strange to say, the Kaffir does not object to this humiliating process.' Events belied this complacent belief.

No one could possibly deny that some whites on the mines took part in IDB. All white workers denied that this gave the companies a right to subject them to the same rigours as blacks: 'It would be a disgraceful and degrading thing if they should be compelled to disrobe in the searching house, and so lowered in the sight of the natives.' The mine companies pointed out that the principle of searching was already accepted by English workers at the London docks. This comparison did not please the overseers. As the *Daily Independent* said, 'Neither can overseers be rightly compared with dock labourers. Actually dock labourers occupy the position of our Kaffirs.' The companies had their way, in principle at least. The Proclamation of 1883 authorized the searching of all employees on the mines, black and white. The white miners formed an Association for Protection of those who objected 'to be searched like a common Kaffir'.

In October the overseers went on strike for a week. A great torchlight procession of miners toured the four mines with a band playing. Property was damaged and one mine manager beaten up. After three days the mine owners seemed to capitulate and said that it had never been their intention to force white men to be strip-searched. The strike ended with, as a prior condition, a promise of no victimization. But two of the companies, the French Company and Baring-Gould, proceeded to dismiss strikers. And then in March 1884 fourteen white workers at Bultfontein were made to strip. Two thousand workers petitioned the governor of Cape Colony, Sir Hercules Robinson, by telegram asking him to suspend the stripping regulations. The companies replied by announcing a lock-out for any who refused to obey the regulations, and on 24 April men chosen at random who refused to strip were sacked.

This action precipitated the great Kimberley strike. It lasted for more than ten days amid considerable disorder. A large crowd marched on the Kimberley Mine where the French – the most obdurate of the mining companies – had stood by an armed guard. In the resultant mêlée six demonstrators were killed. This broke the strikers' nerve and they drifted back to work.

Rudd said that the whole of the fields were in a frightful state, and Rhodes lamented to Merriman that 'The suicidal mania is seizing the community here . . . the doctors say it is almost like an epidemic.' The whole town was as hysterical, nervy and near to despair as ever, perhaps more than ever as the salvation of some great reorganization of the mines now seemed at once so near and so unattainable. Towards the end of the year another episode showed the mining town at its most hysterical and at the same time most sordid.

Almost as soon as Hans Sauer had begun medical practice in Kimberley he had been plunged into an acrimonious controversy, the 'great smallpox war' as he called it. Sauer had detected, as he thought, an outbreak of smallpox but this was categorically denied by a group of doctors headed by Jameson. Sauer suspected the basest motives: a rumour of smallpox would keep black workers away from the mines. His suspicions were confirmed when he was refused permission to inspect the mine compounds. Sauer threatened to report the case to Cape Town where his elder brother J. W. Sauer was a member of the Cabinet. After that he was allowed in to the compounds and confirmed the smallpox epidemic. This made him so unpopular that he needed an armed guard.

Legislation was necessary and, on Sauer's urging, the Public Health Act was soon passed. This 'excellent law made me the most unpopular man on the Diamond Fields. High Society turned its back on me, and as for the ruck of the population, it simply spat when I passed.' Sauer was accused of corrupt motives in return, but at least he had had his way.

The intermittent crises, peaks and troughs which afflicted Kimberley were the effect as well as the cause of the broader process. As the 1890s had worn on they had meant for the mines a series of violent oscillations up and down, but with a general depressing trend. Where in 1873 1.08 million carats had been mined and sold for £1.62 million, in 1877 1,765,000 carats were sold for £1.65m. From just under 30s a carat the price had dropped below 19s. In the worst slumps – or the best buyers' markets – the price had even gone as low as 10s. It was reckoned that under the best management in the most favourably situated part of the mines the cost of production exceeded 15s. a carat and by general agreement the optimum price was around 30s. This was never touched and held for any length of time from the early 1870s to the mid 1880s. The figures for the middle six years of the 1880s speak for themselves. From 1883 to 1888 the annual production of diamonds increased from 2,264,768 carats to 3,565,780. The average price per carat was in 1883: 20s 4¾d; 1884: 23s 2¾d; 1885: 19s 5¾d; 1886: 21s 6d; 1887: 21s 6d.

It took no profound understanding of diamonds to diagnose the complaint. One man learned in the trade summed up the problem. C.J. Posno, a member of a well-known family which had played a dominant role in the diamond business in Europe and in South America, was now chairman of the London and South African Exploration Company. He wrote to Merriman in 1884 to complain that 'the diamond industry was suffering from overproduction. Diamonds are not like a crop of vegetables that must be got and sold ... if you will, forgive me the expression, the Kimberley mine had been ruthlessly destroyed, not intelligently worked.' The implication of intelligent work was not hard to seek. Free competition among rival diggers or rival firms had tended to maximize output with the resultant lowering of prices. Only co-operation or combination could mean controlled, which was to say reduced, output. Free rivalry might have benefited the consumer who bought diamonds more cheaply. It certainly benefited the broader trade: in Amsterdam the 1870s and 1880s are still remembered as the *Kaapsche Tijd*, the Cape time, when the industry leapt out of recession, took on more employees and greatly increased their wages. It did not benefit the primary producers and they looked to the obvious remedy of amalgamation.

By 1884 it was clear that no scheme of amalgamation would work unless backed by European capital. Despite the thrill which the discovery of the diamond-fields had sent out and the pull they had exerted on future seekers, they had not immediately attracted serious investment. This was partly because of the great depression which began in 1873. However, it also reflected a view that diamonds were by no means a good risk for the investor or banker in London, Paris or Berlin. The market was notoriously unstable and in the early years there was no way of telling how long the mines would last.

Slowly the depression began to lift. The long-term prospects of the mines grew brighter. European investors began to look with keener interest at the Griqualand fields. Another scheme for combination was put up, backed by Schroders of London.* Still the time was not ripe. The contestants in the struggle for mastery at Kimberley had several more years in which to husband their resources, to prepare their power bases, and to jostle for position.

They were jostling not only in the way that antagonistic financial groups do, but also in a more literal sense on the two mines, De Beers and Kimberley, divided up into their squares like great chessboards. For several years it was as if the contestants were playing some exotic

---

* And almost certainly by Rothschilds as well: a financier called Gansl who came to the fields was probably representing the great house.

game to gain control of each mine, or at least to block the others. The participants almost saw it in those terms. Frederick Philipson-Stow looked back on the struggle many years later. He had combined with Robert English in the De Beers Mine in 1878. Within a few years his firm, now Stow, English and Compton, 'had practically secured the key to the De Beers Mine. We had succeeded in cutting off the firms of Rudd, Rhodes, Graham, Alderson and Dunmore from the East side of the Mine, thereby preventing any amalgamation of East and West [thus] holding the best position strategically . . .'. What now ensued was indeed an amalgam of three board games: chess, GO, and 'Diplomacy'.

Some players in the game had troubles of their own. Barney Barnato had by now around him a tight-knit clan: his brother Harry and their three nephews, the sons of the landlord of the King of Prussia, Joel Joel, and of Barney's sister Kate: Isaac – always known as Jack – Woolf and S.B., Solomon or Solly. Woolf had come out first and learned the diamond business at his uncles' knees. Before long Barnato was paying the boy a handsome £50 a week. Money well spent: Woolf Joel had a sober, acute mind for figures which Barney with all his imagination never really possessed, and he put some order into Barnato's business affairs. Within three years Woolf was joined by his two brothers. On a visit to London Barnato persuaded his reluctant sister to allow the boys to come and join what was now a family business. The youngest, Solly Joel, also showed flair for judging diamonds. He had a passion for stones and like Beit would fall into a reverie turning a diamond in his hand and looking deeply into it. He was the dandy of the family, too well-dressed for the diamond-fields, a sensualist who loved scents and silks. Jack was no dilettante, but a serious young man who applied himself assiduously to learning the business. With the Joel boys in Kimberley, Harry Barnato spent more and more time in London looking after the family interest there and, with Ludwig Lippert, starting the first informal market in diamond shares on the pavement between Hatton Gardens and Holborn Viaduct. Barnato Brothers were going from strength to strength. In 1884 they bought a block of six central claims in the Kimberley mine for an unprecedented £180,000. The proceeds were shared among the family. Before Woolf Joel came of age that November he was a millionaire.

It was not a happy birthday. Some time before, and partly because of his prolonged absences in London, Barney Barnato had resigned the chair of the Barnato Diamond Mining Company, which had been taken over by Dr William Murphy, another of the doctors on the diamond-fields who had moved into the larger world of diamond mining. Murphy was a curious choice. He was a close associate of J.B. Robinson and no friend of the Isaacs–Joel connexion. Nor was he much loved in Kimberley. 'The Devil himself did not know what a baneful thing he invented when he sent into this world spiteful Dr Murphy, who in his life had been known to serve only two masters – Satan and J.B. Robinson.' At all events Barnato lived to regret the appointment.

At a special shareholders' meeting in October 1882 Murphy attacked Barnato and called the Company Manager, J.H. Pippin, a liar. Pippin resigned. Barnato, refusing to accept his resignation, called – and packed – another meeting. When asked to explain the discovery of an unrecorded diamond, Pippin said not very plausibly that it was left as a 'lark' on Murphy. When he saw the mood of the meeting running against him Murphy lost his temper. His two chief characteristics were irascibility and bigotry. He referred to the shareholders as 'the tribe and supporters' of Barnato. 'Are you not the tribe, the chosen tribe?' The chairman of the meeting was David Harris, who gently said 'we are not here to discuss religious matters'. Murphy's removal was passed without opposition.

His revenge came soon. In 1884, following Rhodes, Robinson and Rudd, Murphy stood for the Cape Assembly. In the course of his campaign he again rounded on the Jews, in what one paper called 'a vile and slanderous attack upon a section of the community by which he endeavoured to arouse religious animosity'. It was also a foolish attack as the Jewish community in Kimberley was large and powerful. In the election Murphy was trounced.

Following the election Jack Joel and some friends accosted Murphy and pushed him about. The sequel was unlucky. Possibly at Murphy's prompting, Jack Joel's dealings were investigated by the Detective Force which enforced the new and fierce IDB legislation. They found three ten-carat stones which had not been registered in Joel's books and for which he could give no explanation. He was tried, acquitted, rearrested on a second charge, and jumped bail. Towards the end of May 1884 he left Kimberley surreptitiously and for good. Back in England he became known as Jack Barnato Joel and began a new life. He married and saw a good deal of his family: the mean East End street where he had grown up had now been demolished, the King of Prussia with it, and Isaacses and Joels went in their new prosperity to live as neighbours in St John's Wood.

By the time of his death more than fifty years later Jack Joel, 'barely scaped from judgement', had been utterly transformed, and was not only a millionaire, the chairman of the great family goldmining business 'Johnnies' as well as a large shareholder in De Beers, but a respected citizen, a successful race-horse owner, and a Justice of the Peace in Hertfordshire.

Before Jack decamped, Barney Barnato had tried to have his trial stopped, first by intervening with Robinson, then chairman of the Diamond Mining Protection Association and, when that failed, by bribery. John Fry, the chief of the Detective Department, was not the man to be bought. He persevered with the prosecution and left Joel with only one escape. Barnato, meanwhile, determined to destroy Fry. If policemen could not be bought then newspapers could. In October 1884 a new paper, the *Diamond Times*, began publication. It was edited by R. W. Murray and ostensibly owned by Woolf Joel, though it was an open secret that Barney Barnato held the purse-strings.

The paper launched a campaign of denigration against Fry. This harrying, together with some influence which Barnato may have wielded at Cape Town, resulted in Fry's suspension and then dismissal. Incautious as ever, Barnato greeted Fry in the moment of victory and told him that he could have kept his job if he had dropped the case against Jack Joel. Shortly afterwards Fry was travelling on the same cart as F. J. Dormer, the editor of the *Cape Argus*. Dormer described the Joel case, plainly accusing Barnato of attempted bribery. Barnato sued for libel, as did Woolf Joel whom Dormer had confused with his brother Jack. Barnato lost his case, Woolf won £10 damages – and the £10 was collected within an hour in a whip-round at Dutoitspan.

This episode demonstrated several things. Even in a rough age the standards of financial morality on the diamond-fields were low. Barnato, in particular, was prepared to take any short cut available. He was utterly unprincipled when need be: ready to make a working alliance with Murphy, his bitter enemy of a moment ago and a violent Jew-hater. But he was not only unscrupulous; he was neurotic. He was highly-strung and always potentially off-balance. He resorted easily to heavy drinking, he became hysterical and the hysterics only abated when he was able to get his revenge. But shocking as the scandal was, it showed finally that anything could be lived down, as the later career of Jack Joel proved.

The spirit of the mining camp lived on in the financial morality of Kimberley; the town itself had changed. By 1885 it became the first town in Africa to have electric streetlights. The Kimberley Tramways Company served the municipality with horsedrawn trams and two years later the Victoria Tramway began a service between Kimberley

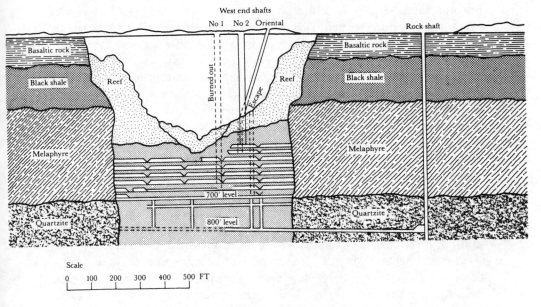

Section through De Beers mine looking West

and Beaconsfield; the line was electrified in 1890. On 31 July 1886 Kimberley railway station was at last opened.

Now respectable families were living in Kimberley in something like the bourgeois comfort they might have known in Europe. Lionel Phillips's fortunes had dipped sharply when, on holiday in England, the young mine manager had lost his considerable savings of £20,000 in a share crash. On his return to Kimberley in 1883 he started again and worked his way up as a haulage contractor, until he was in a position to marry Florence Ortlepp in 1885. The newly-weds lived in a little corrugated-iron house in Beaconsfield with garden, a servant, a horse and trap. Florence read George Eliot from the public library and magazines from England: the *Queen* for fashion notes, and *Truth*, which Lionel recommended as more serious reading.*

Even when they had little money such a couple lived comfortably. Every form of labour was cheap. There was another side to Kimberley. The mass of blacks who had flooded to the fields had not very obviously gained from the prosperity. Their condition struck visitors as degraded. Lady Florence Dixie, an English journalist, was taken round the diamond town and was not much impressed. The general

---

* He might have had second thoughts about it a few years later. See chapter 14.

hospital for non-paying white patients was sordid enough but worse
was the 'out-house set aside for Kaffirs'. It 'more resembled a barn, as
indeed it was, than anything else; and the miserable aspect of the poor
sufferers therein was pitiable to behold. Many lay on mattresses on
the ground, where the cold draughts swept over them night and day,
and on which they were stretched out, helpless to move or assist
themselves.' Not only fleeting visitors were shocked. Merriman wrote
to his wife, 'of course no one expects that the face of Kimberley will be
changed in the twinkling of an eye, but I do maintain that if any
attention was paid to Christianity . . . it would be impossible that the
conditions of Native life there should remain as they are, a scandal
and a disgrace to anyone whose moral sense is not blunted by the
habit of looking at them as mere working animals'. It was maybe
because he thought in those terms that Merriman never became a rich
man, or took part in the great amalgamation which was coming.

The tide seemed to drift inexorably in the direction of amalgamation: by 1885 there were only seven companies and three private
holdings in the Kimberley Mine, eleven and eight at De Beers.
Amalgamation was so much in the interests of the inner ring of
magnates that it is surprising that the mines were not amalgamated
sooner. The reasons for delay were largely personal. Barnato gradually sensed that he could not conquer all the mines – in the first place
De Beers, as well as his own Kimberley – on his own terms. He put up
a fierce fight which lasted almost until the end of the decade and
which distracted the magnates from the new gold-fields.

Yet there was something of shadow-boxing about this great
struggle. Both sides, Barnato as well as Rhodes and Beit, guessed
what was happening and wished only to satisfy their *amour propre*
by not surrendering too early. There may even be more to it than that.
It may even be that Barnato was in Rhodes's pocket from an earlier
stage than events suggested. Certainly in 1883 Barnato had sold a
group of claims in the De Beers mine to Rhodes who badly wanted
them for their position. A ruthless tactician and real enemy would
have held on to them.

At first these were only three among several men with their eyes on
the big chance. Merriman offered his services as mediator with Cape
banks. But apart from his disqualification for the role of mining
magnate evinced by his tender-hearted worries about the condition of
blacks, he had made the further grave error of allying himself with
J.B. Robinson. On the previous occasion when Robinson had tried his
hand at amalgamation, nothing had come of it, partly because of his
personal unpopularity. But that was nothing compared with the

financial rocks on to which his career had now drifted. Landfalls in the Kimberley mine severely damaged his operations and by the middle of the decade he was to outward appearances one of the richest and most powerful of the diamond magnates; he was in reality near to disaster, deeply in debt to the banks, which were further eroding his position – perforce and not from malice – by selling off his shares which they held as security.

In due course Robinson managed to clamber back from the brink of financial extinction but for the moment he was in no position at all to take part in any large scheme of amalgamation, and had to tell Merriman so. Further humiliations were to come. Robinson had resigned his seat in the Cape Assembly ostensibly because of deafness, in reality because of his financial plight. In May 1886 a by-election was held and Robinson stood again, a gesture of hope that his position was now rescued. He was defeated at the polls.

That left Barnato and Rhodes, with Beit inconspicuous but all-important in the background. Rhodes had one more lunge at amalgamation, floating a scheme for a large Trust Company which would be set up by an exchange of shares among holders in the existing companies. Nothing came of this, and Rhodes was persuaded by Stow that the best plan was to consolidate control of the De Beers Mine. This they did for the next four years, buying up more and more properties.

The hardest company to crack was the Victoria, whose owners showed no inclination to amalgamate. But the Victoria was a public company, not an individual enterprise or firm. Its shares were quoted in London and it was there that Beit bought and bought, acting in close collaboration with Rhodes. They thus bought 6,000 shares for a total of £57,000, which brought almost the whole of the De Beers mine within the De Beers Company.

Now that amalgamation of the De Beers Mine was all but complete, the balance of power on the diamond-fields shifted. Barnato had taken advantage of Robinson's shaky position to gain control of the Standard and kept its name when he amalgamated it with the Barnato Diamond Mining Company in early 1887. Several months later the enlarged Standard finally joined with Baring-Gould's Central. Francis Baring-Gould kept the chair of the expanded company, but it was predominantly owned by Barnato and his family. Barnato signalled this amalgamation with the cacophony of dynamite blasts around the rim of the mine. It was such a row that people thought that 'the Transvaal Navy had suddenly appeared and was bombarding dear old Kimberley'.

He behaved as if he was celebrating not only a successful battle but the closing stages of a victorious war, though this was far from over. Soon the new Central was paying 36 per cent dividends and Barnato appeared more powerful than Rhodes. Appearances were deceptive and Rhodes had crucial advantages. He had already effectively amalgamated De Beers, while Barnato, though the most powerful figure at the Kimberley mine, was not all-powerful. Independent companies remained in the mine, especially W.A. Hall, which Rhodes tried but failed to acquire, and the French Company.

With the French Company Rhodes was in a much stronger position as the company was not even paying dividends at the time, but it was of crucial importance as its claims were strategically placed, running across Kimberley north to south, and intersected in turn by the Central. Patently Barnato needed to control the French Company, but here Rhodes's hidden, overwhelming advantage came in: his understanding with Beit. Though Beit did not own the French Company, he played a dominant part in the Porges partnership with its large French Company holding, and he kept Rhodes *au courant* with all dealings.

The struggle grew fiercer. Barnato was able to step up output and push prices down. Rhodes was enormously rich on paper but was short of cash for the battle and could not realize liquid assets without weakening his hold on his own mine. Once more Beit was the key with his excellent connexions in European financial centres. The comparatively easy first step was to raise £750,000 from a French and German syndicate, in return for De Beer shares which could be dispersed in Europe and not be a potential threat to the control of the company. With battle obviously imminent the price of French Company shares had risen sharply and a total of £1.4 million would be necessary for the purchase. The answer was Rothschilds, where Beit provided the introduction, helped by Porges's own relationship by marriage to Rodolphe Kann, a Rothschild connexion and another example of the great interlocking financial cousinry.

The day before Barnato amalgamated the Standard and Central, Rhodes and Williams sailed from Cape Town. They arrived in London on 27 July. The urgency of their business was obvious to everyone. De Beers was only paying 12 to 16 per cent dividends, against dividends twice or thrice as large from Central. Rhodes's pocket was far from bottomless but Rothschild gave the satisfactory answer. If Rhodes could buy the French Company, Rothschild could raise the million pounds. Rhodes left for Paris where once more Beit had done the groundwork. He had persuaded Porges that the sale of their company was necessary.

The deal seemed to be done, but Barnato already owned a fifth of the French shares. Now he announced that he would pay £1.7 million – £300,000 more than Rhodes – for the company. Both Rhodes and Beit returned to Kimberley where a large shareholders' meeting was held on 21 September. The chair was taken by Barnato who asked – on the face of it not unreasonably – 'how on earth' those who took the decision in Paris to sell the Company at such short notice could have taken into consideration the value of the property. When Rhodes tried to cajole him, Barnato high-mindedly claimed to be acting in the best interests of the other shareholders. Rhodes said that he would outbid Barnato as long as necessary. But this was bluff and they both knew it.

The famous consequence was a brilliant deceptive gambit played by Rhodes. Or was it? Rhodes suggested that he should buy the French Company as arranged, and then immediately sell it to Barnato for Kimberley Central shares. By passing the shares on to Barnato he would fool the other into thinking he entirely controlled Kimberley: when in fact by acquiring a fifth of the Central shares Rhodes would have inextricably got his hooks into Barnato. So the story has always gone.

The transaction was announced to the Central Company shareholders on 3 November 1887. Barnato was in high good spirits and continued to struggle with Rhodes for some time yet. In February Rhodes declared all-out war from his new position of strength. To buy the Central would take £2 million which Rhodes had not got. He met Beit at Poole's Hotel in Cape Town that February. Beit said, 'Oh, we will get the money if only we can get the shares.' As soon as Rhodes began buying in earnest, so did Barnato, thereby giving the game away: he did not have a controlling interest in his own company. Every share that came on the market was fought for. In the middle of February, Central shares were £14. Then they rose to £20, to £30, to £40, and touched £49. Every time the price spurted up it would then drop back, as Barnato and his allies sold in the hope of buying back when the price dropped further. The long purse of Beit snapped them all up. By March Rhodes had three fifths of the shares.

Now that Rhodes had established by far the richest mining company on earth he wanted it to be the most powerful. He wanted the new De Beers Consolidated Mines to be given formal power by its trust deed to further imperial expansion to the north. For Rhodes's career now had four faces. He was a triumphantly successful monopolist in Kimberley. He was a leading politician at Cape Town. He was a budding financier on the Rand gold-field. And he was just beginning his adventures in the north, in Zambezia. He wanted no limits to the scope of his combined operations; he wanted to be free to swivel from one foot to the other, to move funds at will, to use De Beers income, if he wished to, for his greater schemes.

A final twenty-four-hour meeting took place between Rhodes and Barnato at Jameson's house. Barnato was determined not to give way and to restrict the object of the Company to diamond mining. At last, exhausted, he relented: 'You have a fancy for building an Empire in the north, and I suppose you must have your way.' They were almost home. Not quite: some Kimberley Central shareholders challenged the legality of the take-over. The case was heard before the Cape Supreme Court. Counsel for the shareholders spelt out what the Company could do. It could 'increase its capital as and as much as it wished, it could acquire any asset, it could move its headquarters anywhere, it could deal not only in any minerals but . . . annex a portion of territory in Central Africa, raise and maintain a standing army, and undertake warlike operations.' He was emphatic: 'Since the time of the East India Company, no company has had such power as this.'

The judge found for the plaintiffs, but Rhodes was not one to have triumph snatched from him at this last moment. He and Barnato owned a clear majority of Central shares. They had a simple, drastic expedient. On 29 January 1889 the Central Company was put into voluntary liquidation. On 18 July an historic cheque was drawn upon De Beers Consolidated Mines in favour of the liquidators of Kimberley Diamond Mining Company for £5,338,650, and it was done.

For all the extraordinary powers that De Beers possessed, it had a more immediate function to perform. 'Overproduction' stopped. There was no more feverish mining to get whatever diamonds diggers could find, to sell at any price, no mutually destructive selling battle between Rhodes and Barnato. In the year of victory, 1889, De Beers reduced the output of diamonds by 40 per cent. The consequence was gratifying: the average price rose by half from some 20s to 30s per carat.

There were losers as well as winners. Rhodes, Barnato and Beit had hung on till the end. Others had been driven out in this prolonged, ferocious struggle: Lewis and Marks early on, J.B. Robinson as the battle moved to its climax. The brothers George and Leopold Albu were not total losers on the diamond-fields. They held a number of claims in the two Beaconsfield mines, Dutoitspan and Bultfontein, but these two southern mines were overshadowed by the Kimberley amalgamation. In 1890 De Beers finally and bloodlessly acquired them both. George Albu had by then moved on to Johannesburg, where he prospered.

Losers in a different way were the men who worked on the mines,

Kimberley Mine, 1873

'Colesberg Koppie', 1872

Paul Kruger

Cecil Rhodes

J. B. Robinson

Barney Barnato

Kimberley Mine

Kimberley: ropeway in the 1880s

Kimberley: 'boys' coming to the mines

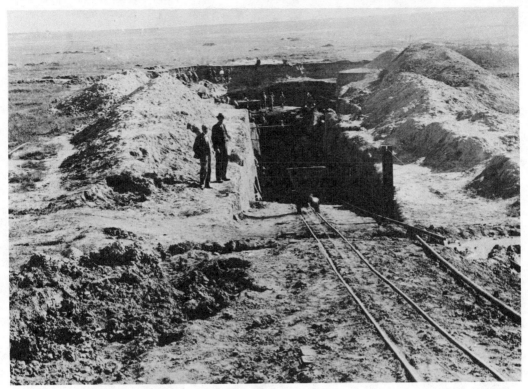

Entrance to a diamond mine

Diamond dealing stand at Kimberley

Kimberley tramway

Washing for diamonds with a 'whim'

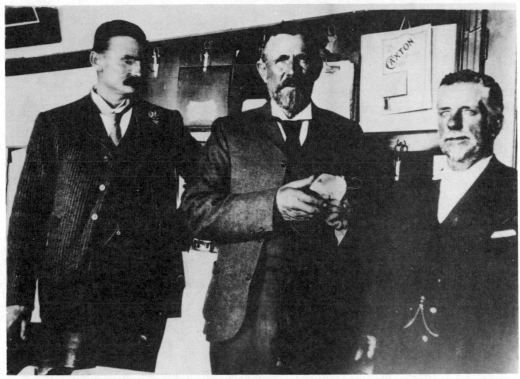

Thomas Cullinan (*left*) and W. McHardy, General Manager of the Premier Mine, (*centre*) with the Cullinan Diamond

De Beers compound at Kimberley

Louis Cohen

Lionel Phillips

Julius Wernher

Alfred Beit

black and white. The white workforce was sharply reduced and for a time Rhodes was so unpopular that when he visited Kimberley his colleagues thought it best to provide him with a surreptitious police escort. The blacks' fate was rather different. There was usually work for them on the diamond-fields and at higher wages than paid elsewhere, certainly than on the new gold-fields. But there was a price for this. They had been broken, controlled by pass laws, compounding, closer and closer control, and more and more exhaustive means of searching. They were all now kept on compounds, De Beers's 'monastery of labour', as Rhodes's doting admirer Flora Shaw put it. Finishing work, they were 'stripped perfectly naked and compelled to leap over bars, and their hair, mouth, ears, etc., carefully examined – not a particularly pleasant duty for the searchers when the thermometer stands at perhaps 100° Fahrenheit in the shade'.

This examination finally drove the miners to rebel. In April 1887 they refused to submit to use of the speculum, an instrument for anal examination of workers as they came off shift. They came out on strike: the first black workers' strike in South Africa's history, but to little avail.

The spoils went to the victors. Four men had taken the closest part in the amalgamation negotiations: Rhodes, Barnato, Beit and Philipson-Stow. They all became life governors or permanent directors of the company. The industry was rescued, as Rhodes had promised. De Beers became and remained the dominant force in diamonds. Kimberley ceased to be the dangerous, exciting, colourful place which Louis Cohen immortalized and became a quiet, efficient company town. All that remained was to tie up the last loose end and formalize an arrangement by which the diamond syndicate sold what De Beers produced. In 1890 sole marketing rights were granted to a group in London of whom the leading firms were Wernher Beit & Co., Barnato Bros. and A. Dunkelsbuhler. The monopoly was now perfect.

Other rewards were to come. Rhodes, Merriman, Rudd and Robinson had already begun their political careers. They were now joined in the Cape Assembly by Barney Barnato. This was the source of legend. It was said that as part of the amalgamation deal Rhodes promised Barnato membership of the Kimberley Club and a seat in the legislative assembly. Barnato had been deliberately excluded from the Club, not because he was a Jew – several Jews were members – but because he was so rough and disreputable. But membership of any club was never of much importance to Barnato.* In any case

---

* Even more absurd is the story that Rhodes asked Barnato to show him a bucket of diamonds: he could have seen a sack of diamonds any time he liked.

Barnato could have had a legislative seat before, if he had wanted it. In the Cape Colony of the 1880s any millionaire could buy his way through an election.

Barnato fought Kimberley in November 1888. He drove around the town in a splendid carriage drawn by four horses in silver harnesses. Even by Kimberley standards his campaign was garish and his meetings rowdy. He was escorted by a local rough called Charley who played trumpet fanfares and beat up hecklers. At one meeting the enemy claque prevailed and Barnato was about to take off his coat and go down into the crowd himself when he was restrained by a colleague on the platform, Rhodes's shady associate Rutherfoord Harris. His candidature shocked Merriman who complained that 'Men are being put forward for election who, if returned, would be a disgrace to any society, and it is quite possible that we may see the spectacle of the dupe on the Breakwater and his employers in Parliament' – a not very oblique reference to the old charges of IDB.

Barnato fought back. Always smart and quick-witted. He now became an effective speaker with sharp repartee and a lucid grasp of facts. He replied to the personal attacks by emphasizing his career as a self-made man grown rich through flair and hard work. Rhodes lent his full support. At one election meeting he took the chair and declared that 'Mr Barnato has been accused of being devoid of honour. If he is good enough to be a co-director with me, he is good enough to represent us in Parliament.' The election cost Barnato a great deal. On the evening of 13 November it was announced that Barnato had headed the poll: 'Kimberley was swimming in whiskey.' Barnato later looked back on that election as one of the greatest of his personal triumphs against the odds. Others took a dimmer view. A friend wrote to Merriman: 'I see Barnato heads the poll at Kimberley. *O tempora! O mores!*

On his first appearance at the legislative assembly Barnato confirmed the fears of staid parliamentarians. He entered the Chamber with a cigarette in his mouth and then made a dumb show as though he had forgotten to put it out. None the less, he was re-elected in 1895 (when Rutherfoord Harris was chosen as the other Member for Kimberley) and remained a Member until his death.

And so the heroic story of Kimberley died away. Those who were in at the death had greatly increased what were already large fortunes. But the length of time which the struggle had ultimately taken was a problem for Rhodes, Beit, and Barnato: it had diverted their attention from the new gold-fields on the Rand. They were all of them soon to make up for lost time.

7

The Corner House

In May 1887 Hermann Eckstein wrote to J. B. Taylor: 'Probably you will have heard that I am going to take charge of Messrs Porges interests at Witwaters Randt.' He had been in South Africa for five years, managing the Phoenix Diamond Mining Company in Dutoitspan and then joining in the Barberton adventure on behalf of Beit who had formed a high opinion of him. Eckstein was just turning forty, a few years older than Beit, and he was about to get married to Minnie Pitt, the prettiest girl in Kimberley he called her and photographs suggest that he may have been right. Together they set off for the north. The place they reached was still little more than a camp site. All along the outcrop of the reef was a hive of activity reminiscent of early Kimberley, gangs of black labourers hacking into the ground, primitive processing works scattered here and there.

The town (scarcely yet deserving the name) to the north of this line of mines saw another arrival at much the same time as Eckstein. After his hopes of making a fortune in diamonds had come to nothing, Louis Cohen had established himself as a journalist of sorts, writing in the *Dutoitspan Herald* as 'Majude' (the Bantu version of Jew). A collection of pieces was published under the title of *Gay Young Creatures*; Cohen's sauce and spice is rather diluted by wearisome Victorian facetiousness, as the titles suggest: 'Silly Marky's Konsistency or A Gay Young Creature Gone Wrong'; 'Mrs Gay Spanker'. Then after more travels he and his family had arrived in Johannesburg. Cohen set up as a stockbroker, first at Simmonds Street in 1888, where his tongue-in-cheek telegraphic address was 'Majude'. He still wrote for the papers, now for the *Standard and Diggers' News* which ran frivolous, gossipy pieces or verses like 'A Christmas Carol or The Children of the Chains': ('Between the Chains' was the extraordinary financial street-market in Simmonds Street on the block where at either end Commissioner and Market Streets were closed off by chains). In his occasional pieces and in his later *Reminiscences*, funny and unreliable though they were, Cohen was the true poet of early Johannesburg.

He described the Reef as he first saw it. From afar the ridge of white water appeared as 'saffron clouds of red golden dust settling on a hotch-potch of what looked like moving dirty sand dunes . . . a huge flame-formed crucible of molten opals'. Nearer to, he found, was a 'laggard place . . . with half-deserted canteens here and there; shabby new-built general stores . . . a few mild-mannered Indian itinerant vegetable traders, some transport riders, meek-faced store-keepers, greasy Malays, sour-looking Boers, drunken loafers and thirsty fossilers, a couple of Barons from the Almanac of the Ghetto, ruined gamblers, cunning smousers, and a bushel of toffs from Poland, destined to oust the Boers and become the aristocracy of the land'.

That was the place which greeted the Ecksteins. They bought a house in De Villiers Street. Every stick of furniture had to come from Cape Town and Eckstein had to ask Taylor to find wallpaper to match as it was impossible to get in Johannesburg. Far from home and nostalgic, he called the house Hohenheim, the name of his home town near Stuttgart.

When he arrived to open the offices of H. Eckstein & Co., the first feverish scramble for property, the jockeying for position along the line of the reef, was dying down. There were still hundreds of diggers working with picks and shovels but they were a breed dying as soon as they were born. The scale of mining gold on this reef even where it outcropped made the small digger an impossibility: no one could grow rich on the proceeds of milling small quantities of banket to sell at a fixed price. Eighteen years earlier men had made small fortunes with a lucky bucketful of diamond-studded soil. Besides, in the early days on the diamond-fields there had been no capital with which to undertake a monopolistic amalgamation. Capital accumulation had now taken place and capitalists had come on to the new gold-field financially strong enough to swallow any small fry.

The incoming financiers were aided also by the Gold Law of the Transvaal, which helped those with ready capital to buy up farms and claims, and as mining rapidly became more sophisticated they had the resources to import the necessary plant and achieve vital economies of scale. As early as March 1887 thirty-three wagons reached Johannesburg carrying stamp batteries, great crushing mills made by Fraser & Chalmers in the United States and Sandycroft & Co. in England, with pestles weighing 850 to 1,000 lb dropped by rotating cams on to the ore to smash it up for treatment. The capitalists could afford the cyanide used to treat the crushed banket, not to say arrange for the great quantities of water needed – 2,000 gallons for a ton of ore – to be brought to the treatment sites. They could afford the fuel. As luck had it coal was soon found at Boksburg and a 'tram line' was hurriedly driven there.

The Kimberley magnates might not only have established themselves severally in a dominant position on the Rand, they might also have brought about early amalgamation or combination there on the Kimberley model. The failure to do so was partly personal. The first financier at Johannesburg was J.B. Robinson, now more obstinate and inhuman than ever. His hatred of Rhodes was beyond sanity and ruled out any co-operation between the two. But despite the fact that Beit was openly a confederate of Rhodes's he managed to work with Robinson for a time. Beit's young representatives Eckstein and Taylor, however, lacked his own superhuman patience and subtlety. The young men tried to get on with J.B.R. in informal partnership but, as Taylor said, Beit knew that Robinson was a man whom it was impossible to work with and that a breach would have to come. When it came J.B.R. made what he thought was a very shrewd deal. He sold out of the Robinson Syndicate for £250,000 and moved on to the West Rand, with considerable success. He was mistaken all the same. In the years to come the Robinson property was to earn Wernher and Beit at least £100 million.

This property was the beginning of the Corner House's vast empire on the Central and East Rand. And it was this that Beit and Porges asked Eckstein and Taylor to come in and manage, 'on the ground floor' as the phrase went, on the most advantageous terms. Beit soon realized that he would have to move to the 'Randt' himself as soon as Kimberley was wrapped up; one of Taylor's tasks was to build a house for him. The young men's work was only in part to manage claims on the gold-reef. There was some gold being milled, but not much: 23,125 ounces in 1887. In any case, for several years, only a minority of 'gold-mining companies' produced any gold at all. Just as important was to be in *in situ* as the group's agents to take part in the struggle for power and position by buying claims and dealing on the vigorous new Johannesburg stock market. For years the greater part of the correspondence among members of the Corner House group consisted of news of share dealings, and a very large part of their income came from market operations rather than from gold-mining. They were at once accumulating yet more capital on a charging bull market, and gaining interests in the numerous mining companies which had been floated.

To run the affairs of this boom town the early Diggers' Committee was soon replaced by a Sanitary Committee which remained the municipal authority for the best part of ten years. At first it consisted of no more than a Sanitary Inspector with a Scottish cart and four oxen. Even its conducting of this nominal duty was at first unsatisfactory. All too often the cart was late on its errand, or tipped its load over in pot-holed streets. Epidemic disease was rife. But the town was changing and growing. It was squalid in some ways, advanced in others. The financiers liked creature comforts but wanted business efficiency even more. It was typical of early Johannesburg that in 1888, long before main sewers were even thought of for the city, the first telephones had been installed. Other amenities took people's minds off the dirt and the dust. By that same year there were 147 licensed canteens in the city.

In 1888 ten times as much gold was mined as the year before. By October the following year the Robinson was crushing 726 tons of ore to produce 3,551 ounces of gold. The mine was wonderfully easy to work, rich ore coming from a wide reef which was plain sailing to mill and treat. The investment spurt on the Rand turned into a boom with shares rocketing upwards from August onwards. The bubble swelled, puffed up also by less well-authenticated, sometimes plainly fraudulent, news of crushings.

As if to crown the boom year Barney Barnato at last arrived on the Reef. At first glance he was no enthusiast for the gold-mines and only reluctantly agreed to send an agent to Johannesburg for Barnato Brothers. Then he went back to Kimberley and attended to diamonds and his political affairs. When he returned to Johannesburg in November he found the Stock Exchange a scene of mad excitement 'men taking their coats off and shrieking like maniacs – fortunes were made and lost in hours'. Robinson shares went up to £80, a former clerk of J.B. Currey's made £20,000 in a day on the Exchange.* Barnato made a sentimental gesture of intent towards the mining town by contributing to the building of its first synagogue. He spoke at the Theatre Royal in Market Street. His audience heard that he looked forward to Johannesburg's becoming the 'financial Gibraltar of South Africa'. He put his money where his mouth was, telling his nephew Jack Joel and his brother Harry, both in London, that he would be investing heavily on the Rand. Perhaps 'other people's money where his mouth was' would be apter. At all events in the new year he was back, in on all sorts of acts. He trod the boards again, playing one of his favourite parts, in *The Ticket-of-Leave Man*.

* Nor did he lose it in traditional fashion: the clerk was W. St John Carr, later Sir William and in 1901 first mayor of Johannesburg.

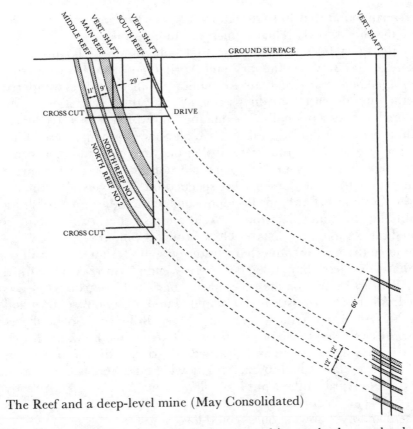

The Reef and a deep-level mine (May Consolidated)

At the end of February the boom neared its peak, the total value of Rand shares touching almost £25 million; but Barnato was gravely ill. He was treated by another old acquaintance from Kimberley, Dr J.W. Matthews. When Matthews told him how serious his condition was, his only answer was 'what a devil of a fight there will be over the chips'. Most people thought the illness was a result of riotous living. Barnato's conduct was more wild, manic-depressive and self-destructive than ever, his manners as gross. He and Lilienfeld had a fight on the floor of the Johannesburg Exchange and he was refused entrance until he had apologized. A little later there was 'a beautiful row' at a newly opened club. Hans Sauer lost his temper and there were several fights, it was reported. 'Barney and Dormer got hideously drunk, vowed eternal friendship and slept together God knows where locked in each others' arms.' When Dormer's wife went looking for him she found him in a hole in a brickfield where Barnato had deposited him. It was an odd way for a great financier to behave. Nor had Barnato's reputation for honesty improved. In Johannesburg

Merriman found that Barnato 'and I regret to say Sievewright* seem
to be the chief rascals. Their names absolutely stink.'

Quite apart from Barnato the first Johannesburg boom was a
rickety affair. Between January and April 1889 bank advances rose
from £300,000 to more than £1 million. Some bankers wondered
whether any potential crushings could justify the speculation: more
deplored the exceptionally lax manner in which Stock Exchange
dealings were conducted in the new city. Share bargains were all on
margin, most of them 'time bargains' or on 'time call', in other words
gambling with money the gamblers had not got, to a greater extent
even than with most stock-market speculation. The general manager
of the Standard Bank in London condemned the Johannesburg
bubble but he could not stop it. At the top of the boom Robinson
floated the Randfontein Estates Gold Mining Company. In what was
to become the best Johannesburg tradition, out of a total of 2 million
£1 shares, no less than 1,809,000 were 'vendor's interest'† retained
by J.B.R. in return for the seven farms on which the company was
based. All the shares issued were snapped up and soon rose to £3 and
then £4. Along with them, of course, rose Robinson's own shares
which he could sell in what quantities he liked, when he wanted.

Early in 1889 a friend had warned Merriman that a crash was
coming. The bubble had to burst: 'Knowing ones here say the fiasco
will come in April – others put it off till autumn. *Nous verrons*!' Already
the banks had begun calling in loans. When the crash came it was
worse than any 'knowing ones' could have guessed.

On 19 March Taylor telegraphed urgently to Porges's offices in
London: 'Following is strictly private. Percy Company, Main Reef.
Below level 120 feet reef changed from banket to quartz, blue, hard,
no free gold but 10 dwt in pyrites. Sunk 50 feet and there is no
change.' It was just like the striking of the blue ground at Kimberley
two decades before. Until then the banket of the outcrop had been
child's play to mine. It was cut by teams of black labourers with picks
and shovels, the ore crushed in stamp batteries, the pulverized rock
mixed with water as slime, the slime passed over copper plates coated
with mercury. The particles of gold in the slime formed a thick

---

* James Sievewright, then a political colleague of Merriman's; operated in the Cape and
Transvaal, as politician and financier; a director of Barnato's 'Johnnies'; implicated in
corruption in April 1893 but backed by Rhodes in Cape cabinet, precipitating Merriman's, J.W.
Sauer's and Rose Innes's resignations; as Merriman had written three years earlier, 'woe to the
law whose Princes are thieves'.

† The proportion of the capital raised (as cash or in shares) which went to the owners of the
claims.

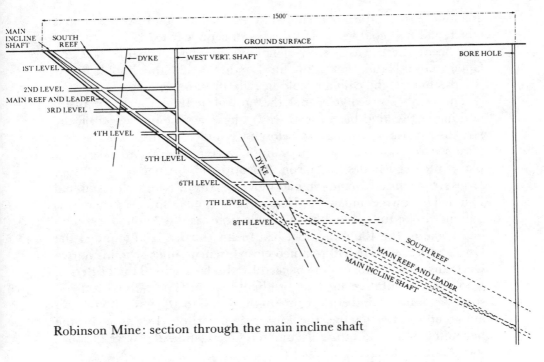

Robinson Mine: section through the main incline shaft

amalgam with the mercury which was scraped off the plates and refined, the mercury being reused. This primitive metallurgical treatment recovered less than 70 per cent of the gold, but as long as extraction was so cheap that did not matter. In the everyday phrase, it was a gold-mine: rich and easy.

Pyritic ore was anything but a 'gold-mine'. It transpired that the reef changed character as it went down. Taylor's news was not private for long, as the same discovery was made at one mine after another: the ore which had been struck on the Percy was found at the same level all along the Reef. The news rocked Johannesburg. The golden dream had ended. There was no answer in sight to the conundrum of pyritic ore. It could not be treated with mercury. As well as gold, it contained iron, copper, lead and zinc, which ate up the mercury without extruding gold. Where the banket reef had been milling at an ounce of gold to a ton of ore or thereabouts, the pyritic ore showed only half as much. The gold was there, but it was quite unpayable, impossible to mine at a profit. Several mines closed and share prices collapsed. One share which early in 1889 had been sought after at £50 was unsaleable at £5 before the end of the year. By March 1890 the total market value of gold shares had dropped by more than 60 per cent. By the beginning of 1889 there were 25,000 whites in

Johannesburg; a third of them left in panic. The South African banks which had had cold feet about the boom soon learned how right they had been: there was a rash of failures among the banks of Cape Colony and Natal. Far away in London when the bubble burst Rhodes too felt the chill and sold heavily in Rand shares.

The crash seemed to signal the end of Johannesburg. It meant nothing of the kind but instead was to have a profound effect on the structure of Rand mining. As before at Kimberley, those with strong nerves came out on top. New names were appearing to take their places besides Rhodes, Robinson, Barnato and the men of the Corner House. Sigismund Neumann had moved on from Kimberley and had founded his own company, S. Neumann & Co. He saw that there was a corner to be turned and he bided his time, as did Adolf Goerz who had come to the Rand in 1888, just in his thirties, and founded the Goerz Syndicate. He had retained close financial links with his native Germany: his syndicate was backed by the Deutsche Bank of Berlin, whose representative he had originally been. The Albu brothers, squeezed out at Kimberley, came to the Rand in 1887 and bought an apparently derelict mine, the Meyer & Charlton. It was to be the beginning of a great future. Neumann, Goerz and Albu were to show something of Beit's own foresight.

It was Beit and his colleagues, however, who stood head and shoulders above the other entrepreneurs: they assumed that there must be a solution to the pyritic problem and that the Rand's future was limitless. The composition of the Corner House group was changing. At the end of 1889 it took on its new name or names. Jules Porges had remained a curiously shadowy figure even when playing a vital part in the final Kimberley drama. He had been godfather to the first development of the Rand but now, still in his early fifties, decided to retire. At the beginning of 1890 J. Porges & Co. ceased to be, and was succeeded by two new partnerships, Wernher, Beit & Co. in London and H. Eckstein & Co. in Johannesburg. Porges walked off the stage of financial history as discreetly as he had entered, living in happy – and enormously rich – retirement until his eighties.

Under the new dispensation four fifths of the Johannesburg firm's profits went to the London partners, Wernher, Beit, Max Michaelis and Charles Rube. Rube was a contemporary of Beit's and former manager of the French Company. Michaelis was a German Jew, born in Saxony, who had come to South Africa for a family firm and moved to the diamond-fields. He dealt with the partnership's diamond interests, which were still very important even as the gold-fields were opened up.

In London it was 'Mikki' Michaelis's function to look after other partners who were visiting England, or their wives, such as the difficult and demanding Florence Phillips. The Michaelises entertained first at a large and sumptuous house they had rented in Gloucester Square, then at a house at Walton-on-Thames. They entertained out also: Florence Phillips was surprised to be taken to dinner parties at the Savoy or the Berkeley. Hotels were becoming respectable for the first time, certainly among the glittering *haute juiverie* which throve on the fringes of London society, enriched by the South African connexion; the sort of people who might be met with the Michaelises: Beit, the Robinsons, Wolf, the Hinrichsens; or for that matter at Mr and Mrs Julius Wernher's house in Porchester Terrace, where one dinner party included the Rothschilds, the Neumanns and Isadore Dreyfus.

Although the division of spoils between London and Johannesburg was unequal, the gold reef city was still the place for a young man to be and the Corner House the group to work for. Lionel Phillips had lost more than one fortune before he left the diamond-fields and he had nearly lost more than that: on the mine he was managing he was travelling one day in a car or glorified bucket which ran down to the foot of the mine when the supporting hawser snapped and he narrowly escaped with his life. This was an augury: Phillips's luck had turned. In 1889 Beit offered him a job on the Rand as general adviser in mining matters at £2,500 a year, 10 per cent of the profits from one mine which he would oversee. Beit told Phillips that he would probably soon be offered another job by Rhodes to work in Matabeleland; he could truthfully say that his own offer was more attractive. Phillips took it.

The Phillipses and their two infant sons left for the Reef in a hired coach, a maid and two nannies squeezed inside with them and the Griqua servants on the roof. Exhilarating as the gold-mines were financially, the view before them did not gladden the eye. Phillips's first reaction was as gloomy as Cohen's. 'Anyone who has gone through the horrors that attended the growth of Kimberley can sympathize with the feelings that would surely oppress one at the thought of going through a similar experience.' Sprinkled among the corrugated iron houses was the odd stone building. The streets were sandy tracks torn up by coaches and ox-wagons. The dust was unpleasant in May, the South African autumn; in October, when they arrived after a winter of drought, it was intolerable. Vegetables were almost unobtainable, prices were sky-high. Butter cost 7s a pound, eggs 4s a dozen, sugar 4s 6d a pound. The Phillipses settled in to make the best of it in their new little four-roomed house at 27 Noord Street, next door to James Taylor.

Apart from its makeshift look, the gold-reef town as Phillips found it was plunged into depression. If only those who had fled Johannesburg in panic when pyritic ore was struck had known. There was an easy answer, one which had in fact already been found. Such refractory ore was already being successfully treated on the Australian and North American gold-fields by a chlorination process. Pulverized ore was dissolved in chlorine which then precipitated the gold. This process was tried on the Rand with only partial success. But by then a greatly superior process had already been discovered in Glasgow. Two doctors, the brothers Robert and William Forrest, and a chemist, J.S. MacArthur, collaborated as speculative scientists. One of their projects was the treatment of gold ore, and together they solved the problem.

One of the pioneers on the Rand, a Scotsman called John Jack, visited Glasgow with a sample of Rand ore which he took to MacArthur for treatment. The process worked. The essential discovery which the ingenious trio had made was that gold has an affinity for cyanide. Crushed ore dissolved in a solution of cyanide of potassium precipitates its gold on to zinc shavings from which the gold can then be recovered and refined.

Although the process was patented in 1887, it was not until June 1890 that a plant was set up on a Rand mine called the Salisbury. The experiment took two months to conduct and was a triumphant success. More than three tons of ore was processed. Free-milling concentrates from the outcrop and tailings all yielded up their gold. At its best the process achieved an extraction rate of 90 per cent, higher than had ever been known with mercury treatment. In November an agreement was signed between the Robinson Gold Mining Company and the Gold Recovery Company for the treatment of ten thousand tons of hitherto useless tailings. The treatment produced thousands of ounces of gold. In 1891 the old amalgamation treatment was still used for 98 per cent of gold produced; by 1894, with the introduction of the MacArthur–Forrest process, the figure had fallen to 70 per cent.* There was a coda to the story. The mine companies paid royalties at first to MacArthur and the Forrests, but then decided that the payments were too high and took them to court. After protracted litigation the inventors' patent rights were overturned. The Randlords gave nothing away.

---

*   Even then most gold still came from free-milling outcrop ore.

When the recovery began it only consolidated the strength of the Corner House. Just as at Kimberley, a slump meant that the weak fell and were absorbed by the strong. The Corner House* not only increased its holdings along the Main Reef, but did more than that. 'Acting with great circumspection and secrecy', they were acquiring blocks well to the south of the outcrop in what was called the deep-level zone.

When the Reef was found in 1886 it was clear enough that it was more than a freak running along the surface. As it was mined it went deeper, at an angle, until it was impossible to mine it from the surface. Presumably it ran deeper still, although no one could possibly know how far. But if it did continue, and at an angle, then the gold could be reached once more. It was only necessary to walk south away from the reef and then dig deep enough to strike the Reef.

Once again success bred success for the Corner House. Jules Porges had first persuaded the Rothschilds to invest in De Beers, and then in the Rand. Rothschilds wanted the best expert advice before proceeding with amalgamation, and it was for this reason that a succession of brilliant American mining engineers was brought to South Africa; many of them had learned their trade at El Callao, a Venezuelan mine controlled by Rothschilds. Gardner Williams went first to Barberton but then was lured to Kimberley by Rhodes's offer of the general managership of De Beers. Then came Hamilton Smith, H.C. Perkins, Hennen Jennings and Thomas Mein.

These men brought with them an expertise which up to then had been quite lacking on the South African mines. That had not mattered at Kimberley, where the art was to control the market price of diamonds rather than to extract them as economically as possible. It had not seemed to matter at Barberton, where gold-mining was little more than a means for partaking in a great gamble. (In fact, the deplorable management of many of the East Transvaal mines contributed to the speedy burst of the bubble there.) It did not seem to matter at first, either, on the Rand. But once it was seen that the Rand was no gamble, no small freak, but a vast reef of gold which must be extracted at the lowest possible cost, mining with the greatest skill and the most up-to-date techniques was essential.

The Reef laid down by that retreating sea millions of years ago is the faintest, thinnest of layers firmly encased in the earth's surface. The geology of the Witwatersrand can be compared to a telephone directory. The gold-reef is one page in the middle of the book. But the

---

* Eckstein and Taylor were the original partners in H. Eckstein & Co. from 1887; Philips became a partner in 1893.

page is crumpled and torn, the book is broken in parts – the geological faults which fracture the Reef. The 'page' needs to be located and removed. Imagine finally that the gold itself is no more than the commas on that page and the exhaustingly difficult task the mines faced is clear: physically exhausting for the labourers in the mines, demanding of the greatest intelligence and skill on the part of their managers. Along with the conquest of pyritic ore the application of the more rational techniques was the next step in the development of the Rand. Mein managed the Robinson mine and worked it up to hitherto unknown efficiency. Hennen Jennings was consulting engineer to the Corner House, and played an essential part in its fortunes with his far-sighted advice.

As early as January 1890 Phillips wrote to London that 'The more I see of the Rand the more I become convinced that deep levels are of the utmost importance to the parent companies. Mr Jennings will, I think, concur with me.' So did Beit, who wrote three months later to stress the importance of the deep levels. On the other hand, in his deranged perversity, Robinson denounced the deep levels as a fraud. Eckstein wrote, 'His tirade against deep levels generally and without exception is nonsense.' Even J.B. Taylor initially doubted that gold could be found 1,000 feet south of the outcrop by sinking shafts as deep as 70 feet. Phillips backed his own judgement. In those dark years the Corner House bought the properties which would become among the most famous mines in Johannesburg: Jumpers Deep, Nourse Deep, Glen Deep, Rose Deep, Village Deep, Geldenhuis Deep. Optimism and daring were well rewarded.

As Johannesburg pulled out of its first slump the development of the deep levels was still in the future but when it came could only aggravate the problems which already faced the mining industry. If a few believed that there was limitless gold deep down, anyone could see that mining would be horribly expensive. The deep levels might provide large quantities of ore but of very poor quality at a high cost. To cut ore from rock deep down and bring it to the surface meant huge expenditure, on labour and on materials, and it was the question of costs which loomed over the mines and their controlling groups, the Corner House above all. 'Materials' meant machinery for head-gear, stamp batteries, pit props, and not least dynamite. It was possible to mine the outcrop with pick, shovel and sweat, but mining deep underground could only be done with explosives.

Once the first deep-level mines had been dug, the pattern of mining underground became what it has remained since on the Rand. A shaft is sunk, itself no easy business. The best shaft-sinkers from all over the mining world were soon brought to Johannesburg, but even with their skill, sinking was still painfully slow and expensive. One experienced shaft-sinker complained that cutting through the rock of the Witwatersrand needed a third as much explosive again as he would have used in Australia. Then a horizontal drive is made away from the shaft at the appropriate level and then from the drives stopes are cut upwards into the Reef. The stopes are cut by men drilling deep holes in which blasting charges are placed which, when ignited, punch out a further section of the gold-bearing rock.

The mining companies would have taken a keen interest in the cost of dynamite even if it had been comparatively cheap on a free market, but it was nothing of the kind. Dynamite was one of the commodities which were supplied in the Transvaal by government-leased monopoly or concession under a policy adopted by the Pretoria government soon after retrocession in 1881. There were several concessions and several concessionaries with whom the Corner House reluctantly dealt, but two loomed larger than any other over the story of Johannesburg in its early years. Wernher had indeed expressed the exasperated, though not very lively, hope that his group might 'become independent of such unreliable parties as N. & L.'. These were the enigmatic figures of Alois Nellmapius and Edward Lippert.

8
Concessionaires and Wire-pullers

All that had happened at Kimberley happened also at Johannesburg, only faster, more violently, more dramatically. Newcomers – 'outlanders' to the Transvaal Boers – poured into the golden city which burgeoned and spread north from the outcrop and its row of mines. So did thousands of blacks seeking work on the mines. For the Republic, its burghers and its president, the finding of the main Reef was a godsend – and a dire threat. Gold rescued the Transvaal from bankruptcy, but looked to destroy the Republic from within. Wolseley's prophesy had been far-sighted. By the beginning of the 1890s it was obvious to all that, unless they could be stopped, these . Uitlanders must soon outnumber the Transvaal Boers and take over the country.

Less obvious than the political conflict between Boer and Uitlander, but in reality far more important, was the economic conflict between Kruger and the Randlords (to give the mining magnates the evocative name coined for them by the British press, whose imagination they had caught). As none of them needed reminding, goldmining was a unique and exacting business because the price of its end product was fixed: the eternal (or so it seemed) gold standard of 84s 11¼d per fine ounce. As 1888–9 had shown, it was possible to make a fortune merely by speculating in mining shares. But after all, the *raison d'être* of the mines was ultimately to produce gold and every penny of costs must be counted.

The Transvaal had not prevented the Uitlanders from coming in and despite the Johannesburgers' complaints it could be argued that the miners were generously treated. But the mine-owners wanted to mine on their own terms. They wanted to arrange their supplies of materials and labour as cheaply as possible. It was this which set them on a collision course with Kruger's Republic. The first conflict came as a result of the Transvaal's policy of monopolistic concessions, the policy which Kruger called 'the cornerstone of the independence of the Republic', and he may have been right. The protagonists on one side of the conflict were those two unreliable parties, Nellmapius and Lippert.

Throughout the vagaries of Nellmapius's career Kruger had stood by him. The opening up of the Rand found him potentially the most powerful man in the Transvaal, controlling the concessions in dynamite and liquor. The liquor concession was sold to Sammy Marks,* but Nellmapius's continued ownership of the dynamite concession was awkward for Beit and Phillips as they were concerned in joint mining ventures with him in the East Transvaal. As he observed Nellmapius's prestigious manoeuvrings, Phillips admitted, 'I don't know whether Nelly is honestly working for us or not, but I have grave doubts about it.'

The Corner House was not long embarrassed by Nellmapius. He had already exploited his friendship with Kruger to the limit. Then he sold the dynamite concession as well. In his last years the limelight barely flickered over him, and his reputation as Kruger's and the Republic's evil genius faded. Those who associated with him did so as reluctantly as Wernher. Even those who had a soft spot for him, as Merriman did, never trusted him. His death in July 1893 was a relief to many people.

A year later Phillips was still clearing up the pieces. Kruger was said to be angry with the Corner House because of its treatment of Nellmapius – 'the best friend the Republic had ever had' Kruger claimed – and his widow had been left with nothing, while the Corner House 'had made a large fortune out of his services'. This was scarcely true, but Phillips assured a correspondent in Pretoria that Mrs Nellmapius was a prime concern to his firm which had intended all along to show her the utmost consideration. In the event she was released from some of the obligations to the Corner House which Nellmapius had incurred.

The name of Nellmapius was invariably linked with that of another power behind the throne in Pretoria, Edward Lippert, who had a close original link with the Corner House as Beit's cousin. But his career had made him one of the most equivocal figures in South Africa and dealing with him had become an embarrassment. He lost heavily in the first crash of 1889–90, but was not slow in finding ways of making money once more. Eckstein had never liked or trusted him, and Beit gradually reached breaking-point with his cousin. The Corner House practised a deliberate policy of conciliation, which

---

* Whose rotgut, distilled at Hatherley was still sometimes known as 'Mapius'.

worked with most of the rival groups; it could not work with men as constitutionally obstreperous as Robinson or as devious as Lippert.

As relations between the Uitlanders and Pretoria grew ever more tense Lippert attempted to act as mediator, but he was unsuited to the task. He could be agreeable when he wanted to, with pleasant manners and considerable charm. His attractive side had been on display in Johannesburg as soon as he settled there in 1890. He bought a part of the farm Braamfontein and built on it one of the first great Randlords' mansions. Not a tree had grown on the Rand a few years before. Now thousands were planted and Lippert planted most energetically of all. The plantation of trees on his estate was known as Lippert's Plantation. Later he renamed it Sachsenwald. It was not principally a decorative or ecological adornment: the mines needed wood for pit props. The pretty little forest remained. At the time of Teutonophobic anglicization during the Great War it became Saxon-wold, which name the Johannesburg suburb bears till this day. On his estate Lippert lived with his highly intelligent wife Marie, authoress of remarkable travel letters.

He attracted shoals of enemies. The reason for this – and his true disqualification for the role of go-between – was his position as a dynamite concessionaire. Through negotiations with the Pretoria administration and with the Volksraad – negotiations in which bribery always played a large part – he obtained the explosives concession for his Zuid-Afrikaansche Maatschappij van Outplofbare Stoster Beperk, the Explosives Company. There was a moment when the Corner House might have joined hands with Lippert in the dynamite concession. If this had happened, the history of the Rand might have gone very differently. Instead they fell out. Lippert's conduct made closer dealings impossible.

The encumbrance of Lippert's dynamite monopoly grew more insupportable. The monopoly was continually attacked by the Chamber of Mines from the moment the Chamber was founded, from its first annual report which rehearsed the complaints about Lippert's dynamite: 'its want of uniformity and the frequency of partial explosions; . . . its want of power; . . . the loss of time and injury to health caused by its fumes'. But the quality of domestically manufactured dynamite was only part of the problem. Lippert had a sixteen-year concession to manufacture, but not to import, dynamite as a finished product; his materials for the manufacture had to be imported.

He wanted to do a deal with one of the European explosives cartels, Nobel or the French-Latin Trust. These were both enterprises which operated with great duplicity and unscrupulousness, Nobel distinguishing itself by its superior cunning. Each was eager to exploit the South African market by selling as much dynamite at as high a price as possible, and the Transvaal Government's concessions policy was obviously enticing.

The chief complaints about Lippert's dynamite concerned its high price and poor quality. The quality suggested that it was, as it was supposed to be, manufactured in the Transvaal, but events showed otherwise. Towards the end of 1889, agents of Nobel's in Cape Town attacked the Explosives Company. It was alleged that the company was not importing merely guhr – clay from which dynamite was made – but finished explosives. In Cape Town harbour the ss *Baron Ellibank* was searched and dynamite was found, but no action was taken. Early in the following year the company itself asked the Transvaal government to investigate an alarming spate of mine accidents. The government did nothing.

More serious for the mining companies than the quality of Lippert's dynamite was its price. The Explosives Company was selling 50 lb cases of dynamite at up to £7 10s a case; the Chamber negotiated with John Grice, Nobel's agent in Natal, for imported dynamite at £5 a case. The government stonewalled on negotiations, and Nobel made another offer of £4 15s a case, at which the Explosives Company offered to reduce its price to £5 2s 6d. The government's Dynamite Commission, to no one's surprise, pronounced that Transvaal-made dynamite was safe and the Dynamite Commission backed Lippert's concession yet again. The British government entered the dispute to complain that British firms trying to export to the South African Republic suffered unfair discrimination.

But although the Transvaal government brushed aside almost all criticism, it felt obliged to appoint a new commission in April 1892. The evidence of increasing accidents was irrefutable. Lippert felt a cold wind and offered terms to the Chamber of Mines. If they made peace with him he would cut his prices. But relations with the Chamber – which effectively meant the Corner House, which in turn meant Phillips – had by then deteriorated too far. On 8 May 1892 and following a tip-off to the Chamber the government investigated another cargo of illicit dynamite on the *Highfield*. Though the Chamber's intelligence system was very effective, it did not discover that Nobel were double-dealing by having entered into negotiations with the Explosives Company. Lippert was on the ropes. He suggested that the government itself should assume the dynamite monopoly, using the Explosives Company as its agents. But then in July it was announced that the *Highfield*'s cargo had indeed been finished dynamite. As had been so long and so widely suspected, Lippert had been importing dynamite from Europe surreptitiously to resell it at a profit margin of 200 per cent. No wonder that a Johannesburg wag adapted Madame Roland's words: 'O Lippert, E! What crimes are committed in thy name!'

The Explosives Company's contract was revoked in August, and an agreement with the French, German and British consuls allowed for the importation of 15,000 cases by each of their countries. The government bided its time and waited for the Volksraad to create a new monopoly. The Volksraad in turn moved at the leisurely pace of its burghers, horsemen and graziers unused to political manoeuvring, suspicious of the Uitlander though susceptible to his inducements. There was constant pressure from foreign governments and trust companies for the right to import more explosives. This hiatus created a severe shortage of dynamite and stimulated the resentment of the mining companies.

In September the government was again granted a state monopoly. This could be assigned to agents, and on 25 October it was assigned to the South African Explosives Company. The terms were that the company would make a maximum charge of £4 15s a case. It would pay the government a royalty of 5s a case, and in addition a fifth of its profits would be made over. The company also undertook to build a new factory within thirty months. This deal went ahead despite the counter-offer made by the Chamber of Mines, which had said that it would manufacture dynamite and give the government 5s a case and half the profits. The Chamber attempted to come to terms with Nobel for an agreement that would have undercut the Explosives Company. These negotiations at least succeeded in bringing about Lippert's belated resignation from his totally invidious position on the Executive Committee of the Chamber.

Throughout 1893 the Corner House achieved little more in the matter of explosives. But in Europe all unknown to Phillips the rings which controlled the dynamite trade were combining to form a new company, 220,000 of whose shares would be owned by Nobel and 25,000 by Lippert. The new company was registered in May 1894. Chamber and Corner House had been bamboozled. All the other parties in this elaborate quadrille stood ranged against the interests of H. Eckstein & Co. And so the vital dynamite issue lay unresolved.

That was only one of the questions which the Chamber of Mines made its business. This institution took in everything which touched the mines in the 1890s. It first saw life in late 1887 when its membership was individual rather than by companies. From the beginning its history was troubled and fissiparous. A group led – of course – by Edward Lippert objected to the first Chamber's composition and proposed forming a mining union. At all events the Chamber collapsed with the first great boom of 1888–9. As the *Diggers' News* put it, 'prosperity killed it even as depression had given it birth'.

When the Chamber was re-formed in 1889 as the Witwatersrand Chamber of Mines it was on a different footing. The leading houses now openly became its backbone: Rhodes's Goldfields, Barnato's 'Johnnies', above all the Corner House. Twenty-one out of 150 companies quoted on the Johannesburg Stock Exchange were represented. The sort of problem which a collective voice could address itself to was seen later that year when drought led to food shortage, which made it difficult for the mine companies to feed their workers. An appeal was made to the government in Pretoria, through the mining commissioner, Jan Eloff, a Transvaaler who had fought as a very young man in the Anglo-Boer War of 1880–81, whose voice carried weight. A prize scheme was set up that spring by which the first 250 wagons carrying 6,000 lb of food to arrive received a bonus of £20. They began to arrive in early November.

The Council of the Chamber was replaced by an Executive Committee of eleven which met monthly. Henceforth there was no more important body of men on the Rand. Critics of the Chamber were numerous. It was charged with monopoly rule by the big groups, which was true enough. Its foes were very wrong, however, when they saw all the large magnates – Corner House, Robinson, Rhodes, Barnato, Goerz, Albu – as a monolithic bloc; so events were to show. The first President of the reformed Chamber was Hermann Eckstein. After a couple of years he was succeeded by Phillips and it therefore became an unwritten tradition, like the Habsburgs' title to the Empire, that the presidency would be held by a man from the Corner House. The honorary President of the Chamber was Kruger himself, but this was never an honour which weighed heavily with him.

From its first annual report the questions which preoccupied the Chamber were the same: the dynamite monopoly, railways, food supply, and perhaps the most important question of all: native labour.

Although the Transvaal as a whole had a large black population there were few Africans living on or near the Witwatersrand when gold was found. The nearest sizeable community was north of Rustenburg, seventy miles distant. Blacks were soon coming to the

mines to earn money to pay for cattle, but this casual to and fro of
nearby tribesmen did not satisfy the appetite for labour in the mines.
Quite soon extensive recruiting of labour on the periphery of the
Transvaal had begun. Once again the co-operation of the government
was required. Phillips hoped 'the Government will take this matter in
hand. Our principal supply of niggers comes from the North Transvaal and East Coast', that is, Mozambique, where there was a
recruiting station at 24° South. Places were naturally needed on the
migratory road for the men to shelter and buy food. Most important
was the need for a safe passage back for those men with the money
they had earned; some were being cheated of it by Boers en route,
Phillips said, and stressed the importance of safe passage to the supply
of 'cheap and plentiful labour. . . . As our work increases so we shall
want more labour.' These last words indeed summed up decades of
the history of the Rand.

This demand for more labour was a result of the peculiar circumstances of the Rand fields. The price of gold was fixed, the price of
other necessities such as dynamite was high, or hard to control, as was
white labour. A steady stream of skilled miners was coming to the
Reef, 'hard-rock' men from England, especially Cornwall and the
north-west, and from Australia. The mines needed their skill with
modern mechanical rock drills and with explosives ; these miners
could demand and get high wages.

The blacks were a quite different case, unskilled labourers working
first with picks and, as the mines spread underground, with hand-
drills. They dug the mines with sweat and muscle. But, although
South Africa was still blessed with its great raw material of cheap
native labour, the problem was how to collect and use it ; and above
all, since it was the only outlay of the mines which could be controlled
and reduced, how this could be achieved. In the first years wages rose
until in 1890 the average pay of natives was as high as 63s a month,
exclusive of keep.* A concerted effort by various companies led to an
agreed scale by sixty-six companies with the result that in the course
of three months the average rate was reduced to 41s 6d. Still the
problem remained, and from its inception the Chamber directed
much energy to the 'question of the reduction of native wages'.

The problem was twofold : annual shortages of labour and high
wages. The Chamber's first annual report complained that 'the
supply of Kaffirs is at present totally inadequate', and that 'a steady
rise of wages all round . . . is adding a very heavy additional expense'.

---

*   A (white) junior mill hand was paid £20–25 a month.

The two aggravated each other. When labour was in seasonal short supply different companies would compete for it with high wages. A labour shortage and competition between mine managers repeatedly drove up black wages, it was claimed. Several mine owners were enthusiastic about a pass system, but this had to be weighed in the balance, and for the moment most managers were against it : 'any system by which additional restriction might be placed upon the Kaffir . . . might have the effect of creating a bad impression in the Kaffir Districts . . . and so operate to retard the influx of Natives'. At Kimberley the mines had broken their blacks' spirits by compounding, searches and passes. On the Rand, the problem was different and more intractable : how to lure labour to the mines while holding wages steady or even lowering them.

In August 1890 the Chamber agreed with the mine managers on a monthly maximum wage of 40s. The agreement was theoretical: average wages were still running at 63s 4d per month. But by October a general reduction to 48s 10d had been effected. For two reasons, however, it was difficult to make these reductions stick. Firstly, much as companies and individual mine managers wanted to reduce costs they still more wanted to keep their mines milling, and in 1891 a severe shortage of labour led to the defection of several companies from the agreement. Secondly, there were important mine owners who were always opposed to any agreement on maximum wages – not out of kind-heartedness towards the labourers but through obstreperousness and resentment of the other companies. Mine owners could use black labourers' wages much in the way that Fleet Street press lords like Beaverbrook used printers' wages as a means of cutting their competitors' throats. Chief among those rebels, it need hardly be said, was J.B. Robinson.

Things looked brighter for the owners during the course of 1891 but August 1892 again found Phillips complaining that 'Our niggers are getting paid far too much.' By this time a pass system had at last been put into action to control the movement of labour and to prevent the 'theft' of labour by one mine from another, which had become widespread. By May 1892 the number of black labourers on the Rand had increased to 19,000. The Chamber formed a Native Labour Department in recognition of the importance of the question, which co-operated with Pretoria over the pass system. The Republic was quite amenable in that matter, but a thorough operation of the pass laws was difficult because of lack of funds and general inefficiency.

In all its activities, the Chamber of Mines bore witness to the central economic problem of the Rand. Most people were slow to see that the problem was diametrically opposed to Kimberley's. The magnates, however, very quickly saw that in the case of diamonds the price was variable and had fluctuated wildly in the two decades before amalgamation. The price of gold by contrast was fixed, and the problem was to control variable costs. In each case the answer was combination or cartelization but with opposite ends in view. In Kimberley the object was monopoly in the strict sense, a market condition with a single seller who can then dictate the price: upwards. On the Rand the object was monopsony, the market condition when there is a single buyer who can dictate the price: downwards. This was the aim towards which the magnates conspired in the critical matter of labour. It was not necessary to amalgamate the mining companies if there was another institution. What De Beers Consolidated was to Kimberley the Chamber of Mines was to Johannesburg.

By 1892 the Rand was well on the road to recovery and the Corner House men both in Johannesburg and in London were forming a plan of campaign: the launching of a great new mining company. Much of that year was spent in making arrangements for the purchase of vendors' interests and for the flotation which was to come. In the same month Beit wrote from London to agree on the name of the new company, Rand Mines Ltd. It was to be the culminating achievement of H. Eckstein & Co.

Yet the man who gave the firm its name did not live to see the flotation. The personnel of the Corner House was changing rapidly. J.B. Taylor had announced his intention of retiring, still a young man. He had married Miss Mary Gordon of Pietermaritzburg in March 1891 not long after his thirtieth birthday, as he had already made a respectable fortune and they both craved the life of the English countryside. Eckstein was also headed for England, to join Wernher, Beit at the start of 1893. He handed over the Presidency of the Chamber to Phillips, resigned his other Transvaal directorships, and the family sailed north. In London Eckstein supervised the last details of the new company. The directors were to be Wernher, Phillips, Harry Mosenthal, Neumann, Rudd and Eckstein himself. The strain took its toll and he was in poor health when he left to spend Christmas in Stuttgart. On 12 December he wrote to Phillips, 'I am not very strong as yet but am picking up.' After Christmas he was back in bed, doctors continually attending him. On 16 January he died in his young wife's arms.

Rand Mines was Hermann Eckstein's memorial. Until then the Corner House had not been directly concerned with mining operations

Each mine was floated as a joint-stock company with its own directors and its own manager. The Corner House kept control through share ownership but more importantly by dominating the boards of directors. These mining companies were never what would later be called wholly-owned subsidiaries. That indeed would defeat the whole object of the exercise which was to raise capital by flotation, by the sale of vendor's interest, in some cases by 'booming' the stock.

The 'group system' had quickly come to dominate the Rand, with each mine controlled by one or other group. Often there was keen rivalry between the groups. When Robinson was involved, any rivalry was sure to be very bitter. And yet the operations of the mining groups were very far from a Hobbesian state of all against all, or indeed the *laissez-faire* economist's dream of pure, collectively beneficial competition. Kimberley had been cartelized, to the great benefit of the cartelists if not of those who bought diamonds. Despite early attempts at amalgamation, and persistent subsequent rumours, the Rand did not lend itself to cartelization in that way. And yet the houses had numerous common interests expressed through the Chamber of Mines. Beyond that, the Corner House practised a conscious policy of conciliating the other houses.

The flotation of Rand Mines first of all raised cash. The company was floated in February 1893 with a nominal capital of £400,000 in £1 shares. Here indeed was 'a gold-mine'. Those who came in on the ground floor were bound to make a very great deal of money through the issue, though even then just how much could not be guessed: within thirty months, in mid 1895, these shares stood at £45. H. Eckstein & Co. contributed 1,300 claims to the new company as well as the controlling interests in five existing companies. This vendor's interest was compensated with 200,000 shares for the partners who all made substantial, new private fortunes from the flotation: Beit, Wernher, Michaelis and Rube in London; in Johannesburg Taylor, Phillips and Friedrich Eckstein, Hermann's younger brother, who had come to Johannesburg in 1888 and now succeeded him.

Next, the flotation greatly increased the scope for 'conciliation'. Eckstein had always tried to persuade the London partners of the use of working alliances. As he argued, the business 'is a peculiar one and by excluding others from good bargains now and then a good thing will slip by me'. Now, others were let into the good bargain.

After the vendor allotment, 27,000 Rand shares went to the Rothschilds, divided between the London and Paris firms. Four members of the Diamond Syndicate which Wernher had founded divided 8,400 shares. The engineers who had come to the Corner

House by way of Rothschilds were also rewarded: Hennen Jennings received 6,000 shares, H.C. Perkins, 4,000, Hamilton Smith 3,000 and E.G. De Crano 2,000. Rodolphe Kann, Porges's kinsman and the Rothschilds' connexion took 3,000, Ernest Cassel, the German-born financier and crony of the Prince of Wales, took 6,000.

The largest allotment of all outside the Corner House was 30,000 shares to Goldfields. This was the supreme example of the policy of keeping potential allies sweet. The friendship of Rhodes and his connexion was essential. In the same way, 10,700 shares were divided among Abe Bailey, Carl Hanau and Sigismund Neumann. There was a final allotment which for the most part was taken from the partners' own allocation, for what has gently been called public relations purposes. Shares were allotted at par to J.G. Kotzé, the chief justice of the Transvaal and to J.G. Leyds, the state secretary, who both received 200. What was remarkable was not so much the 'insider trading' as the way in which the greatest financial power in the Transvaal was in effect working to turn the Republic into a company stage. Therein lay the seeds of terrible conflict.

Within the Transvaal and despite the hostility of the government the mining industry had many advantages, not least its press influence. The *Star* had begun life in Grahamstown before it was brought to Johannesburg. It was consistently favourable to the mining houses, which owned it. That was scarcely surprising as its editor, Francis Dormer, had received a par allotment of 200 Rand shares. The position of its rival, the *Standard and Diggers' News*, was more equivocal. It was generally hostile to the Randlords and sympathetic to both the poor whites and the Pretoria government, which from time to time subsidized it. In 1894 it briefly displayed an unwonted sympathy towards the Corner House but that was easily explained. Phillips told the London office: 'Lippert was paying at the rate of £1,000 p.a. He declined to continue. They came to us and we have agreed to give them £83 6s 8d per month so long as they support the firm's influence in the place. This is strictly private.'

The Rand financiers could buy the South African press and they doubtless bought the Paris press when it suited them. The London press was less biddable. From the discovery of gold the attention of English newspaper readers had been focused on the Rand. London papers watched the magnates' doings with a fascinated but far from uncritical eye. Even *South Africa*, a weekly magazine founded in 1890 to gratify the growing interest of the investing (or punting) public in 'Kaffirs', was by no means sycophantic. Still less were the serious London financial press: papers like the *Economist*, the *Statist* and the

Financial News took a close interest in the wheelings and dealings of the Rand houses, however discreetly they tried to wheel and deal.

A year before Rand Mines was launched the *Financial News* had expressed its dissatisfaction with South African mine shares policy. The financiers, so the paper noted, were taking full advantage of the distance between London and Johannesburg. When companies were floated it had become the practice to hold in reserve certain portions of the authorized issue in order to procure further capital as needed: so far so good. If the companies had their headquarters in London those shares would be offered *pro rata* among existing shareholders either at the current market quotation or at another fixed price. 'But such an arrangement as this is, apparently, thought unsuitable by the Johannesburg wire-pullers.'

The Crown Reef Company was one of the most successful on the Rand: its dividends had risen from 12 per cent in 1888 to 50 per cent in 1891, and stayed at that level for several years. Its shareholders had been asked to sanction the issue of 14,000 shares which had been held in reserve. 'Messrs Wernher, Beit & Co., who are a good deal mixed up with many of these Rand mines', arranged proxies and in effect packed the meeting at which it was agreed that 4,000 shares should be sold at £4 17s 6d to Rudd, who 'appears as usual to have been acting for Mr Rhodes'. Then Rudd was to buy another 5,000 at £5 in June and the balance at £5 5s in December. 'It is impossible to imagine any justifiable reason for the Directors preferring to play into the hands of the coterie for whom Mr Rudd acts', the *Financial News* thundered.

Eighteen months later another deal came to light when a group of claims had been sold to Eckstein's by the Robinson Deep Company, all of whose directors were connected with the Corner House, in such a way that property of unknown value could be explored with shareholders' money. The *Standard and Diggers' News*, in the days before it had been temporarily tamed, complained that in view of 'the discredit such transactions bring upon the Rand, many must heartily wish that these gentlemen's genius were not so fertile, or that they would transfer their interests to some other country'. A London paper explained the transaction in detail, the profits accruing to 'the great Eckstein–Beit combination' whose confederates might 'realize a secret profit of 2,000 per cent in the space of twelve months . . . with all the opportunities for stock-jobbing on the bear side if the claims had not proved worth acquiring', finally lamenting that 'the morality of the financial magnates of Johannesburg is at such a deplorably low ebb'.

The financiers could shrug off criticism. After the successful flotation of Rand Mines, those who had designed it were cushioned by still further wealth. Lionel Phillips, now also a rich man, went to London to join Florrie who had bad-temperedly left Johannesburg for a long absence. In the late spring of 1893 the Imperial Institute of South Kensington was opened by Queen Victoria. There was a 'Cape Court' decorated with diamonds and attended by South African courtiers in the form of Beit, the Julius Wernhers, the Max Michaelises, the Lionel Phillipses, and the Barney Barnatos among many others. Back in Johannesburg work was under way on the Phillipses' new house. It was to be called Hohenheim, like the little four-roomed house in Noord Street which they had taken over from Eckstein, but on a very different scale. It was two miles further north, on the farm Bramfontein, which Eckstein owned. The new residential area was named Park Town and was developed by a company of which Phillips was a director. Several of the Randlords were to build their houses on or about Parktown ridge, but few as grand as Hohenheim – and none so extravagant. It soon became known as 'Phillips's Folly' and was a painful burden to him.

Beit had more immediate and more serious worries. His cousin Edward Lippert had a younger brother William who had also come to South Africa from Mecklenburg and had joined the Union Bank in Cape Town. In a period of financial squeeze the bank was near insolvency and William Lippert tried to cover with securities. Unable to provide his own funds, in a panic he made up a bill and added a forged signature. Beit honoured the bill, and another to a total of £150,000. The third time he rebelled and exposed the forgery. In the resulting scandal the bank crashed with much ruin. Lippert fled the country.

He finally returned to face trial, and Alfred Beit had to give evidence against him. Reproached by the judge for condoning the first two forged bills Beit said, 'I thought of only one thing – to save the family.' Lippert was sentenced to seven years' hard labour. It was not thought an excessive sentence. It was a minor episode, but a distressing one, and Beit must have thought that he had suffered enough from his Lippert cousins.

Kimberley had been benevolent godfather to the Rand from the beginning as it was diamond capital which started the big groups in Johannesburg and opened up new mines. When diamonds had first been found by the Vaal they had had no purpose except decoration. It was during the very decades when Kimberley was dug and amalgamated that a 'useful' purpose was found for diamonds. As well as large and fabulously valuable gemstones the four great pipes yielded up masses of diamonds too small for jewellery. Could not some use be made of them, the hardest of all minerals?

There could and there was. Drills had advanced in design so that, adding section after section of shaft, a powerful engine could sink a bore hundreds or even thousands of feet down through the earth's surface; with a bit harder than rock itself, bores might be drilled anywhere. Quantities of tiny diamonds packed together formed a hard edge or tip for cutting and drilling tools. So it was that on 5 June 1893 Kimberley and Johannesburg, diamonds and gold, came together with physical yet almost poetic congruence. On the Rand Victoria, a second-row deep mine 4,100 feet south of the outcrop, drilling with a new diamond-tipped bit, the main reef was struck 2,391 feet down. The golden city's future was certain.

9
The Young Burgher

Most of the great diamond men had moved to Johannesburg by the early 1890s, but one was an absentee gold magnate. Despite Rudd's 'unbelievable stupidity' on the Rand and Gardner Williams's lack of enthusiasm, Rhodes had soon got to grips with the gold-fields. But he did not base himself at Johannesburg, which was, along with diamonds in Kimberley, only one of his centres of activity, the others being expansion in the north and politics in the Cape. Events must be retraced during the late 1880s, when Kimberley was amalgamated, his career in Cape politics blossomed, and his Zambesian schemes began.

His uniquely ambitious plan of action was, he liked to think, inspired by a distinctive, odd and in truth weirdly immature philosophy which he had formed in his lonely days on the diamond diggings and at Oxford. Though brought up as a son of the parsonage, Rhodes soon abandoned his father's faith, confessing to a friend that life on the diamond-fields had not strengthened his religious beliefs.

In this as in so many things he was a child of his times. It was in the late nineteenth century that Europe at last lost its religion. God died and the Sea of Faith ebbed with its melancholy, long, withdrawing roar. Like many contemporaries far more remarkable in their mental scope, Rhodes looked for something to put in the place of religion. Chesterton said that when men cease to believe in God they do not believe nothing, they believe anything, and that could be applied not unfairly to Cecil Rhodes. He concocted a curious pagan mixture, part Marcus Aurelius, part Gibbon, part *The Martyrdom of Man*, Winwood Reade's strange mishmash of history and religion, amateurish but infectiously enthusiastic. That book's message was in large part vulgar Darwinism – so popular at the time – and it taught that suffering is an essential cause of human progress.

This struck a chord with young men growing up in a time of violent change, of hardship and inequality and, as Rhodes entered manhood, of a new idea, the imperial mission. England's overseas empire was growing apace. In the 1870s there persisted in the Liberal party and in the Colonial Office itself a tradition of detachment and isolationism. It was partly moral in inspiration, though it had its practical side. The Little Englanders as they would become known could truthfully claim that it was Free Trade and not Empire which had given England her generations of greatness. But the tide was turning. When Disraeli's second administration came to office in 1874 it was too early for imperial passions to influence the election.* But Disraeli and his Colonial Secretary Lord Carnarvon had set upon a consciously imperial course for the first time. The Queen became Empress of India in 1876 and Disraeli was ennobled. His title, Beaconsfield, appropriately passed on in turn to the mining town where the Dutoitspan and Bultfontein mines stood.

At the same time there was an intellectual current running. It is too easy now to dismiss imperialism as no more than a racket and to suppose that all those who succumbed to its spell concealed sordid motives. There was idealism as well. In 1857 the famous David Livingstone had lectured at Oxford and Cambridge on the theme 'Dedicate your lives to Africa'. Then the greatest English moralist and aesthetic philosopher took up the cause. John Ruskin became Slade Professor of Fine Art at Oxford in 1869 and delivered an inaugural lecture before the University in the Hilary Term of 1870. He spoke of the need for making honest wares, he damned the materialism of the industrial age, he reproved the pursuit of pleasure; these were his habitual themes. And then:

> There is a destiny now before us – the highest ever set before a nation to be accepted or refused. We are still undegenerate in race; a race mingled of the best northern blood. We are not yet dissolute in temper, but still have the firmness to obey ... [England] must found colonies as fast and as far as she is able, formed of her most energetic and worthiest men: seizing every piece of fruitful waste ground she can set her foot on, and there teaching these her colonists that their chief virtue is to be fidelity to their country, and that their first aim is to be to advance the power of England by land and sea ... if we can get men for little pay, to cast themselves against cannon mouths for love of England, we may find men also who will plough and sow for her, who will behave kindly and righteously for her, who will bring up their children to love her and who will gladden themselves in the brightness of her glory, more than in all the night of tropic stars.

* The reasons for the defeat of the Liberal government were domestic, notably its collision course with the licensed trade. As Gladstone said, 'We have been borne down in a torrent of gin and beer.'

To read this inaugural lecture now is to understand that imperialism was not just pillage. There *was* force and fraud and exploitation, and the Empire's bounds were pushed wider in part for economic reasons, but that was not the whole story. In the 1870s and 1880s the 'imperial idea' spoke to exactly the kind of high-minded, energetic and meddlesome young men who two generations later would turn to communism.

Among those young men was Rhodes. He read Ruskin's lecture and in his two years at Oxford became his disciple. Ruskin was added to the odd mixture of influences on Rhodes's mental formation, such as it was. If one considers these influences, it may be that he has been treated with a seriousness which he does not deserve by his many adulators, and even by those who turned against him in hatred, as did Olive Schreiner, who took his estimate of himself at face value. It may be that his mental formation was much more mundane, that his real mentors were not in fact Ruskin, Winwood Reade and Marcus Aurelius but 'the editorialists of colonial newspapers and the cruder outbursts of the *Standard and Diggers' News*, mixed with the finer sophistries of the *Pall Mall Gazette*'.

He left behind damning evidence against the self-promoted legend of the thinker as a man of action. Rhodes wrote several wills. The first, in 1872, left all his then modest wealth to the Colonial Secretary to be used at his discretion for the expansion of the British Empire. Within five years this curious seed had sprouted wildly. His second will was written on 2 June 1877, the day when Rhodes became a Freemason. Masonry was part of the brew fermenting inside him as he wrote, along with Ruskin's mysticism and imperialism and Reade's unknowable God whose providence was worked out as evolutionary progress. The result is a very strange document entitled 'Confession of Faith':

> It often strikes a man to enquire what is the chief good in life; to one the thought comes that it is a happy marriage, to another great wealth . . . to myself thinking over the same question the wish came to render myself useful to my country.

He then speculates as to how that might best be done.

> I contend that we are the finest race in the world, and that the more of the world we inhabit, the better it is for the human race. Just fancy those parts that are at present inhabited by the most despicable specimens of human beings what an alteration there would be if they were brought under Anglo-Saxon influence.

Gold mine vignettes, 1889

Outside Johannesburg Stock Exchange, 1890

Early Rand mine: compound life

Johannesburg: tailing wheel
at Robinson Deep

George Albu

Isaac Lewis and Sammy Marks

Miners on the Reef

Minnie and Hermann Eckstein

Friedrich Eckstein

Alois Nellmapius

Edward Lippert

Headgear

Outside the goldmine shaft

Stamp battery

In a drive

Drilling in the stopes

Adolf Goerz

Abe Bailey

Ernest Oppenheimer

Harry Oppenheimer

He thinks of the power exercised by the masonic order, and when 'I read the story of the Jesuits I see what they were able to do in a bad cause, and I might say under bad leaders.' From these two examples comes his plan:

Why should we not form a secret society with but one object, the furtherance of the British Empire, and the bringing of the whole uncivilized world under British rule for the recovery of the United States, for the making of the Anglo-Saxon race but one Empire. What a dream, but yet it is possible ... [The loss of America to British rule was regrettable.] Even from an American's point of view just picture what they have lost, look at their government, and not the frauds that yearly come before the public view, a disgrace to any country and especially theirs which is the finest in the world. Would they have occurred had they remained under English rule ... think of the countless ooos of Englishmen that during the last 100 years would have crossed the Atlantic and settled and populated the United States. Would they not have made without any prejudice a finer country of it than the low class Irish and German emigrants! All this we have lost and that country loses owing to whom? Owing to two or three ignorant pig-headed statesmen of the last century, at their door lies the blame. Do you ever feel mad? Do you ever feel murderous? I think I do with those men.

He goes on to consider that 'Africa is still lying ready for us, it is our duty to take it.' Then he returns to the secret society 'which should have its members in every part of the British Empire working with one object and one idea, we should have its members placed at our universities and our schools'. A young man once selected would be 'sent to that part of the Empire where it was felt he was needed'. The Confession ends by leaving his goods to S.G. Shippard, the attorney general of Griqualand West and the Secretary for the Colonies 'to try to form such a society with such an object'.

Rhodes returned to South Africa to ponder and expand his scheme. In September he made a more detailed will which he deposited, sealed, with a lawyer in Kimberley who should hand it on to Rhodes's death to Shippard. He in turn was to execute the will with the Colonial Secretary. The entire estate was to be used

for the establishment, promotion and development of a Secret Society, the one aim and object whereof shall be the extension of British rule throughout the world, the perfecting of a system of emigration from the United Kingdom and colonization of all lands wherein the means of livelihood are attainable by energy, labour and enterprise, and especially the occupation by British settlers of the entire Continent of Africa, the Holy Land, the valley of the Euphrates, the Islands of Cyprus and Candia, the whole of South America, the islands of the Pacific not hitherto possessed by Great Britain, the whole of the Malay Archipelago, the seaboard of China and Japan, the ultimate recovery of the United States of America as an integral part of the British Empire, the inauguration of a system of Colonial Representation in the Imperial Parliament, which may tend to weld together the disjointed members of the Empire, and finally the foundation of so great a power as to hereafter render wars impossible and promote the best interest of humanity.

Not surprisingly, Rhodes's admirers have passed by these documents in silence. Their semi-literacy and vacuity – small tribute to Oriel – is striking enough; more so is their puerility. Rhodes was no longer a boy or an adolescent when the Confession was written; he was twenty-four years old, past the age at which those effusions might be forgiven. But then, in a sense, he never grew up. His energy and ambition gave him a masterful personality. He dominated not only abler men but more sophisticated ones too, but there remained something deeply immature inside him all his life.

Just as his adulators wanted to build up Rhodes as a great man, so his detractors wanted to put him down as a bad one. Maybe he was a bit of both and of neither. Rhodes has been compared* to the great totalitarian tyrants of the century into which he barely lived. That is unfair. Whatever may be said of him, Rhodes was not a deranged murderer on the scale of Hitler and Stalin. Yet there is something telling in the comparison. One part of it is this boyishness, this inchoate yearning and love of dizzy speculation. Just as Hitler would indulge in hours of world-historical speculation, talking about vast new empires and great new buildings to be made with his beloved concrete, so Rhodes dreamed beyond ordinary mortal bounds, even beyond the earth.

'Expansion is everything,' he would say. 'These stars that you see overhead at night, these vast worlds which we can never reach! I would annex the planets if I could. It makes me sad to see them so clear and yet so far away.' The same childish musing comes out again and again: the delight in scheming, in making plans, in drawing up lists, in conspiring. The comparison goes further. It is said that great men are almost always bad men, but some great men are also

---

* E.g. by Hannah Arendt in *The Origins of Totalitarianism* (1958).

men of parts. Neither Henry VIII nor Frederick the Great was a good man, but they were both gifted musicians; Bismarck may have been unlovable, but his speeches and writings belong to German literature. In the case of Rhodes, as with the modern dictators, there is a yawning gap between what they did and what they were, between great deeds of conquest and personal insignificance and lack of talent.

Both his contemporaries and those who lived after him pondered all this as they tried to work out what made Rhodes tick. An unmentionable subject, in public at least, during his lifetime was his sexual nature. He was a neurotic man with quirks such as a morbid fear of snakes; he was a misogynist whose shunning of female company was especially notable at Kimberley where most men had mistresses of some sort; he had almost girlish characteristics and a high-pitched laugh; he sometimes drank heavily; and he formed intense friendships with younger men, some of them his secretaries, the most famous being Neville Pickering, to whose deathbed he hastened though in the middle of critical negotiations on the Rand.

All this must lead post-Freudians to speculate that Rhodes was homosexual. The case is inconclusive. His misogyny is poor evidence of homosexualism: everyday observation shows that many homosexual men have close friendships with women. (It is the professional philanderer who actually dislikes women.) Rhodes almost certainly never gave physical expression to his sexual nature. Whatever it was, it was suppressed and became part of the force driving him on. Expansion is everything. Maybe Freudian psychology is less useful here than Adlerian. Where other men are driven by the need for sexual gratification, Rhodes had a very strong taste for power. He had great schemes to accomplish and, as more than one heart attack reminded him, limited time in which to accomplish them.

He turned from making money to exercising power. In his guise as Cape politician Rhodes had spent the period 1882 to 1885 involved in territorial questions over the land to the north of Griqualand West and the west of the Transvaal which he counted as a 'road to the north', his projected satrapy in Matabeleland and Mashonaland between the Limpopo and Zambezi. This strip of land became the subject of much contention. The dispute culminated in a conference at Fourteen Streams in January 1885 where Cecil Rhodes and Paul Kruger met for the first time. Of the two, Kruger was more in favour of a customs union at least between the Cape and Transvaal; he was still in desperate financial straits; the Main Reef had not been discovered.

The road to the north was declared a British protectorate, which meant as far as Rhodes was concerned that it stood open for him. He now wanted to advance a step further and establish a protectorate over Lobengula's Matabele kingdom in western Zambesia. At his urging Sir Hercules Robinson, the Cape Governor, sent a deputation to arrange that Lobengula's foreign relations should be in British hands. Eight months later in October 1888 Rhodes sent his own mission led by Rudd which by one means or another extracted a 'concession' which granted a monopoly for the exploitation of mineral rights in the whole of Zambesia and Mashonaland as well as Matabeleland.

The idea of opening up this new colony by means of a chartered company was not originally Rhodes's but a rival group's. In London Rhodes inveigled these rivals into joining forces with him, and then overbore not only the Colonial Secretary Lord Knutsford but also the Prime Minister Lord Salisbury. Against their judgement, in July 1889, a charter was issued for the British South Africa Company. It was granted powers of government as well as of mineral exploitation in Zambesia.

Salisbury's position was poignant. He had come to office in 1886 after the failure of the First Home Rule Bill and the Liberals' historic split over Ireland. With one three-year Liberal intermission he remained Prime Minister, and Foreign Secretary also, until the twentieth century, so he witnessed the extraordinary upheavals in South Africa: the consolidation of monopoly power in Kimberley, the opening of the Rand, Rhodes's Cape premiership, the attack on Kruger's republic from within and without, the Jameson Raid, Chamberlain as Colonial Secretary in his own Cabinet scheming with Milner as High Commissioner at the Cape and thus with the mining magnates, and finally the outbreak of war. As prime minister he was closely involved in – ostensibly responsible for – the part Great Britain played in South Africa. Yet Salisbury had an old-fashioned Tory's distrust of imperial expansion and an old-fashioned gentleman's contempt for jumped-up adventurers. He disdained the feverish, hysterical imperialism of the 1890s almost as much as a radical or a Little Englander, but he was fated to preside over it.

After winning his charter, Rhodes moved fast. More missions were sent north, and in September 1890 the Pioneer Column which had left Kimberley under young Frank Johnson pitched camp at what would become Salisbury (in some ways an ironical name) and Rhodesia thereby came into being. Rhodes himself visited the new territory for the first time a year later.

In the second half of the 1880s Rhodes's career in the Cape had been innactive. Even he could not conduct a battle as frenetic as that at Kimberley while giving full-time attention to politics. He was known as a representative of the diamond interest, of course, but also in Salisbury's words as 'rather a pro-Boer MP'. The Member for Barkly West with his Dutch-speaking rural constituents was sometimes called 'de jonge burgher'.* He formed a political friendship with J.H. Hofmeyr who led the Afrikanerbond, the political expression of the Cape Dutch. It was Rhodes himself who was to call the Dutch 'the coming race in South Africa', by which he really meant that they were emerging as the dominant force in Cape politics. They were no longer merely despised peasants. After the British defeat at Majuba Rhodes told Hofmeyr that the battle had made English and Dutch respect one another. The Dutch language acquired equal status in the Cape Parliament in 1882, a belated recognition that it was easily the most widely spoken language in the Cape.†

However genuine his admiration for the Dutch may or may not have been, Rhodes sincerely despised the 'English' party in the Cape Assembly which in the 1880s was 'hopelessly divided and incapable. And it had no policy at all beyond that of serving office'. By contrast the Afrikanerbond was united and resolute under the direction of Hofmeyr, 'without doubt the most capable politician in South Africa'. Hofmeyr was no extreme nationalist. He held the Afrikanerbond back from the policy of a united – that is, Dutch-ruled – South Africa, an outcome which was to be achieved in very different circumstances sixty years later. He favoured 'colonialism' which meant the Cape looking after its interests free from British interference, and he favoured Rhodes's northern schemes which would have to be achieved in South Africa itself, with no help from the 'imperial factor', a popular phrase which Rhodes coined. Rhodes would have preferred – not from modesty – Hofmeyr to head the ministry but Hofmeyr declined and on 17 July 1890 Rhodes became Prime Minister of the Cape Colony with Hofmeyr's support. His government was a coalition, with members of the Afrikanerbond as well as Liberals such as Merriman.

Having ascended to the top of the greasy pole, Rhodes pursued his dreams with political power in one hand, economic in the other.

---

* Maybe his nickname was a joke of Merriman's; maybe Rhodes, as he later claimed, had said to Kruger on their meeting, 'You are the old burgher and I am the young burgher.'

† 'Dutch' in the sense of High Dutch was still the official written language; the spoken language in the Cape and the two republics was the Dutch-descended Afrikaans. Apart from the fact that 'Cape Dutch' outnumbered English in the colony, Afrikaans was, as it remains, the language of the 'Cape Coloureds.'

When the Chartered Company had been set up, two of the largest
blocks of shares in it had been granted to the two great mining
companies, De Beers, his creation at Kimberley, and his new creation
on the Rand, Goldfields of South Africa, which was registered on 9
February 1887 and held its first meeting in London on 19 March.
After his initial maladroit performance on the gold-field, Rudd had left
for London where Rhodes pursued him by means of letters. Rhodes
journeyed to the Rand again, where 'I took Witport from the Jews,
and paid them £1,500 for a half'. This frenetic activity was matched
by Rudd who now made up in the City of London for the indecision he
had shown on the gold-fields. Rhodes and Rudd had learned the
lessons of Kimberley well, and applied them well.

At the first meeting of their new company, Rudd said that he could
give no details about the properties which the company controlled or
of their likely value. All the directors of the company apart from
Rhodes and himself – the joint managing directors – were more or
less ignorant on the subject and had not the practical knowledge of the
Transvaal which he had himself. In effect the company was to be
floated on the reputation of two men. As a sign of good faith there
would be no vendors' interest. This was in striking contrast to the
other new gold-mining companies. Of the remaining nine South
African companies floated in London that year, the vendors' interest
was usually well over 70 per cent: £350,000 of £450,000 in the case of
the Great She Gold Mining Company, £55,000 of £60,000 with Gold
Fields of Apollonia.

On the face of it Rhodes and Rudd were generous towards their
new shareholders, but they asked for something in return. The two
were to receive three fifteenths of all profits on account of their
Founders' shares, and a further two fifteenths in respect of their
position as joint managing directors. They would turn over to the
company any properties they acquired in the Transvaal, but their
remuneration was to run for an initial three years and could be
changed only by a vote of three quarters of shareholders. The
readiness with which this arrangement was agreed to testifies to
Rhodes's prestige. It also set the tone for the future of South African
mining company promotion, despite the Barberton fiasco. Investors
privileged to take part in these ventures had to understand that the
interests of the magnates on the spot would come first.

The flotation of Goldfields of South Africa was a brilliant success.
Within its first week 70,000 shares were taken; by the end of April
94,000 in London; and by the end of October all 250,000 of the £1
shares had gone. The flotation was no more than a beginning. More

and more stock was sold to a willing public. As vendor share interests in mining properties were acquired, they were resold to the investors. Goldfields bought 202,034 £1 shares in the Luipaards Vlei Estate mine for a mere £67,000 before reselling them. In the space of six weeks during 1889 the company sold more than £300,000 worth of Rand shares.

The impression that this gave of Goldfields as one of the leading companies on the Witwatersrand was misleading. For one thing Rhodes and Rudd had, through ill-luck and bad judgement, not bought wisely in the scramble for mining properties. One of the farms which Goldfields bought on the West Rand was on poor grade ore, another was on a geological fault, a break in the Main Reef series. In any case Rhodes still believed that his financial base lay in diamonds. And the capital raised through Goldfields was substantially returned to Kimberley. By December 1887 Rhodes and Rudd had spent £57,000 out of unused capital and profits from gold share dealings on De Beers and Kimberley Central shares. Although Kimberley capital was to help launch the Rand the relationship worked backwards also. The first Rand boom paid for the Kimberley amalgamation. By the end of 1891 the investment portfolio of Goldfields contained gold shares to the cost of £33,000; its diamond shares cost £277,000.

The different strands were coming together: Kimberley, the Rand and the north. Rhodes had ensured that he could use De Beers funds for his northern expansion. He was now using Goldfields' capital – money raised from the flotation rather than from any gold as yet mined – for De Beers, and also for imperial ventures to the north. In 1889 120,000 new £1 shares were issued 'to develop the Matabele concession lately obtained', and he bought further shares in the British South Africa Company. Rhodes's various interests could not now be easily unravelled. He raised capital for the Chartered Company by selling watered stock, in the hope which he held for the rest of his life of finding a 'second Rand' in the north. The pioneers who took off for Rhodesia went not as farmers but as would-be gold-claim owners. They were deluded. So were those who bought the more or less fraudulently promoted Chartered stock.

The intricacy of these dealings puzzled Rhodes himself and sometimes enraged him when, as it seemed, power was being removed from his central grip in the name of immediate financial expediency. When Rudd found it necessary to sell some of Goldfields' De Beers interest in order to pay a dividend, Rhodes was furious. 'I think the GFSA have behaved disgracefully. I am thinking of resigning but shall await your decision . . . I do not think that I shall attend the yearly meeting' of Goldfields. And he never did, not once in his lifetime.

For all this petulance the floating of Goldfields had done well by Rhodes. He had more cash in hand, more capital as well as income to spend. Even his irritation had its productive side. He had been like a man gambling at four tables at once. Now, in his exasperation at Goldfields, he could take his eyes off one table and for the moment give his full attention to the Cape government. As prime minister of the Colony a visible change was coming over Rhodes. In 1890 he still had some vestige of a liberal reputation. It did not last long. A series of executive and legislative actions lost him whatever support he had enjoyed among enlightened South Africans.

Many people who had backed Rhodes recoiled at the legislation he now introduced. There was a Master and Servant Bill giving employers rights of corporal punishment over employees (white and black respectively, of course), which was familiarly known as the Strop Bill, or the Every-Man-to-Wallop-His-Own-Nigger Bill. This horrified Olive Schreiner who, up to then, had been a keen admirer of Rhodes. Thenceforth she put her considerable energy and reputation into denouncing Rhodes, even after his death.

Once he had achieved his office in the Cape, Rhodes dreamed more than ever of a great South Africa federated under his rule. On one question he was prepared to humour, even to entice, the Boer republics: 'native policy'. The Cape still kept up its policy of colour-blindness. A good few 'Coloureds' had the vote, so did a small number of the many black Africans who now lived within the expanded borders of the Cape Colony. Rhodes's attitude to the blacks was always paternalistic at best – he often spoke of Africans as 'children' – and he had no wish to let the Cape tradition stand in the way of warmer relations with the Dutch. He introduced a Franchise and Ballot Act to raise the qualifications for the vote so as to reduce the number of Coloured and black voters. He launched the phrase 'equal rights for all civilized men below the Zambezi', but what he originally said was 'all white men', an appeal to the Boers, and only a visit by a deputation of Coloureds from Kimberley led to the half-hearted change of phrase.

The most important piece of legislation, as concerns the light it shed on Rhodes's career, was the Glen Grey Act, which regulated the lives of 'Bantu' dwellers in the eastern Cape. It conveniently disenfranchised them, and introduced a labour tax (though this was not fully imposed). It was even more convenient that Rhodes, who jocularly defended the tax as 'not slavery but a gentle stimulus' was both mining magnate and politician. The mines, first at Kimberley and then on the Rand, utterly altered the relationship between white-ruled South Africa and its black peripheries. Until the middle of the century black and white had largely ignored each other. What contact there had been was, despite intermittent wars, often amicable and co-operative. Now all of a sudden white South Africa needed the black man as labourer on the mines. Many new laws and regulations provided that gentle stimulus for tribesmen who might otherwise have lived in contented idleness, by reducing the size of the lands blacks owned and thus their capacity for self-sufficiency, by demanding the payment of hut or labour taxes in cash which could only be earned on the mines.

Rhodes's term of office in Cape Town had some happy results for the colony, most of all in the realm of agriculture. He introduced American vine stock resistant to phylloxera; he introduced the ladybird to counteract a pest which attacked citrus trees; he pushed through a measure for the compulsory dipping of sheep against scab, despite obscurantist Boer opposition; and he improved the breeding of horses and cattle. Not least, he was the first man of his age to take a serious interest in Cape Dutch buildings, the beautiful and distinctive eighteenth-century houses of Cape Town and the surrounding Cape hills which had been badly neglected in the nineteenth century. Rhodes began their re-evaluation and restoration, starting with his own house at Groote Schuur. If he had governed an entirely agrarian country his ministry might be looked back on a golden age. As it was, there were many men and women even at the time who found his rule more and more sordid.

Although Rhodes had much criticism to bear, he was garlanded also with much flattery. That was no accident, as he had come to control a sizeable portion of the South African press through his various companies, the 'Rhodesian press' as J.A. Hobson was to call it. The jingoist and Tory section of the British press took him at his own evaluation. He was only one of many colonial politicians but when he came to London in February 1891 he was saluted as a world eminence. At first he supported the Liberals: on that visit he made a secret donation of £5,000 to the Liberal Party. Three years before he had given money to Parnell and the Irish Party; Merriman facetiously wondered whether he wanted 'a seat from Parnell, or is some leading shareholder in the French Company a Fenian?', but there was a plan behind it. Rhodes wanted to keep Irish members at Westminster if Home Rule was granted, in view of his fanciful dream of a greater imperial legislature.

To describe Rhodes as merely an absentee magnate on the Rand is, however, misleading. He was physically absent from Johannesburg and the money raised through Goldfields was put back into diamonds and Zambezia. The Chartered Company did not acquire the Rudd concession at first but that came in 1893 when the United Concessions Company, controlled by Goldfields, sold the concession to the Chartered Company for £2 million, once more enriching Rhodes and his friends. Rhodes had recruited to his personal service Dr Jameson who had paid several more visits to Lobengula to try to keep him sweet. Despite his absence Rhodes was closely concerned in the Rand: it was more than merely owning a large gold-mining company. Just as important was his intimate connexion with Beit, Father Joseph to his Richelieu. The two worked hand in hand in both diamonds and gold. Not all of Rhodes's other associates approved of the connexion. From London the severe Stow, whose relations with Rhodes were anyway always tense after the amalgamation, disapproved of attempts to rig the diamond market surreptitiously and warned Rhodes not only against Barnato but also Beit 'and the speculative element'. That only showed Stow's ignorance. Rhodes and Beit were inextricably bound up. Rhodes had a personal account with H. Eckstein & Co. and operated to a great extent through the Corner House. It was not an accident that the Corner House and Goldfields followed the same paths, for instance in moving into the deep levels. The two groups were intertwined. Just how much Rhodes owed to Beit is impossible to estimate; but financially it was a great deal.

The connexion with Beit was especially important to Rhodes once his northern venture had proved an economic flop. The Matabele were bloodily crushed in battle in 1893, a victory which further enmeshed in Rhodes's toils the British High Commissioner, Sir Henry Loch, boosted Dr Jameson's opinion of his own military abilities, and sent Chartered stock higher. But the gold in the north, the second Rand, was not there, as Rhodes finally conceded after visiting Rhodesia with his mining adviser John Hays Hammond in September 1894. And the British investing public were not the gullible mugs that the Randlords sometimes took them for – not quite. Rudd warned Frank Rhodes of the very strong feeling among Chartered shareholders about the managing director's remuneration and Davies in London reported 'a general complaint with the public that they are not allowed to get in on what is termed the "ground floor"'.

In South Africa the political outlook for Rhodes grew bleaker. He had hoped for some form of federation, to be dominated of course by him; but Kruger would not play. He tried to win Delagoa Bay, the

Rand's natural link with the sea, for the British Crown. That came to nothing. Then he tried to persuade Kruger, to the old burgher's horror, to seize Delagoa by force. Nothing happened. Rothschild considered buying Delagoa from the Portuguese but nothing came of that scheme either. As the gold-fields financially strengthened Kruger's government the possibility of federation seemed to slip away and with it Rhodes's hopes of dominating South Africa. The pressure was tightening upon him, both politically and economically.

All of this Beit saw with his attentive eye. He scarcely needed Phillips to tell him in June 1894 that Kruger had developed a deep fear that 'foreigners (of whom he regards Rhodes as the head) will gradually buy the whole country and oust the Boers'. That was a process which Kruger could scarcely have stopped without closing the mines. It could have done Rhodes's work painlessly. But it needed time, the one thing Rhodes did not have. After a severe illness in 1891 he felt more than ever that he might not live to be fifty. If he was to play a great stroke it must be soon. And so he began to make other plans. Beit told Phillips in the last month of 1894, 'I had a very confidential talk with Rhodes and the gist of it is that he will not stand for long Kruger's government.'

10
All the Conspirators

By the new year Rhodes was far from alone in his wish to be rid of the turbulent President. Ever since the Uitlanders had begun to flood into the gold-reef city they had ruled against the rule of the Transvaal authority as represented by bothersome petty officials and 'Zarps', the police of the South African Republic. Johannesburg bubbled with resentments justified and unjustified, grievances real and imaginary. The Uitlanders' dislike of Kruger's Republic had grown – and had been carefully nurtured. In 1892 the Transvaal National Union was formed to publicize the Uitlander cause and promote their claims, which on the face of things were plausible. Writing more than thirty years later, one of the Uitlander leaders, the American mining engineer John Hays Hammond, listed their grievances.

There was no sewage system or clean water supply in Johannesburg; barely one per cent of money spent on education in the town was spent on Uitlander children, that is to say on education in English; the Uitlanders paid for the city but had no say in running it; the mining industry was harassed by government monopolies, especially of dynamite; the railways were in the hands of another monopoly, the Netherlands South Africa Railways Company, 'whose shareholders were entirely German, Dutch and Boer', and whose freight charges were so extortionate that rather than use the last stretch of the line controlled by the Company it was cheaper to unload goods coming from the Cape on the Free State side of the Vaal, reload them into ox-wagons to cross the drifts, the Vaal fords, and bring them on to the Rand; the liquor distilled by the liquor monopolists was poisoning the Kaffirs at the mine, rendering them dangerous at work, or unfit to go to work; Kruger and his cronies leant heavily on the Transvaal judiciary, which was not independent or impartial; finally Pretoria expected the Uitlanders to do military service.

Each of these grievances told a different story and some affected one section of the Uitlander community more than others. Foes of the mining magnates could say that the grievances were factitious, no more than a weapon for agitation. Some were trivial. The National Union agitated much on the franchise question. Before gold was found it was quite easy for white incomers to obtain citizenship and the vote in the Transvaal. Then the influx of men to the Reef meant that Uitlanders would one day – perhaps soon – outnumber the original burghers. In 1890 the Volksraad raised the qualification for the franchise to fourteen years' residence, when of course most Uitlanders on the mines had been there for only a few years. A petition requesting an easier franchise was rejected by the Volksraad in contemptuous language.

But although the claims were sincerely promoted by some of the National Union leaders, they did not bear much examination. Even as 1895 drew to its close, quite a few observers thought that the Uitlanders' conduct showed that they were not sincere rebels: they were too busy making and spending money to take politics seriously. One acute observer said that the franchise issue was nonsense. Barnato compared the South African Republic to a limited company whose shareholders were the burghers. Why should they agree to a watering of the stock? (A metaphor taken from financial sharp practice was especially vivid for him.) Merriman agreed. He disliked the obscurantism and intransigence of 'Krugerism' but at the same time he saw that the mass of Uitlanders had no intention of making the Transvaal their home.

Even the agitators said as much, Hammond himself admitting that many were on the Rand only temporarily, and in private Phillips said even more bluntly that most Uitlanders did not 'care a fig' for the vote. It was a useful stick to wave at Pretoria, no more. So also in the matter of education: Barnato said that it was unreasonable to expect a Dutch government to treat its own language as foreign. Some of the grievances were real enough. Sewage and water were grave problems – as Barnato, owner of the Waterworks Company, knew all too well – and if Kruger had been wise he would have drawn several agitating teeth by granting some degree of municipal self-government to Johannesburg. But in truth, franchise and local government were unimportant: there were other grievances, economic ones, of quite a different order and very real indeed, not for the mass of Uitlanders but for the mining companies.

There was provocation enough on both sides in the early months of 1895. In his parading of the independence of the Transvaal Kruger persistently nettled the Uitlanders. On 27 January he attended a 'Kaiser Kommers', a banquet in honour of the birthday of the Emperor William II of Germany. He spoke of cementing the close ties of friendship between the Republic and Germany in inflammatory language. Then in April the Transvaal government tried to commandeer a group of Uitlanders to go on commando, to serve in one of the primitive expeditions against outlying black tribes – not a formality at a time of black unrest and insurgence.

The Uitlanders refused. They said that it was not for the country which refused them the vote to ask them to fight for it. The 'commandeering crisis' brought relations between Pretoria and the Uitlanders to boiling-point. The High Commissioner, Sir Henry Loch, came from Cape Town to Pretoria for talks with Kruger. Now it was time for provacation on the other side. Loch was met at Pretoria station by a rowdy mob of Uitlanders who unharnessed the horses from his carriage and dragged it through the streets singing 'Rule Britannia' and carrying a portrait of Kruger with a Union Jack hanging over his face. This was a gross display in a foreign capital. In the Cape parliament Merriman – now on the back benches – asked whether Loch had undertaken his meddlesome errand on the advice of Rhodes. Told that the High Commissioner was acting independently, Merriman asked why he was paid £3,000 a year if it was not to have some control over his actions? Rhodes ignored Merriman's sniping. The young burgher's plans to remove the old burgher were now moving apace.

First there was ground to make ready both in London and in Johannesburg.

> The stock was sold, the Press was squared.
> The Middle Class was quite prepared.

– or so they needed to be. Rhodes had to take control of the Uitlander movement. At first the Randlords had stood aloof from the National Union, or rather, as Phillips put it, given 'a little pecuniary assistance without our name appearing'.

At the time of the attempted purchase of the *Standard and Diggers' News* Phillips had paid his firm's money into 'C. Leonard's special account'. Charles Leonard was a lawyer turned politician from the East Cape who had moved to Johannesburg. He was soon in the thick of Uitlander agitation and became President of the National Union. He was sincere and principled, typical of the sincere and principled men who did not quite realize what they were in for when they supped with Rhodes.

Another man who was meant to take part in the squaring of the press was Francis Dormer, editor of the Johannesburg *Star*, the mining house's tame paper before and after the unsuccessful attempt on the *News*. He was no innocent, as his par allocation of Rand Mines shares showed, but in his way he, too, was a man of principle and, when he learned of the plot which was taking shape and the part Beit expected him to play, he resigned rather than comply. A more pliant editor took his place and another link in the chain was forged.

Controlling the South African press was a simple task for Rhodes compared to bending the British government to his will. And yet this was essential if he was to depose Kruger. On his visit to the Transvaal at the time of the commandeering crisis Loch met Phillips. They talked about whether Johannesburg might rise against Pretoria and if so for how many days it might expect to hold out. The meaning was clear: if the 'imperial factor' intervened and an outside force came to the rescue, how long could it afford to take? On his return to Cape Town, Loch communicated this potential course of action, a 'spontaneous' Uitlander rising in Johannesburg to be followed by British military intervention and then his own arrival to impose a settlement. The dying Liberal government which had been in power since August 1892 lived up to its dying traditions and refused to countenance such a course.

Then events moved Rhodes's way. Lord Rosebery's ministry (he had succeeded Gladstone as Prime Minister in 1894) staggered from one misfortune to another until in June 1895 office slipped from his nerveless fingers. In the Tory government which succeeded, again under Salisbury, the brilliant and sinister figure of Joseph Chamberlain became Colonial Secretary. He was the dazzling shooting star which had shone over British politics for twenty years. Starting as a municipal politician – something quite new in England – and fortified by a personal fortune, this 'boss' of Birmingham had moved on to the national stage in 1876 as an extreme radical. His radicalism was indeed of a modern kind for which other words might have been found in the twentieth century: he was a populist, appealing to the 'little man', a statist, and now an imperialist. He had helped to break up the Liberal Party by his opposition to Irish Home Rule, and he had joined with Salisbury as a 'Liberal Unionist'.* In Salisbury's new administration Chamberlain was back in office for the first time since 1886. He was dynamic, ambitious, unscrupulous. Here was a man after Rhodes's heart, a spiritual ally where it mattered in London, and here was his chance.

---

* 'Oh, they count as Tories. They dine with us. Or come in the evening, at any rate.' Lady Bracknell, *The Importance of Being Earnest*, Act I.

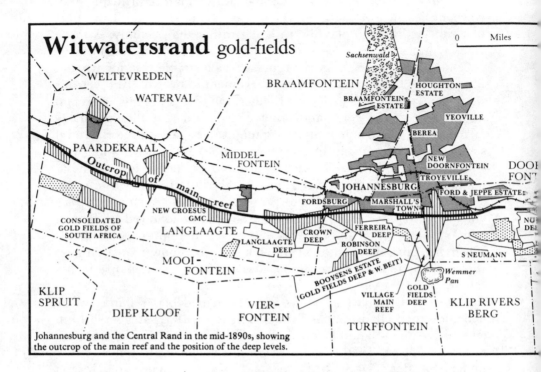

Johannesburg and the Central Rand in the mid-1890s, showing the outcrop of the main reef and the position of the deep levels.

The Vaal basin gold-reef is a thin sheet, shaped roughly like a saucer, which lies underground. Its edges come upwards and break surface or outcrop. The richest of these outcrops runs along the crest of the Witwatersrand at some 6,000 feet. This simplified mining map of Johannesburg and the Central Rand in the mid-1890s shows the outcrop. That is where the gold-bearing Reef or banket was first detected in 1886 in the open, and mined in rudimentary fashion straight into the ground as it ran.

The mynpachten *to which landowners were entitled was, as can be seen, mostly staked along the line of the outcrop, where it was at first supposed the richest mining would be. North of the outcrop grew the built-up estates which together came to be known as Johannesburg.*

Before long the shape and nature of the banket Reef was better understood. Although there are odd auriferous patches to the north, the Reef proper is to the south; if one imagines that the map is shaded below the outcrop, then all of that land

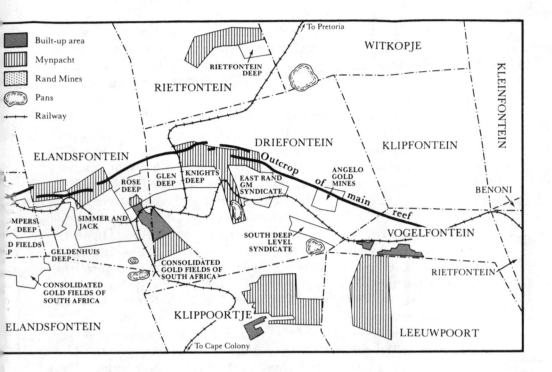

holds auriferous banket at some depth beneath its surface. Further mining properties were thus staked to the south and acquired for exploitation. At first these deep levels were only a few hundred yards from the outcrop. Then, further 'deep deeps' were staked.

Their advantage was the great riches they held below ground; their disadvantage, the very heavy cost of sinking shafts and opening up the mines, all of which costs had to be laid out before the deep levels began to yield gold. The owners of these mines at once had the greatest expectations for the future and were most concerned with counting the cost, exaggerated as it seemed by the policies of the South African Republic, of opening the new mines. A significantly high proportion of the deep-level mines were owned by Rand Mines as well as other companies controlled by the Corner House and Wernher, Beit in London, or by Rhodes's Consolidated Gold Fields of South Africa — the prime movers behind the 1895 conspiracy.

To complete his plan from outside the Transvaal Rhodes needed a jumping-off point for an attack on Johannesburg. On the Vaal to the south the Transvaal marched with her sister republic, the Orange Free State. The northern border on the Limpopo facing Rhodesia was too far away from the Rand. That left the western border, which since 1885 had adjoined the British protectorate of Bechuanaland. It was from there that a raid would have to be launched, but that meant careful arrangements.

The Uitlander movement seemed also to be in Rhodes's hands, but in fact there were difficulties in Johannesburg. To the outside world the Randlords appeared as a unity, a single group with common interests and aims. How very far that was from the case was soon to become apparent, as the year 1895 revealed what insiders knew already, that the dozen or so great mining magnates were deeply divided among themselves. On one side stood Rhodes and his constant ally Beit, which meant all of the Corner House. They were not themselves a great monolithic block always moving as a group, or even as two. There were fierce dissensions within the Wernher-Beit-Eckstein partnerships, usually between London and Johannesburg, and similarly within Goldfields between Rhodes in South Africa and Rudd in England.

For all that, in important economic and political matters, Rhodes and Beit and their groups acted with a common purpose. They now had another ally. After those early days in Kimberley, when Trollope had met him, George Farrar had moved on to the Rand. His first business had been selling machinery but he soon turned to mine promotion. By 1891 he was a leading figure, on the board of nine Rand companies, and the managing director of the Anglo-French company. He had begun to work in partnership with another daring, even reckless, promoter, Carl Hanau. In 1892 they put their initials together and formed H.F. Syndicate.

The Syndicate supplied £90,000 working capital to five mines and in 1893 Farrar floated East Rand Proprietory Mines which in turn acquired the Syndicate's assets. ERPM controlled the largest single block of mining ground on the Rand, but it was a shaky operation. Its subsidiaries were located on a rough section of the Reef where a dyke or fault broke up the pattern of the reefs near the outcrop. Farrar was not a serious or far-sighted mining financier as was Beit. Until 1895 these subsidiaries had produced less than £12,000 worth of gold, although that did not stop ERPM shares from quadrupling during the Kaffir boom. Farrar showed little inclination to direct his personal profits from market operation into mining. In 1895 the Corner House still had substantial holdings in the H.F. Syndicate, but within a few years they sold out.*

Nevertheless Farrar was one of a comparatively small group of magnates within the plot which Rhodes and Beit were forming. Barnato was conspicuously not in the plot. He was Rhodes's and Beit's partner in De Beers but that required no active collaboration; indeed they sometimes acted individually and secretly on the diamond market. In his dealings with Rhodes Barnato was irascible, self-pitying and wheedling by turns. He complained bitterly to Rhodes that he had been left out of a Goldfields ground-floor issue, despite his 'profitless support' of Rhodes in the past. Or he could write ingratiatingly, attaching a press cutting with that day's race card: 'I think you know how strong my belief is in you when I saw you entered for a heavyweight handicap, see enclosed, knowing you have carried some heavy penalties in your time.'† But Barnato would always have been excluded from an intimate conspiracy with Rhodes because of personal incompatibility, unreliability – and a difference of long-term interest.

That was equally true of J.B. Robinson. He spent most of his time now in London at his house on Park Lane and watched with cynical detachment as 'Afrikander patriots imported from Holland found themselves confronted by sturdy British imperialists from the ghettos of Warsaw'. He thought and was happy to say that the agitation against Kruger was factitious, the grievances exaggerated, and any attempted coup doomed to failure; and if his hatred of Rhodes coloured those views they were not far wrong for all that. In any case J.B.R. like Barnato had shown that it was possible to make a great deal of money effectively working with the Pretoria government. Like Sammy Marks he had learned to speak the Taal and made it his business to get on with the Transvaal leaders.

The real culprits of Hammond's complaint about the liquor monopoly were Lewis and Marks, also well outside Rhodes's plot. The Hatherley distillery and its output showed once more that there was no easy explanation of motives on the magnates' part. To begin with the mine-owners had been happy enough with a supply of cheap liquor which helped to lure 'boys' to the mines and kept them docile once there. It was only when the level of drinking among black miners reached the point where it incapacitated a significant number of workers that the managers and owners of the mines had second

---

* Wernher complained that Farrar was 'as much discredited in Paris as Barney' and that he had been 'lining his own pockets'.
† The horse named Cecil John Rhodes won at 6–1.

thoughts, and agitated for prohibition or control. Thus for the moment Lewis and Marks became effective antagonists of Rhodes and Beit.

The liquor monopoly illustrated the conflict of interests among the magnates; and it illustrated a similar conflict on the other side, among the Boers. As the story goes, when Paul Kruger had heard of the first gold finds on the eastern escarpment he had cursed 'the element which brings more dissension, misfortunes and unexpected plagues in its trail than benefits' and begged to be spared from it: 'I tell you today that every ounce of gold taken from the bowels of our soil will yet have to be weighed up with rivers of tears.' But by 1895 he had no escape from the element, or the tears. He could not live with the gold-mines, which were fated to change and disrupt his Republic, but he could not live without them. The old rural Transvaal of Kruger's dreams was by now just that, a dream. The 'new' South African Republic had developed symbiotically with the gold-mines.

Boer and Uitlander seemed to hate each other, but they needed each other. For all their contemptuous behaviour, the ruling class of the Transvaal had done well from the gold-fields. It was Boer landowners who had sold or leased their farms to mining companies and estate companies; it was they who sold their provisions to the mining town, not least the fodder essential in a horse-powered economy; and pressure from farmers prevented the Johannesburg tramways from being electrified as early as they might have been. It was they who grew the grain which Hatherley distilled into rotgut. It was they who benefited from the fog of petty corruption which enveloped Pretoria through the concessions system and general bribery (Kruger himself was not immune). The Randlords recognized this. As well as squaring the press, they tried to live with the system.

The mine-owners complained that government and justice in the Transvaal were corrupt and partial – a complaint which took a good deal of nerve for them to make. They joined enthusiastically in the wholesale bribery during Raad elections. Phillips told Beit, 'I think we *can* influence the election with a fund of £10,000, but it *must be subscribed secretly*. If Kruger finds out that there is £10,000 at the back of the progressives' – the loose group of Volksraad members around P.J. Joubert who were favourably disposed to the mining industry, many with good personal reasons – 'he will put up £20,000 out of the secret service funds (all of which comes out of our pockets).' He added demurely that 'those members who take money much prefer to take it from the right side'.

Besides that, the Corner House had placed several senior officials of the Republic in their pocket, including judges. In his book on South Africa, Trollope had mentioned a 'boy judge', Johannes Kotzé, who became chief justice of the Transvaal at twenty-seven. Kotzé had done his best to preserve judicial independence, a losing battle under Kruger's presidency. Their first clash had come in 1886 when Kruger had pardoned the egregious Nellmapius after his conviction for theft. In 1893 Kotzé stood for President, to come a humiliating third.* His final battle with Oom Paul was yet to come. But Kotzé was not quite the angel of integrity that his story suggested from the outside. He had been on the list of friends for the Rand Mines par issue; and he received even more straightforward favours. Early in 1895 Phillips recorded that Kotzé owed the Corner House £4,289, adding with as straight a face as he could manage, 'we had helped him out of pure good nature, without any ulterior motive'.

The truth was that most of the magnates did not object in principle to the Republic. Robinson claimed that the terms on which the gold-fields had been thrown open were 'more generous than would have been obtained anywhere else in the world [and] it was this generosity that had made it possible for the Uitlanders to accumulate great fortunes', and in private Beit and Rhodes might have admitted that there was truth in this. And a co-operative independent country had advantages over a British colony.

Despite Kruger's rhetoric, the conduct of his Republic had often shown that it was not fundamentally hostile to the mines. It was just incapable of meeting all their demands. All the traditional backwardness and incompetence of the Boer Republic outweighed rational avarice. The pervasive corruption of the Republic was not even a sensible or lucrative policy from Kruger's point of view: only about an eighth of the Explosive Company's huge profits made their way to the state revenue in Pretoria. But short of handing his country over for the Corner House to run there was a limit to what more he could do to help the mine-owners.

The conflict was insoluble. Kruger did what he could to help with Gold Law and pass laws; the magnates' grievances remained, and were real enough in their own terms. A partial control over black labour, a hope to see the dynamite monopoly mitigated – these were not good enough for mine-owners who counted costs in pennies despite an enormous capital outlay. All Kruger could think of doing to shore up the Independence of his Republic was to keep the concessions policy going

---

* Kruger polled 7,854 votes, Joubert 7,009, Kotzé 81.

and to form a counter-group of independent capitalists. This only brought him into fiercer conflict with the Randlords.

Or rather with a particular group of them. All the leading magnates were by now immensely rich. In late 1895 the *Mining World* assessed Beit's shareholdings at £10 million, Wernher's at £7 million, Robinson's at £6 million, Rhodes's at £5 million and Barnato's at £4 million. So far their income had come not only from market operations but also from the outcrop. That was just as true of the Corner House as of Barnato. It controlled only ten of the seventy-nine mines which produced gold between 1887 and 1895; but those ten produced 32 per cent of gold mined and paid 45 per cent of the dividends. There was a far deeper difference between those mine-owners who were and those who were not inside the plot to depose Kruger in 1895. What set the plotters apart was not superior wealth, or indifference either to outcrop mining or to market manipulating – far from it – but a commitment to the future and to the deep levels.*

The economy of the Rand resembled a race meeting, in more ways than one. The mines were the runners, the groups the trainers. On a racecourse there are two distinct economies: the formal economy in which horses pay entrance fees to take part in races in the hope of winning and collecting prize money, and a secondary or submerged economy – betting – from which some trainers have been known to make the greater part of their income. So it was in Johannesburg. The formal economy was the digging and crushing of gold and ultimately the paying of dividends by profitable companies. The secondary economy was stock-market operations, with the 'trainers' backing 'horses' – their own or others'. In this sense every one of the Randlords was to some degree a gambling trainer, all of them until the mid 1890s making a larger part of their income from the stock market than from gold mined and sold. Beit, Wernher and Rhodes were distinguished by the fact that they had horses which they were running to win. They had great expectations on the deep levels. On the other side were J.B.R., Barnato and the Joels, Goerz and Albu, all of them manipulators whose almost sole preoccupations were holding operations and playing the market.

A pure gambler does not care whether it is Derby Day or a down-country meeting. Each is the opportunity for a punt. So it was for Barnato. To Rhodes and the Corner House it mattered very much how the race went. The deep levels so far did not produce money, but

---

* This is the theme of a famous article by Professor Geoffrey Blainey, 'Lost Causes of the Jameson Raid', which provoked a lengthy conversation; see Bibliography p. 286.

only absorbed it, and on a huge scale. Before it was completed in
March 1896 the Geldenhuis Deep cost £328,000, and that was a
comparatively shallow mine. In 1895 two mining engineers, F.H.
Hatch and J.A. Chalmers, estimated that to open a 'deeper deep level'
would cost £650,000; to open two second-row deeps would cost as
much as the entire gold dividends of the Rand for 1894. And still the
rewards had not come. The first producing deep levels had had to
close temporarily in February 1895; the Champ d'Or Deep ran up
debts of £100,000; Geldenhuis was not making a profit five months
after it began crushing; up to June 1895 no deep-level company had
paid a dividend.

No wonder that those who had banked on the deep levels were so
worried about the return on their investment; no wonder that they
resented even the smallest financial impediment. Mining gold on the
Rand was always a matter of maximizing output while minimizing
costs. What was true of outcrop mines was far truer of the deep levels.
Monopolies and restrictions made no difference to a speculator whose
Simmer & Jack or City & Suburban shares had just risen another 20
per cent; but they bore peculiarly heavily on the deep-level mines
whose development and working costs were bound to be burdensome
even if the mine-owners could have arranged their own costs.

Although the Gold Law was in some ways as generous to the
capitalists as Robinson claimed, it had certain unintended consequences which still further vexed the deep-level mines and set them
apart. When gold was found on a farm it was proclaimed as a
gold-field and prospectors could stake their claims; the owner of the
farm had the right at proclamation to reserve a tenth of the farm as a
mynpacht-brief. Naturally the owner chose what seemed the most
lucrative part of his farm: a mining map shows how each *mynpacht* ran
along the line of the outcrop. Once reserved the *mynpacht* was sold at
high profit to mining companies. These were the first companies to
mine large quantities of gold and might have been expected to bear
the burden of whatever taxation was imposed. They did not. By mid
1895 the three richest companies had been charged with nominal rent,
little more than £100. This extraordinary exemption could only
impose in effect another burden on the deep-level companies.

There was one further source of friction: the *bewaarplaatsen*. These
were the sites on each farm which had been nominated by claimholders or companies for non-mining purposes. They were used for
water reservoirs or tailing dumps. An area could be held as a
bewaarplaatsen for a sixth of the rent due on a mining claim, and so
naturally that was how the companies held these sites. Naturally also

they chose them to the south of their mines, before it was realized how far the Reef ran. When that was learnt, it was too late: the law forbade mining under *bewaarplaatsen*. The government invited applications for mining rights under them in September 1892 but then dragged its feet. In the background lurked Edward Lippert who wanted to control the sites. It was one more straw to bear, another reason they might wish to change the regime.

Every one of these hindrances nagged at Beit, Phillips, Rhodes and Farrar. The future was theirs if they could only take it. If Kruger and Krugerism could be done away with a truly golden age for the mines could dawn. After Kruger, what? Was it to be a new kind of state, or under quasi-imperial control, like Egypt, as Phillips suggested, or a 'reformed republic' under honest young burghers, or a new domain of the British Empire for Rhodes to present to Queen Victoria? By not answering this question early and precisely the conspirators were storing up trouble.

11
Beautiful, Bountiful Barney

All through the long, dry South African winter of 1895 Johannesburg quivered with excitement and anticipation. Investors had slowly recovered their confidence in the Rand after the slump of 1889–90. A gently rising market accelerated until it became the 'Kaffir boom' of 1895. The market was quite mad, Beit said. Robinson shares climbed to £9, Simmer & Jack to £16 10s, City & Suburban to £25. All of Europe was buying Kaffirs. Not quite all of Johannesburg: for someone to buy, someone else has to sell. The great magnates who had so adroitly let themselves in on the ground floor could choose to take profits as the boom went on, or to sit tight, or to do both.

The Corner House did both. The partners of Wernher, Beit in London and their South African colleagues, skilfully used the bull market to sort out their vast stock holdings, winnowing wheat from chaff. The chaff could be sold off, extremely profitably, and the cash used to consolidate their holdings in the deep levels. Just as had happened in the first Rand crash of 1889, and before that so many times at Kimberley, Beit with his financial genius showed that the really adroit operator could weather both boom and slump and emerge stronger than before.

Although they were closely allied, Rhodes's case was different from Beit's. He benefited greatly from the roaring Kaffir boom by selling heavily, stockpiling cash for the schemes he was brewing. In London the shady figure of Rutherfoord Harris took up his part as Rhodes's go-between to Chamberlain. Harris was another of the doctors who had come out to Kimberley and moved from medicine to business. He was taken on as Rhodes's confidential agent and had gone with Jameson on a mission to Lobengula; then he became secretary of the Chartered Company; then he had become Member for Kimberley in 1894, alongside Barnato. Now he had to undertake a mission of peculiar delicacy for which he was in some ways qualified and in others not, for though an unscrupulous intriguer he was a clumsy one.

He met Chamberlain in London on three occasions between 1

August and 5 September. His object was twofold: to arrange the transfer of territory from the Bechuanaland Protectorate to the Chartered Company, and more vaguely, but most importantly, to compromise Chamberlain in Rhodes's plot. At the first meeting there were no officials of the Colonial Office present; Harris was escorted or sponsored by Lord Grey who had become a director of the Chartered Company – not to say a complacent stooge and figurehead – having met Rhodes through the conceited and odious muckraking journalist W.T. Stead.

At this meeting Harris tried to say 'something in confidence'. He was cut short by Chamberlain who later said that 'it was unnecessary and undesirable that I should receive from a private friend details of the proceedings which I could not afterwards act upon without a breach of confidence'. Harris of course knew about Rhodes's plans to station troops of the British South Africa Company on the border and attempted 'guardedly' to bring the question up. According to his own later account, Harris managed to say that if a rising took place in Johannesburg 'of course we shall not stand by and see them tightly pressed', before Chamberlain abruptly ended the interview. Harris's fly had landed short, but he had still got Chamberlain to rise. It was only a matter of time before the Bechuanaland strip was settled.

Then came the stock market crash. The bubble had been fantastic – the total market value of Kaffirs was reckoned at more than £151 million – and it burst with a bang. The day it happened was not as violently dramatic as Wall Street's Black Friday in October 1929, but it can be dated just as precisely: 28 September 1895 was the Day of Atonement. On the great Jewish holy day of Yom Kippur the Johannesburg stock exchange closed.* Business sensibly slackened on the leading bourses of Europe. It was the perfect moment for a stock market coup. That day enormous blocks of Rand shares were thrown on those markets where dealing was taking place, principally the Paris Bourse. The price of shares plummeted.

No one could say who was responsible for this great bearing of the market, but it would, at least, have been hard to execute without the knowledge of the Corner House and of Rhodes. Rhodes had sold heavily throughout the summer and autumn of 1895, both his own Chartered stock and probably gold shares also.

Rhodes's critics were later to accuse him of being the man behind the September crash. He had deliberately beared the market, they

---

* Something which shocked the radical writer J.A. Hobson when he visited the Rand four years later.

said, in order to make a killing after his political schemes for the
Transvaal had been successful. In the view of Hobson, Rhodes 'was
able, by his political genius, to give a temporary cloak of political
significance to adventures which were *au fond* operations of the stock
markets'. Ironically enough that was to romanticize events. Of course
there was a conspiracy under way, but it was at once larger and
simpler than Hobson knew. His nose for skulduggery did not deceive
him but his radical passions did. In any case, the conspiracy theorists
presupposed, as they always do, a higher degree of control over events
by individual men than really existed.

As events took their course, Rhodes himself made just the same
mistake in exaggerating his influence. He instinctively over-estimated
his control of events, as Kruger stepped up his war of nerves with the
mine-owners. He had been exacting a steep duty on goods coming
into the Transvaal from the Cape or Natal with the scarcely disguised
object of destroying this trade in favour of his own Netherlands
railway now operating from the Rand to Delagoa Bay. To avoid the
duty, as Hammond had complained, traders used the railway as far as
the Vaal and then unloaded, and crossed the drifts or fords at
Viljoensburg to cover the last two score miles or so to Johannesburg
by ox-wagon. At one point 120 wagons had been crossing the drifts in
a day. On 1 October Kruger closed the drifts. It was a deliberate
defiance, both of the Uitlanders and of the British government. The
new Colonial Secretary was not slow to act. Troopships on their way
to India were ordered to call at Cape Town and make a show of force.
Chamberlain dealt sharply with the Cape Prime Minister also. He
asked Rhodes whether in the ever more likely event of armed conflict
between the British Empire and the South African Republic the Cape
would play its part and provide a fair portion of the fighting force.
Chamberlain was determined that any coming war should not be
conducted entirely at British cost.

For the moment things were running Rhodes's way; not Barney
Barnato's. Since his arrival on the Rand he had set the pace, if not the
tone, in brashness of behaviour and in brazenness of company
promotion. Despite the fact that he was always quicker to 'boom' a
company than to pay dividends his name still carried its own magic
for the British investing public. As he said himself, people followed his
name as they had Fred Archer's. And as with the great jockey, the
punters who backed Barnato often made a packet. What did they care
about the long-term prospects of a mine so long as they could get in
and then get out taking a profit? These punters understood also that
their interests came second to those of the promoters — at least they

ought to have realized that. No one who had followed closely the flotation of any of Barnato's companies should have been under any illusion.

One of the first flotations was the Eagle Gold Mining Company in February 1889. At a time when no deep-level strike had yet been made this company bought 400 acres of a farm seven miles south-west of Johannesburg. The company was capitalized at £350,000. Of that sum £50,000 was retained ostensibly as working capital. The balance was 'vendor's interest', or in non-technical terms was pocketed by Barnato and his connexion. Shortly afterwards Barnato floated another company, the National; 140,000 of 170,000 shares were retained by the vendors. The first Rand crash held up further company formation until 1891 when these two companies – which in fact lay far apart physically – were amalgamated as the Unified Main Reef Gold Mining Company; of its £55,000 capital, £40,785 was issued to former shareholders. After further and various transmogrifications it emerged in 1893 as the New Unified, with an issued capital of £140,000, but with so little working capital that it had to borrow – at 12 per cent – from Barnato's investment company.

Another remodelling of the company in 1895 issued 56,000 shares for scrip; shareholders got two for five. Part of the share issue was offered at par to the guarantors of the scrip issue, who were of course the Barnatos. They held a large number of the old shares, which they had allotted themselves at rates much below par.

In addition, Barnato had an assortment of interests: 'Johnnies' owned and managed subsidiaries such as the Johannesburg Waterworks Estate and Exploration Company which Barnato bought in 1889 when the slump had brought it close to bankruptcy; after the Boer War Johnnies sold the Waterworks Company to the Johannesburg municipality for £1 million. Johnnies also owned a stake in the Johannesburg Estate Company, one of the few cases of effective collaboration between Barnato and the Corner House. It was one of the most successful companies which latched on to the expansion of Johannesburg as a residential city.

The central enterprise of Barnato's group – Johnnies and its subsidiaries and Barnato Consolidated Mines – was, however, booming shares. Between them these companies created more than five million shares in the 1894–5 bull market. Barnato's methods are illustrated by the statements of Barnato Consolidated, or rather by the mixture of *suppressio veri* and *suggestio falsi* which they contain. The company claimed to own 'equivalent to about 2,500 claims, some of which are situated in the immediate neighbourhood of companies whose shares represent . . . a value from £20,000 to £50,000 per claim'. Some of these were 'held in conjunction with' Goldfields, Rand Mines 'and other . . . firms or corporations'. In fact, while Goldfields and Rand Mines were almost exclusively interested in the central and east-central Rand, no more than eighty-seven of the Barnato group's claims were on the central Rand, while 700 were in the western, and 1,500 in the far-eastern, Rand. During the boom the Barnato Banking Company was formed with an issued capital of £2.5 million. It bought heavily in mining shares and was caught short by the next crash: Barnato had sold many of these shares to the bank and had to promise to buy them back at the same price. In the event the bank was liquidated in 1896 when Johnnies took over its assets.

All this did not go unnoticed. If some British investors were pigeons, the London financial press was no caged bird. Barnato was execrated in private by almost all his confrères in the mining business and, although he basked in a kind of glory at home, there were critics also. A couple of years earlier *South Africa* had noticed with more than a shade of irony that Barnato was travelling south on the same steamship as one Bishop Alexander and suggested that he ask for an episcopal blessing: 'It may not be as much good as a reconstruction scheme or a deal in De Beers or a put-and-take of Jagers; but a bishop's blessing on a millionaire ought to be worth something because of its rarity.' Other papers were less oblique. They noticed that even as the Kaffir boom took so many of Barnato's companies rocketing upwards, only three of those companies were actually paying dividends in 1895. When he launched the Barnato Bank as well that year the *Economist* thought that a better word than banking might have been found to describe it.

With true *chutzpah* Barnato ignored this sniping. To those who did not read the financial press he seemed at the height of his glory. Music halls and glee clubs sang:

> I'm beautiful, bountiful Barney
> And Beit may go to pot!
> Beautiful, bountiful Barney
> And Robinson may rot!

He spent more and more time in London, cutting more and more of a dash there, his behaviour more extravagant than ever.

He became a backer of West End shows, gave lavish parties, tipped with reckless generosity, his name headed the list with 5,000 guineas for a charity show organized by the Rothschilds. He was spending £10,000 a year on the turf, sharing the Prince of Wales's racing manager Lord Marcus Beresford, though not achieving any great success with the second-rate horses which Joe Cannon trained for him. It was an age when English racing was still dominated by aristocratic owner-breeders and it was not easy for a newcomer, or upstart, to enter the charmed circle which controlled first-class bloodstock.

His first house in the West End was in Curzon Street. From there he moved to a suite at the Berkeley Hotel. Beit had bought a house in Park Lane, and Robinson soon followed suit, taking a long lease on Dudley House. Barnato followed them and bought a site in Park Lane from the Duke of Westminster for £70,000, where he began to build a new house.

When the charity show was mounted Barnato hoped for a part, but was not given one. Instead he took the Vaudeville theatre for a night and put on *The Bells*, with himself in Irving's part, just like the old days at Kimberley. But London was not Kimberley. The performance was an embarrassing flop. One newspaper acidly remarked that South African audiences must take their pleasures sadly. It was a portent. Barnato's whole life, his financial life included, was a performance which had to be carried off with verve on the player's part and a willing suspension of disbelief on the audience's. Barnato himself knew better than the music-hall singers: there was too much about his companions which did not bear examination. Still, his residential estates in Johannesburg – Yeoville, Berea, Houghton – were dotted with prosperous little houses. His shares soared. He floated yet more companies, moving into the French money market with the London–Paris Financial and Mining Corporation. But of all the bubbles blown by ingenious financiers in that boom his was the most fragile.

The crash of September plunged Barnato into grave trouble. The long-awaited spring rains rescued his waterworks but could do nothing for the rest of his card-castle of newly floated companies. Not only did Johnnies fall by more than half: the new Barnato Bank, an openly speculative operation, was particularly vulnerable and its shares fell from £4 to below 30s, Johnnies dropping from £7 to £3. In his own way Barnato bore responsibility for the crash. Along with Robinson and Farrar, it was he above all who had created the bubble in the first place by flooding the market with rubbishy stock.

Now it was time to make amends. Barnato was a gambler and a manic-depressive: his nerve might fail or pull him through. This time he came through. Throughout the whole of October he frantically bought Kaffirs to hold up the market and pull it back from collapse.

Others sneered. What FitzPatrick called 'the gorgeous philanthropy of Barney', with 'his "three millions" put in to check the slump is not the pure chivalrous disinterested thing that it was supposed to be'. Of course it was not: Barnato himself was the principal gainer by rescuing Kaffirs. But others gained also. Fortune for a moment seemed to smile again on Barnato. By strong nerve he pulled the market round.

Four months before the Kaffir boom burst Barnato had given a dinner party in London for 250 people. Most of his guests were from the sporting and theatrical outer fringes of London society, but the Lord Mayor of London came too: the City might not trust Barnato but it could not ignore him. In fact, Sir Joseph Renals, the Lord Mayor, became a regular diner at Spencer House, the Palladian town house in St James's which Barnato had taken while his palace in Park Lane was being built. He now returned the hospitality. On 7 November, as the drifts crisis was coming to its end, Renals gave a magnificent banquet at the Mansion House for Barnato, saviour of the Kaffir market. Fanny Barnato in tiara, the Lady Mayoress, Harry Barnato and Woolf and Jack Joel joined Barney to hear Renals's inept eulogy: 'one of the few Englishmen – Englishmen by birth and not by adoption – who have contributed to a superlative degree to the great prospects of one of our greatest colonies. Coming to the Transvaal, we find in Mr Barnato the portent of some of the most successful enterprises.' Barnato's reply was vulgarly boastful: 'To those who have placed their money in well-managed companies, I say they do not need to fear.'

Not all listeners might have reckoned 'well-managed' to include Barnato's own companies. In any case the City grandees had stayed away and *The Times* snorted: 'The Mansion House is not the proper place for glorifying a successful operation in a department of the Stock Exchange.' All the same, Barnato had rescued his own position. He knew that the Kaffir market was bound to recover in the fullness of time as the deep levels were opened up, and with the market his own less genuine enterprises. What he did not know about was the plot being hatched between Rhodes, Beit, Chamberlain, Jameson and a group of hotheads on the Rand.

12
The Baffled Band

While Barnato banqueted at the Mansion House his third nephew Woolf Joel was on the Rand and in the thick of things. His experimental prank with rockets fired into clouds had brought a sharpish shower; another kind of storm was soon to break over Johannesburg.

The inner circle of the conspiracy was putting the finishing touches to the plot. Those who guided it, the magnates with the most serious commitment to the deep levels – Beit and Rhodes above all – had left the market operators like Robinson and Barnato outside the conspiracy, although Woolf Joel was an enthusiastic anti-Krugerite Reformer. To effect his plans Rhodes needed the co-operation not only of the British authorities, or at any rate of one crucially placed personage, but also of many Uitlanders on the Reef. Herein lay another conflict of interest and of objective. The National Union represented the voice of the Uitlanders, resentful of obscurantist Boer rule. Its members were typified by Charles Leonard, President of the Union in 1894, high-tempered but honourable. The Union wanted to see Kruger brought to heel and if necessary replaced; it did not wish to see his Republic overthrown by force and absorbed into the greater South Africa controlled by Rhodes. In the following year a Reform Committee was established in Johannesburg. It was effectively dominated by the Corner House, with Lionel Phillips as Chairman and Percy FitzPatrick as Secretary. At least its senior members were aware of the plot that was being hatched. The Uitlanders on the Reef would rise up spontaneously against the intolerable regime and then, spontaneously also, a force would dash to their aid from the Bechuanaland border.*

But with what purpose? After Kruger had been toppled, what next? The failure to establish a generally understood answer to this

---

* Not dissimilar to the 'unpremeditated' intervention of Anglo-French forces at Suez sixty years later.

question lay at the root of the impending fiasco; and at the root of the failure lay Rhodes's disingenuous misleading of his colleagues.

On the same day as Barnato was lauded by Renals, Phillips opened the new building of the Chamber of Mines with a speech violently attacking Kruger. His fellows on the Reform Committee applauded. They too wanted to be shot of the old man – but to achieve a 'reformed republic', not a new satrapy of the British Empire. This was a point of honour with them, and anyway, how could it be otherwise when the Reformers were so cosmopolitan a group? Thirty-four of them were British, but eighteen were South African, along with eight Americans, two Germans, an Australian, a Swiss, a Dutchman and a Turk. Even the 'reformed republic' was open to a wide variety of interpretations. The mass of Uitlanders disliked the Boers, but disliked the Randlords also; the skilled miners from Cornwall and Australia, who found their anti-capitalist feelings expressed in the *Standard and Diggers' News* – these men looked for something like the 'working man's paradise' of which Kimberley diggers had once dreamed. Some of the magnates looked for a mining company-state, rather of the sort which the Union Minière briefly realized two generations later in Katanga under Tshombe; that was perhaps the prospect Phillips meant to hold out when he spoke of putting the Transvaal under international control in the manner of Egypt.

If Charles Leonard was typical, however, the Reformers wanted something much more literally close to what the phrase 'reformed republic' meant. They had reckoned without Rhodes's capacity for deceit. The fact was that by now Rhodes was obliged to speak with a forked tongue. Rhodes and Leonard met. Was Rhodes preparing for a South African federation by *coup de main* and did his Chartered Company stand to profit by a rising? Rhodes insisted that the answer to both qeustions was No, and as Leonard listened to these mendacious replies he was, in his own words, 'gradually drawn under by the singularly magnetic personality of Rhodes'. He was to look back remorsefully on the way in which Rhodes had played on his weaknesses 'until retreat was all but impossible'.

Meantime Rhodes was giving precisely contrary assurances. He was not, he let it be known to those who mattered, 'going to risk my position to change President Kruger for President J.B. Robinson'. In reply to a telegram from Rutherfoord Harris in London asking, 'We have stated positive that results of Dr Jameson's plans include British flag. Is this correct?' Rhodes replied, 'I of course would not risk everything as I am doing except for British flag.' The Union Jack would fly over the Rand as soon as the rising began. Following this exchange Chamberlain returned from a holiday on the Continent and at last instructed Robinson, the High Commissioner, to place the strip of Bechuanaland Protectorate bordering the Transvaal under the administration of the Chartered Company, which would take over the troops and equipment of the British Bechuanaland police from the imperial authorities.

The Resident Commissioner at Pitsani at the southern end of the strip was to be Dr Jameson. He had been busy 'in the north' for several years, spending on one occasion three months with Lobengula (when he much helped matters by successfully treating the king's gout), then serving as Administrator of Mashonaland, then provoking and winning the Matabele war. The war had been waged with considerable ferocity; after its conclusion, Jameson had removed large acreages from Matabele hands and had administered the territory, replacing at Rhodes's wish a less ruthless official. He was the man for Rhodes, and the man for the hour. In the mean time, that other doctor, Rutherfoord Harris, was in London, where he had helped to frustrate an appeal by the Bechuana chief, through English missionaries, against the transfer. With the negotiations successfully concluded he and Beit left for Cape Town. The next day, 7 November, Kruger reopened the drifts.

So things stood when Sir Hercules Robinson wrote a lengthy despatch to Chamberlain on the situation in Johannesburg. In Cape Town, the High Commissioner's imperial Secretary Sir Graham Bower tried to clarify Robinson's position and urged caution. But events now had their own momentum. Jameson went to Johannesburg to see the conspirators. A date for the Uitlander rising was provisionally fixed at 28 December and the Reformers gave him an undated letter which asked him to ride to their rescue.

> Thousands of unarmed men, women and children of our race will be at the mercy of well-armed Boers, while property of enormous value will be in the greatest peril. . . . We guarantee any expense that may reasonably be incurred by you in helping us, and ask you to believe that nothing but the sternest necessity has prompted this appeal.

So far was this last sentence from the truth that the letter was in effect extorted from the Reformers under duress. It was signed by Leonard, Lionel Phillips, Frank Rhodes, John Hays Hammond and Farrar.

In London other allies of Rhodes were at work. The manager of *The Times*, Moberly Bell, had given his heart to Rhodes and the Uitlander cause and was doing all he could to glamorize the projected rising. He recruited two other enthusiasts on the paper, the special correspondent Captain Younghusband and the colonial correspondent Flora Shaw. She was the first woman journalist to rise to a high position on an English newspaper; none of her successors can since have wielded so much power, or done so much harm.

On 10 December a telegram went from 'Telemones' in London to 'Veldschoen' in Cape Town, that is from Miss Shaw to Rhodes: 'Can you advise when you will commence the plans, we wish to send at earliest opportunity sealed instructions representative London *Times* European capitals; it is most important using their influence in your favour.' Flora Shaw advised bringing the rising forward to 16 December, Dingaan's day, when the Boers celebrate their ancestral victory over the Zulus at Blood River in 1838. But, although Chamberlain had now given his assent for Robinson's plan to deal with the Transvaal after the rising, things had begun to go wrong. Rhodes had some other date in mind, early in the New Year. Bell in London counselled haste and Rhodes relayed this advice to his brother Frank in Johannesburg: any further delay of the 'meeting' or the 'flotation' would be most unwise. Still nothing happened. Far from Johannesburg's being united and eager to revolt, it was riven by dissent and paralysed by indecision.

Then distant events took a hand. The United States was indeed far away in the 1890s. The Republic's own story was unfolding dramatically within its borders but it played no part in the affairs of Europe or the empire which Europe was building, except when that empire was on the American continent. Although – or perhaps because – the United States was still largely an economic and cultural dependency of England, anti-British sentiment was strong and the anti-British card a useful one for an American politician to play. President Grover Cleveland was in the penultimate year of his second term. He was a Democrat and his party was in the throes of a savage internecine controversy, one which was highly relevant to the Witwatersrand: the 'money question'. Since the United States had adopted the gold standard it had come under attack from advocates of bimetallism. These 'silverites' denounced the 'crime of '73', when Congress effectively demonetized silver.

In one respect they had been objectively right: there was a shortage of gold in the world economy in the 1870s and 1880s, although that problem was in process of being resolved by the mines of the Rand. The silver cause – an expression of southern and western agrarian populism against the business interests of the east – remained vigorously alive and found an orator of genius in William Jennings Bryan. In March 1895 in the 'Appeal of the Silver Democrats' he called on rank-and-file Democrats to seize control of the party, a campaign which was to reach its climax the following year at the Chicago convention when Bryan denounced New York, the great cities, big business, and the gold standard: 'you shall not crucify mankind upon a cross of gold'.*

So, in late 1895, Cleveland was facing a fight for his political life and needed to strike popular attitudes. Not for the last time in a presidential election did foreign affairs present him with a case. There had been a protracted dispute over the border between British Guiana and Venezuela in which the Venezuelans had skilfully sought American support. Dingaan's Day passed and on 17 December Cleveland sent a message to Congress which was in effect an ultimatum to the British: an American commission would be appointed to define the boundary and its decision would be imposed on Great Britain in the name of the Monroe doctrine, if necessary by war.

This greatly complicated the situation in Johannesburg. The many Americans and Germans in the Reform movement had always been in an awkward position. They surely would not fight to incorporate the Transvaal in the British Empire. Kruger had been for some time flirting with the Berlin government, in defiance of the 1884 Convention, and boasted of these dealings. The German community were citizens of a country which was hostile to the British Empire. Now the Americans on the Rand were potentially also at war with the British.

This did not calm the more intemperate spirits on the Rand. At the Old Barbertonians' Dinner on 18 December FitzPatrick made a violent speech. Kruger's aim was to ruin the mining industry and to drive out the accursed Uitlander, he said, but the latter would not tolerate their continued ill-treatment any longer. His audience cheered him to the echo and sang 'Rule Britannia'.

In London the Permanent Secretary at the Colonial Office, Sir Robert Meade, wrote to Chamberlain on 18 December suggesting that 'the Uitlander movement' should be postponed for a year or more. In reply, Chamberlain said that the rising should either 'come *at once* or be postponed for a year or two at least'. Edward Fairfield of the Colonial Office was even more vehemently for delay. He wanted the rising postponed indefinitely. He had to argue the case with

---

* He was nominated by acclamation – and went on to defeat at the hands of William McKinley, the Republican gold-bug, who pushed through the 'crucifying' Gold Standard Act of 1900.

Rochfort Maguire, who had known Rhodes from their Oxford days and was now another of his shadowy go-betweens. Maguire was strongly against postponement. On 20 December he sent a telegram to Cape Town. It is one of the 'missing telegrams' but its gist can be inferred from the telegram which Beit immediately sent from the Cape to Phillips in Johannesburg: 'Our foreign supporters urge immediate flotation.' Three days later Rhodes signalled to Jameson on the border: 'Company will be floated next Saturday.'

For Jameson this was the supreme moment. He was to dash to the rescue of Johannesburg. It would be at once a noble, gallant and tender mission – to rescue the thousands of unarmed men, women and children – and a fierce passage of arms, seeing off the brutal and overbearing Transvaal burghers. For this dream to come true the rising would have to have taken place successfully, otherwise a mission of mercy would be no more than a squalid filibuster,* and Jameson's troopers had to be a match for the 'well-armed Boers'.

Neither condition was fulfilled. In London Maguire now used several misleading if not downright dishonest arguments for immediate action. One was that the capitalists were rapidly losing control of events in Johannesburg. There he spoke more truly than he knew. No one was in control of events on the Rand. The 'flag dispute', unresolved because of Rhodes's shiftiness, now broke out again. Many of the would-be rebels made it clear that they would not rise under 'English bunting', the Union Jack. Early on Christmas morning Leonard was visited by Farrar who expressed himself in the strongest possible terms: 'I have induced every man who has joined me and who is helping me in this business to go in on the basis that we want a reformed republic.'

Two Reformers, Leonard and Frederic Hamilton, editor of the *Johannesburg Star*, left to see Rhodes with that message. Rhodes knew what to do and yet, for a moment, the great man of action was inert. He could have sent an unambiguous and categorical signal to the border ordering Jameson to stay put. He did not. On the evening of 28 December he summoned Bower and told him the full story. The rising, which Rhodes had been financing, had 'fizzled out like a damp squib' in part because of the flag issue; Jameson would do nothing. Bower was overcome by relief and went home to spend a carefree Sunday. His peace of mind was short-lived. Just as he was going to bed that night Rhodes's coachman dashed up with the news that Jameson had crossed the border.

---

* Here in its sense, as the word was continually used at the time, of a piratical adventurer, or his adventure.

The rest of the tragi-farcical story scarcely needs retelling. That afternoon Jameson had read out the 'letter of invitation' to his force though omitting crucial words which it contained, 'should disturbance arise here'. They all sang 'God save the Queen' before setting off from Pitsani across the Transvaal border. Having once hoped for fifteen hundred men Jameson rode with less than four hundred and one day's rations: they would surely pick up more *en route*, and anyway the struggle would be over in a trice.

Jameson thought he enjoyed an element of surprise. That was an astonishing delusion. The conspiracy in Johannesburg, the comings and goings on the border, the furious telegraph traffic – all had been watched by Kruger. As he said later, he waited for the tortoise to put its head out to lop it off. Jameson ordered all telegraph wires to be cut. The trooper given this task had anticipated victory and in his drunken stupor neglected to cut the wire to Pretoria – a microcosm of the Raid. The High Commissioner also knew what had happened and so did the British Agent in Pretoria. They sent messengers begging Jameson to turn back. He rode on.

He and his men were proud of their bloody victories over the Matabele, Maxim gun against asegai; they were no match for the Boers. They were followed all the way and harassed by sniping patrols. Jameson still hoped to meet up with rebels from Johannesburg. None came. On 2 January at Krugersdorp, more than twenty miles from Johannesburg (and to rub salt into the wound Oom Paul's home town), the Raiders made a sad last stand and then surrendered to a much larger force of burghers.

No rising had taken place in Johannesburg. Within a day or two Edmund Garrett, the young editor of the *Cape Times* and another satellite of Rhodes's, coined the cruel name 'Judasburg', which stuck. But really the Reformers on the Rand were confused and pathetic rather than treacherous. They had been deceived and bemused by Rhodes. They were hopelessly divided amongst themselves. They had certainly begun to get cold feet about the 'polo tournament' or 'flotation' and not without reason. Even the hottest heads had repeatedly advised delay. Hays Hammond was strong for action, but he advised waiting until the arrival of Beit. Phillips was confused, so was Frank Rhodes. Colonel Rhodes, Cecil's younger brother, was a man universally liked, with pleasant manners and competent military skill. He was no Bonaparte or Lenin. Even as the plot approached its climax, his mind was on other things. Jameson came to see him only to find a note from one well-known *homme à femmes* to another: 'Dear Jimjams, sorry I can't see you this afternoon, have an appointment to

teach Mrs X the bike.' No doubt they had an enjoyable ride: that was
the true spirit of Johannesburg in 1895.

Confusion piled on confusion. Hays Hammond heard the news of
Jameson's ride from Sammy Marks, straight from Pretoria and the
horse's mouth. He was appalled but decided that Jameson must be
supported. An emergency meeting of the Reform Committee was
convened as refugees started to flee the town. An overcrowded train
on its way to Natal crashed, killing many women and children, one
tragic episode amid absurdity. On Monday the city was deserted, all
shops shuttered, the 'Zarps' keeping off the streets.

Far away in Paris, Florence Phillips's maid brought her news of the
Raid. She rushed to London where she was first told that Lionel was
in Cape Town, and then made a great nuisance of herself to the
partners of Wernher, Beit as she badgered them for more news and
help. In Johannesburg Phillips now added his bit to the confusion.
Seeing that there was no hope of a successful rising, he tried to
compromise. He led a deputation to Pretoria to see that old friend of
the Corner House, Chief Justice Kotzé. Kotzé talked gently, saying
that some of the Uitlander grievances were negotiable. Then he asked
for proof that the Reform Committee was the true voice of the
Uitlanders; disastrously, Phillips agreed to give him a full list of
members' names.

At the Rand Club the Reformers sat drinking, talking, playing
poker, waiting. They saw the ignominious arrival of Jameson in
Johannesburg, as he was marched round Market Square under guard
before being taken off to Pretoria gaol. But it was not the Raiders for
whom Kruger's vengeance was in store. Sir Hercules Robinson
arrived. Rhodes's original plan was that he would be ordering
Kruger about, but Robinson was taking orders. Kruger told him that
if there was any more trouble in Johannesburg he would shoot the
Raiders. Robinson had to tell the would-be revolutionaries that they
must co-operate or forfeit the sympathy of the British government.
Warrants were issued for the arrest of the Reformers and on the night
of 9 January sixty-four of them were collected at the Club, and taken
off to the 'tronk', the prison in Pretoria. Solly Joel light-heartedly took
with him a couple of silk shirts and a bottle of lavender water.

Even as the Raid collapsed the myth-making began. It had been a
fiasco: it could still be presented as an heroic failure of the sort which
the English have always celebrated, in the tradition of Corunna or the
Charge of the Light Brigade. The Raiders may have been wrong-
headed, may have been ruffians even, but they were inspired by
quixotic chivalry; so the story went.

One of the earliest to tell it was Alfred Austin who was made Poet Laureate that New Year's Day of 1896. Now only remembered as a master of bad verse and unintentional comedy, Austin was a more interesting figure. An amateur if prolific versifier, his real career was in political journalism, as a leader-writer on several Tory papers and then as editor of the *National Review*; a writer devoid of poetic talent who had risen to eminence through political pamphleteering.* The cynical Salisbury found a reward in the laurel. The Raid gave Austin the opportunity for his first public poem, and 'Jameson's Ride' appeared in *The Times* on 4 January.

> Wrong! Is it wrong? Well may be,
> But I'm going all the same.
> Do they think me a Burgher's baby,
> To be scared by a scolding name?
>
> They may argue and prate and order;
> Go tell them to hold their breath;
> Then over the Transvaal border,
> And gallop for life or death.

Austin took the high view of the Raid. They had no choice but to ride:

> There are girls in the gold-reef city,
> There are mothers and children too
> And they cry, 'Hurry up! For pity!'
> So what can a brave man do?

His interpretation won many hearts, but not all. Neither *The Times* nor the cheap jingo press could persuade everyone that the Raiders' hands were clean; there were sceptics as well as sentimentalists. That January Austin's poem was recited from the stage at London music halls, with its stirring last stanza:

> We were wrong, but we aren't half sorry,
> And, as one of the baffled band,
> I would have rather have had that foray
> Than the crushings of all the Rand.

Among those who heard it declaimed from the boards was Louis Cohen. He added, 'Them's just my sentiments ... the crushings were a mere item.'

---

* When reading a life of Austin, Evelyn Waugh remarked, 'A close parallel with Spender'.

13
A Grain of Gold

The Raid left a bitter taste. Whatever the verdict of the press and music-hall, honest and perceptive observers sensed that something far beyond the normal run of public events had taken place. The brilliant young lawyer Jan Smuts had returned to South Africa from Cambridge in 1895, a Boer but an anglophile. He had defended Rhodes's record at a public meeting in Kimberley. The Raid revolutionized his feelings. He was shocked and he felt betrayed, recognizing as he later said that the 'Raid was the real declaration of war in the Great Anglo-Boer conflict'. From a quite different perspective Winston Churchill also sensed disaster. Looking back he dated 'the beginning of these violent times in our country from the Jameson Raid'. Merriman agreed. Years later, on the occasion of Jameson's death, he wrote to Abe Bailey about the man who was by his 'fatal mistake the *fons et origo* of all our misfortunes'.

Before that verdict was pronounced, Jameson had far to go. For him and the other central participants in the Raid time's whirligig was to bring astonishing changes of fortune.

For the moment Rhodes was crushed, but he was not finished; this was his supreme test. Barnato had flatteringly joked about the heavy weights which Rhodes had carried in the past. At the beginning of 1896 he had the heaviest weight of all in the stiffest of handicaps. He needed every resource of energy, resilience and ruthlessness to win through for he was entirely compromised in Jameson's filibuster. He had lost control of events, first setting up the plot, then trying to halt it, then failing to stop Jameson, then hoping against hope that Jameson might in some manner win through. Beit was utterly broken, prostrate with worry and misery. In hysterical penitence he drafted a new will which left £1 million to those who had suffered in consequence of the Raid.

Others closely concerned shared stronger nerves. After he heard that the Raiders had crossed the border, Rhodes tried to get word to Chamberlain, through Flora Shaw, not to disown Jameson: 'I will win and South Africa will belong to England.' The appeal was unavailing. Chamberlain heard the news of Jameson's ride at his home in Birmingham. While Rhodes was fretting, raging at the High Commissioner's proclamation against the Raid, praying for some impossible success, Chamberlain had gone his own way. He told his family, 'If this succeeds it will ruin me', and rushed to London to condemn Jameson to the Prime Minister and the public.

With the news of Chamberlain's disavowal and then of the Raiders' final surrender, Rhodes knew the full extent of his plight. In private he complained bitterly about Jameson's folly: 'Twenty years we have been friends and now he goes and ruins me.' But it was no time for recrimination. Rhodes set about an immediate plan of damage control. He made sure that *The Times* in London received a copy of the 'women and children' letter which was dutifully published and which, as Austin's poem showed, began to alter the public mood. Then Kaiser William II of Germany came to the help of Rhodes and Chamberlain. Always the most cack-handed of monarchs, he had forgotten nothing of the worst examples of his Hohenzollern forebears and learnt nothing of the statecraft that Bismarck might have taught him. On 3 January he sent Kruger an arrogant telegram: 'I express my sincere congratulations that without calling on the aid of friendly Powers you and your people, by your own energy against the armed bands which have broken into your country as disturbers of the peace, have succeeded in re-establishing peace, and defending the independence of the country against attack from without.' Kruger replied enthusiastically. The 'Kaiser's telegram' inflamed opinion in England and did Rhodes no end of good.

Kruger received another telegram of encouragement from Hofmeyr, who had already said that if Rhodes was behind the Raid they were friends no more. The Cape ministry was breaking up, and on 7 January, still refusing to disown Jameson, Rhodes resigned as Prime Minister. On the same day Chamberlain agreed to Hofmeyr's request for a full enquiry into the Raid. Rhodes left for England soon afterwards on a number of vital missions and with the supreme objective of saving his Charter. On 25 January, Jameson and his associates followed. Some Transvaalers had wanted to string them up with or without trial. One enthusiast brought a piece of the wooden gallows on which the British had hanged the Boer rebels at Slagter's Nek eighty years before. That vengeful burgher was not alone, moreover. Besides the jingoist papers in England, there was a current of opinion turning against Rhodes and his lieutenant. The traveller and radical writer Wilfrid Scawen Blunt detested Rhodes and rejoiced that 'the blackguards of the Chartered Company' had come a cropper and been captured, Jameson among them: 'I devoutly hope he may be hanged.' A reflective Tory agreed: A.J. Balfour, then leader of the House of Commons, Salisbury's nephew and his eventual successor as Prime Minister, thought that Jameson was an attractive man, but for all that he thought the Doctor should be hanged.

But Kruger was too *slim* for that. He unceremoniously bundled the Raiders off to Durban and home to England, for the imperial government to deal with – or be embarrassed by. Meantime Kruger held on to his more illustrious prisoners, the members of the Reform Committee.

Having done his work in London, Rhodes returned to South Africa in time for the Reformers' trial in the Market Hall at Pretoria. The leaders pleaded guilty to high treason, persuaded by their counsel that they could 'plea-bargain': by admitting guilt the sentence would be lenient. Whether sincerely meant or not to begin with, the bargain went horribly wrong. At the Doornkop surrender, the Boers had captured a despatch box belonging to Major Robert White, '*De trommel van* Bobby White', containing cipher keys, note books and the damning, because undated, letter of appeal. At the end of the trial the five who had signed the letter, Lionel Phillips, George Farrar, Frank Rhodes, John Hays Hammond, as well as Percy Fitzpatrick, each came before Judge Gregorowski. Their interpreter broke down half-way, but no translation was needed for the words, '*hangen bij den nek*'.

After the verdict Barnato stood at a nearby bar drinking heavily and cursing the injustice of it all when Gregorowski came in. Barnato started to harangue him as a 'Kaffir lawyer', a government hack. 'You are no gentleman,' the judge shouted back. 'And you are no judge,' retorted Barnato. When he had calmed down, and after being carried shoulder high into Johannesburg Stock Exchange, Barnato returned to Pretoria to see the President. He first pleaded gently, then threatened: if the sentences were not commuted Barnato would close down all his mining operations, plunging the whole Rand into economic collapse. Probably he could not have carried out this threat; probably Kruger had no intention of making martyrs of Uitlanders. He was under considerable outside pressure; pleas for clemency came from the American Secretary of State as well as the British government. After a short delay he commuted the death sentences.

Even so the strain on the prisoners was great. Conditions in Pretoria prison were disgusting and the prisoners did not know what would become of their families. On 16 May one of them, Frederick Gray, borrowed a razor and went into a privy. After sitting there a long time he cut his throat. It was a warning to Kruger that he could not play cat-and-mouse indefinitely. Two weeks later the lesser prisoners were released on payment of a £2,000 fine, and in June the ringleaders were also released on payment of £25,000. These huge sums were met by Rhodes, who paid out altogether more than £300,000 in consequence of the Raid.

It was a small price to pay: Rhodes was straining every muscle to recover from the catastrophe. His first success was in London with Chamberlain. The Colonial Secretary did not threaten to remove Rhodes's Charter. How could he, given what each knew about the other? He could not prevent a parliamentary Committee of Inquiry being set up, but he could at least see that it was convened at a leisurely pace. Rhodes was free for the moment to go to his private colony in Rhodesia.

There also disaster nearly struck. The troopers who had so ingloriously ridden into the Transvaal were men of the Chartered Company's gendarmerie. They were no longer in place to do the job of keeping down the blacks, which they had been created to perform. The Matabele had meantime been treated with provocative folly, stripped of land and cattle as white settlers were given farms. They were frustrated, oppressed, and hungry, with cattle disease a new hardship in early 1896. They had lost the traditions of centralized kingly rule thanks to the Company. They were inspired by strange religious impulses, by prophesying spirit mediums, and had learnt how to fight the white man. In the past the Company's troops had beaten Mashona and Matabele on the well-tried principle:

> Whatever happens, we have got
> The Maxim Gun and they have not.

Now the Matabele had learnt not to attack in massed groups and be mown down but to hit and run.

As Rhodes was on his way up country from Beira he heard that the Matabele had risen, and when he arrived in Salisbury he took direction of the war. More than one bird was conveniently killed: he replied to a request from Chamberlain for his and Beit's resignation from the Company's board : 'Let resignation wait. We fight Matabele tomorrow.' The Company's depleted troops were rallied by him and held their own. They made great slaughter, and when one officer had been needlessly merciful of African lives Rhodes told him to kill as many as he could to teach a lesson. A peace was concluded on 13 October. It was during the parleys with the Matabele that Rhodes went walking on his own in the Matopos, and there found the magical vantage-point, the 'View of the World'. This, he decided, would be his last resting-place.

The repercussions of the Raid were not the only topics which exercised Johannesburg in 1896. The first days of January had been a disaster for Jameson and Rhodes ; on Ash Wednesday, 19 February, there was another kind of disaster which struck a different chord at the gold-mines. A train was drawn up in a siding at Braamfontein with a cargo of more than 2,000 cases of Lippert's dynamite ready for unloading and transport to the mines. The marshalling yard was in the middle of the teeming locations and slums where the poor of Johannesburg huddled together. The dynamite had been sitting for three days in the summer sun when an engine carelessly backed into the train and set off its whole load. The explosion flattened acres and left hundreds dead and wounded.

Many Johannesburgers blamed Lippert for yet another crime committed in his name: the explosion was a frightful reminder of rankling grievances. The Corner House was even more concerned with the other great expense: labour. White labourers were highly paid even when 'practically in the position of the ordinary Kaffir labourer', the London office complained. And yet the firm was opposed to a combined movement on the part of employers to reduce the pay of white workers; politically that was dynamite of a different sort.

One answer was to reduce the proportion of white to black, but that too was an explosive issue, destined to become even more so. Another was to import cheaper white labour, and for several years this chimera, as it proved, beckoned the mining companies. The difficulty was the policy and attitudes of the Transvaal government. The white miners on the Rand who had come from Cornwall or New South Wales might be Uitlanders but they did not compromise the racial purity of the Boers. And so, while the government was prepared to consider Italians or Hungarians they 'will not allow Armenians or suchlike at any price'.

Another answer was to increase the supply and reduce the price of black labour. But as Phillips said, 'with a pass system indifferently administered, and the unfettered ill-treatment of the "boys", shortages were not surprising'. Things began to look up in 1897. In April Goerz was in London and said that there was an abundant supply of labour and that an immediate reduction of wages was possible. The trouble was the odd men out. Determined efforts were being made to get all the mine-owners into line, through the Chamber and through the Witwatersrand Native Labour Association, a body formed to recruit labour outside the Transvaal and bring it to the

Rand at the right wages. But two men in particular would not play. Robinson could be relied upon to break up any united front. So too could Albu who was doing his best to raise black wages by not adhering to the scale already agreed upon; not of course from charitable motives but to beggar his neighbours.

These enemies of Rhodes now watched with undisguised amusement as he struggled for his public life. The Cape parliament went to work more quickly than did Westminster. Merriman dominated the Cape Committee of Inquiry but it had an almost impossible task. It was treated with contempt by all the conspirators. Rhodes and Rutherfoord Harris declined to attend and give evidence; other witnesses showed a grand disregard for the truth. On one occasion when the Committee had been struggling all day against a wall of mendacity, one honest witness told them the truth: the legislators rose as one man to shake his hand. The Report of the Committee condemned Rhodes for taking part in a conspiracy at odds with his position as Prime Minister; Rhodes simply ignored it.

London was more difficult. There he had considerable public and political support, but also formidable antagonists. He returned from his Rhodesian triumphs to London to give evidence at the Committee of Inquiry which finally began its work on 5 February 1897. Two of his greatest enemies were on the Committee: Henry Labouchere and Sir William Harcourt. Harcourt, a Whig aristocrat, had long disliked Rhodes, but his dislike was as nothing to the hatred which Labouchere felt and had often expressed in parliament and in the columns of his magazine *Truth*. For 'Labby', every penny of the Chartered Company's profits had been taken from the pockets of British investors or was 'stained with the blood of African natives' while the 'Empire jerry-builder', Rhodes himself, was 'a mere vulgar promoter masquerading as a patriot, and the figurehead of a gang of astute Hebrew financiers with whom he divided the profits'.*

On the other hand, and astonishingly enough, Chamberlain himself sat on the Committee which he had been forced to set up and of whose inquiries he was in effect a key suspect. Sir Graham Bower was not the only one to be astonished and shocked by this: it was as if in a criminal trial a criminal were allowed to appoint his own friends to the jury 'and to finish off by taking a seat on the bench himself. The world would hesitate to say whether it was a joke or the act of a lunatic. . . . The real criminal on trial was Chamberlain. He drew the indictment on the wrong charges and then packed the jury.'

* Verdicts with which several recent, especially Marxist, historians would effectively agree, even if they did not use Labouchere's language.

That was one advantage. Another was that the two foes, the unpacked jurors, went off at tangents. Harcourt interrogated Rhodes skilfully and forcefully, but with all his dislike of him was too much of a gentleman to penetrate his evasions and untruths. Both Rhodes and Chamberlain lied sturdily on oath, and got away with it. For his part Labouchere for all his gifts was too much of a show-off and *flâneur* to mount an effective attack, and he was too obsessed with his theory of a stock-market coup to see the wood for the trees.* Besides, the Inquiry was conducted in an unhelpful political atmosphere. The Tory government was on the crest of a wave, the Liberal opposition in an extremity of fissiparous demoralization, so the outcome was no surprise. When the Committee reported in July it severely censured Rhodes but entirely acquitted Chamberlain of complicity.

To many this verdict was preposterous. Evidence had been, as it were, openly suppressed. The Committee saw forty-four telegrams out of a longer series that had been exchanged in the critical weeks. Rhodes refused to produce the others, and these 'missing telegrams' encouraged the belief that there had been a cover-up of something nasty. Chamberlain's friends, including men of honour, even Salisbury himself, believed his word, credulously it now seems. Chamberlain merely denied some incriminating episodes. He flatly denied that the 'guarded allusion' had ever been made by Rutherfoord Harris. It was his word against Harris's, and Harris was an obvious cad. More than that, there was a well-founded belief that Harris with Rhodes's connivance had deliberately worded the telegrams so as to compromise Chamberlain, and those dirty dealings back-fired. As early as 4 February Rhodes's lawyer Hawksley had mentioned the telegrams to an official of the Colonial Office in what might have been taken to be a minatory manner. In June, when he had seen all the telegrams that were subsequently published, Chamberlain offered his resignation to Salisbury. Salisbury refused and complained to others of 'the monstrous libels which have been invented against Chamberlain and for which proof has to a certain extent been manufactured'.

Again, it was argued in Chamberlain's favour for decades afterwards that his disavowal of the Raid before he knew of its failure was proof of his lack of foreknowledge of Jameson's plan. But the plain truth is that in Chamberlain Rhodes had met a man more decisive, more selfish and less scrupulous even than himself. Chamberlain's immediate condemnation of the Raid showed a master politician acting in almost impossible circumstances. Though he did not then

* This persistent accusation was an interesting case of Freudian projection. Labouchere himself used *Truth* to boost shares in which he had an interest.

know that the filibuster would so soon be humiliated, he knew, or could intelligently guess, that the Johannesburg revolution was fizzling out. Still more, as J.A. Hobson put it many years later, Jameson's actual conduct was too foolish for prediction. Thus Chamberlain could later deny any foreknowledge of the Raid with a certain casuistical truthfulness. In any case, the whole escapade had implicitly been got up on the old principle: if anything goes wrong we will disown you. That was precisely why Chamberlain had so abruptly ended the 'allusion' interview with Harris.

A fall-guy was needed and was found in the person of Robinson's faithful imperial Secretary, Sir Graham Bower, who took the blame, exonerated Robinson and Chamberlain of any complicity, was condemned for grave dereliction of duty and had his career broken. He nurtured a deep and entirely justifiable sense of grievance but he never gave the game away. Bower was the perfect English gentleman and a credit to the Royal Navy with whom he had served for twenty years. He upheld the tradition of the Silent Service. Or almost: before his death in 1933 at the age of eighty-five he had prepared a time-bomb in the form of a memorandum to be released in 1946, fifty years after the Raid, describing it as he recalled it. It was discreetly worded, a low-explosive bomb, but it nevertheless began the final demolition of Chamberlain's reputation.

The full extent of Chamberlain's complicity is no longer in doubt. Even when it was belatedly grasped, a puzzle remained: what had happened in the eighteen months following the Raid? There too an answer lies. Just after the news of the Raid had come through, Rhodes was talking to the young editor of the Johannesburg *Star*, Frederic Hamilton, at Groote Schuur. Rhodes said that he had Chamberlain by the balls.* 'Then he really is in it, Mr Rhodes?' 'In it? Up to the neck.' But the truth was that they were 'in it' together; or rather that Chamberlain had Rhodes in the same grip. That was the background to their talks, which must have been tense indeed, in early 1896. They were engaged in a form of mutual blackmail. Each could break the other, but with strong nerves, as they surely both had, together they came through.

Rhodes sailed for South Africa in April 1897 after his interrogation, but before the London Committee reported, being rightly confident of its outcome. The whitewash was slapped on. Not for nothing was the Committee known as the 'Committee of No-Inquiry', or more wittily as the 'Lying In State at Westminster'. Each of the blackmailers kept

* Presumably his phrase; Hamilton mentions a 'strong anotomical metaphor'; Rhodes's brutal speech often shocked his contemporaries.

his side of the bargain. Each kept to his story. Evidence was cheerfully suppressed. Chamberlain had an additional ally: Sir Robert Herbert was the former Permanent Head of the Colonial Office. On his retirement, in the best traditions of British public service, he joined the boards of various public companies, including the Eastern and Southern African Telegraph Company. He was thus in a perfect position to suppress the copies of incriminating telegrams, and was no doubt appealed to on the grounds that the national interest was at stake.

Rhodes never revealed what he might have of Chamberlain's complicity; and when Chamberlain was exonerated he was able to ignore the calls for Rhodes to be expelled from the Privy Council and to lose the Charter, making a strong speech in Rhodes's defence, including the ridiculous statement that Rhodes had done nothing affecting his 'personal position as a man of honour'.

In retrospect Chamberlain's very defence reeks of imperial arrogance. The distinction made at the time between the hypothetical legality of troops entering the Transvaal on the orders of the Crown to help a rising in Johannesburg – the plan proposed by Loch – and the actual illegality of an incursion by troops of the Company is remarkably fine. Rhodes knew why the troops of his Company were being mustered on the border at Pitsani; Chamberlain claimed not to have known, and though many believed him at the time, nobody will now. Chamberlain admitted that he knew of an imminent rising on the Rand, but he also knew as well as Rhodes did that the potential rising was not popular and spontaneous but engineered by Rhodes and Beit. When Chamberlain was told of the accusations levelled against him by Hawksley, he gave the magnificent reply, 'What is there in South Africa, I wonder, that makes blackguards of all who get involved in its politics?' That was the last word on the Jameson Raid.

The players in the dramatic events of 1895–6 began to drift away from the Transvaal but not yet from the pages of history. Far from being destroyed by the Raid, several of its participants emerged not only unscathed but vigorous for higher things. Not the least of the dubious deeds of Rutherfoord Harris was to insert the date '28 December' on the Reformers' letter of appeal. He made an ignominious showing at the Committee of Inquiry, impudent and shifty by turns. If anything his conduct after the Raid explained why his urgent appeals to Pitsani not to move had been ignored: Jameson thought him an inferior and untrustworthy personage, typical of the hangers-on and toadies with whom Rhodes had surrounded himself, and in this respect Jameson's judgement is hard to fault.

For all that, Harris was elected Member for Kimberley in the 1898 Cape Colony elections. He supported the Progressive Party of Rhodes and Sir Gordon Sprigg and at the same time kept a connexion with the Transvaal. The one wholly useful act of his life was when he chaired the committee which established the School of Mines in Johannesburg, later to become the University of the Witwatersrand. But Harris did not remain long in South Africa. He returned home and was elected to the House of Commons in the 1900 'Khaki election', when it was no disadvantage to have been a satellite of Rhodes's. Another of that odd crew, Rochfort Maguire, did not return to the Commons which he had left in 1895 (he had been an Irish National-ist) although he tried again in 1900 as a Liberal and failed. He became a successful businessman, as a director of Consolidated Goldfields and of the Chartered Company, and played a large part in the develop-ment of the Rhodesian railways.

No recovery was as dramatic as that of 'the Doctor' himself. Back in London he was tried under the Foreign Enlistment Act of 1870, the only part of the book which could be thrown at him in an English court. He was sentenced to fifteen months' imprisonment. In March he was visited in prison in the greatest secrecy by Chamberlain, and was released after four months on grounds of health, a curious reflection on this pioneer in the African bush. At the 'Committee of No-Inquiry' of 1897 he played his part, reading a statement which said that he had acted on his own. He returned to Rhodesia and thence to Cape Town. During the Boer War itself he was caught in the siege of Ladysmith where he was valued as a doctor.

Then his resurrection began in earnest. He was returned as Member for Kimberley in May 1900, after Rutherfoord Harris, and became a director of De Beers. His reception in the parliament was hostile and when he came to make his maiden speech after two years he referred briefly to the 'abominable Raid'. Rhodes's death shattered Jameson, but at the same time it was a release, political and psychological. He took up politics with new energy and by March 1903 became, astonishingly, Leader of the Progressive Party; more astonishingly still, in the following year, after his party won the general election, on 22 February 1904, just over eight years after the Raid, he became Prime Minister of the Cape Colony, and remained in power for four years. When his ministry took office the mainly 'Dutch' opposition not surprisingly regarded him with suspicion. But as always in South African 'racial' disputes between Briton and Boer there was a card to play: the other racial question. Jameson did much to conciliate the Dutch. There was economic support for the farmer, with irrigation schemes; there was an amnesty for the Cape 'rebels' of the Boer War who had fought alongside their cousins from the two Dutch republics.

There was a sop to the Dutch opposition in the Chinese Exclusion Act, and the biggest bribe of all, the 1905 School Board Act. The yellow peril was alive throughout South Africa in the first decade of the century as Chinese-indentured labourers were brought into the Transvaal; but in the Cape where there were few black Africans and fewer Asians the fear of the white, especially the Afrikaans-speaking, population was directed towards the mulatto Coloureds – Afrikaans-speaking themselves and indeed often only to be distinguished from 'white' Afrikaners by a pedantic nicety of pigmentary differentiation. The School Board Act made primary education compulsory for white children while restricting Coloured children to mission schools, and was a most important step in the vertical separation of peoples: it is not too much to call it proto-apartheid. Jameson's term of office was a great portent for the future, when the 'racial' difference between the two white tribes would be settled at the expense of others.

Dr Jim became Sir Starr when he was knighted in 1911. No less remarkably, he was appointed a Privy Councillor in 1907, little more than ten years after he had stood in the dock. In 1912 he retired to England and became President of the Chartered Company. He died in London in 1917 and his body was returned to South Africa to be buried alongside Rhodes in the Matopos Hills. Kipling's poem 'If . . .' is said to have been inspired by Jameson. That December of 1895 he had conspicuously not kept his head when all about him were losing theirs. All the same, he had met with Triumph and Disaster, and he of all men could judge them both imposters.

One other person who also survived the conspiracy with remarkable success was John Hays Hammond. He left the Transvaal and opened an office in London as a consulting engineer. Then, after the Boer War broke out, he returned to the United States, and in the gaudy heyday of American capitalism he flourished even more than on the Rand. He was taken up by the mine promoter Daniel Guggenheim at a salary of $250,000: it was recorded that Hays Hammond's total income at the time – the early years of the century – was more than a million dollars a year. He became a friend of President Taft, might almost have been his Vice-President, represented the President at the coronation of King George v in London in 1911, was on intimate terms with presidents Harding and Coolidge and died aged eighty-one covered with honours – including an honorary degree from his old college, Yale – and enormously rich.

In the months after the Raid Barney Barnato had problems of his own. He tried to play the peacemaker once the prisoners were released. He heard that an acquaintance had commissioned two marble lions to be carved. Barnato made him an offer and when the lions were complete presented them to President Kruger, outside whose little house in Pretoria they have stood ever since. But the London investors needed some other form of peace-offering than carved lions.

As Barnato Bank shares fell towards par he called a meeting for press and shareholders at the Cannon Street Hotel. He arrived elaborately dressed, looking at his notes through his pince-nez, put them down, picked up a mug of champagne, and spoke. It was the old Barney, supremely confident. Investors need have no fear; there had been bad times but they would all come through; 'I tell you that the name of the Barnato Bank will not die out whilst the name of Barnato Brothers lives.' Courage? Recklessness? Insolence? Whichever, it was a bravura performance. But however he might impress others outwardly, within Barnato was close to the brink. The trial, when he generously put up £20,000 bail for Hays Hammond after he heard that the American was gravely ill in Pretoria prison, was almost as much a strain for Barnato as for the prisoners. He was overwrought and drinking heavily.

A mere nine months after that show of bravado in the City, Barnato Bank was hurriedly absorbed by 'Johnnies', still without paying dividends, and carting those who had bought the bank's shares at the top of the market and were inadequately compensated with 'Johnnies' stock. Barnato's name now had a very unpleasant ring to it, in Johannesburg, in London and in Paris. With the Dreyfus affair at fever pitch there were sections of the French press all too happy to gloat over the discomfiture of this grandson of a rabbi.

In London too his enemies were gleeful. There was the story of the young clerk who had tried his luck with the firm's money and whose body was found in a railway carriage with a pistol in his hand and a note: 'Regret I have made a mess of it all. Barnato's Bank is the cause of everything.' It scarcely mattered whether the story was true or not: everyone had heard it. The fantastical house at 25 Park Lane was being topped out. At roof level there was a series of gargoyles or weird sculptures. What did they represent, a friend asked Labouchere as they drove down the Lane. Barnato creditors petrified while awaiting settlement, Labby cheerfully replied. Barnato himself had lost all interest in the new house. It was a year after the Johannesburg slump and he observed Yom Kippur devoutly for the first time in his adult life, fasting for twenty-four hours and keeping the full appointed time in synagogue. He was out of things, increasingly in deep melancholy, absent-minded.

He went to see off Solly and Abe Bailey who were sailing for South Africa, and on impulse went aboard to travel as far as Madeira, and then decided to complete the journey, in increasing despair. Others might sneer at him, but only he knew how precarious his position was. He reviewed his mining operations on the Reef with despair and made a revealing admission to another financier. He was determined to cut down expenditure and stop all unprofitable work. 'As he says, he cannot create new shares now and therefore it is his money that goes.' Barnato was running out of gullible punters, and skills with which to gull them.

In Johannesburg his spirits revived for a time. He took more interest in his new house which was being built in Berea, Barnato Park, than in 25 Park Lane, but he was highly and increasingly unstable, his instability aggravated by drink. An unfortunate incident took place at Vereeniging where the Barnatos were changing trains: a clumsy customs official mishandled his rifle which fired, wounding Fanny in the leg. This preyed on Barney, on top of his other obsessions.

He was still sane enough to take part in the parliamentary session in Cape Town in early 1897, and took Merriman's vituperative attacks on him in good part. His old love of the theatre cheered him. The great music-hall singer and *comédienne* Marie Lloyd was in Cape Town; Barnato fêted her, gave her a bracelet of gold and diamonds and made her promise to come to the new theatre he was building in Johannesburg. But by the end of May he had relapsed. He was delirious, sometimes raving, deluded; at night Fanny found him counting imaginary banknotes or trying to claw diamonds from the walls.

Solly Joel was summoned and rushed to Cape Town where he booked a passage and persuaded his uncle that he must return to London. The *Scot* was sailing on 2 June. It would be full of important personages, including Rhodes's successor as Cape premier, Sir Gordon Sprigg, all on their way to celebrate Queen Victoria's Diamond Jubilee. Barnato must go too. He had an appointment: a house-warming party in Park Lane on 22 June, Jubilee Day. They all left together, Barney, Fanny and Solly. On the voyage Barnato seemed to calm down. He drank little and was cheerful company with children. But he thought of London and shareholders weighed heavily on him.

On 23 July the ship was south of Madeira. After lunch Barnato made Solly walk round and round the deck at exhausting pace. The younger man collapsed into a deck-chair. He heard Barnato ask the exact time, and replied, 'Thirteen minutes past three'. The ship's fourth officer W.T. Clifford was sitting near by. A moment later he heard a commotion and a cry and looked up to see Solly Joel clutching at something over the side of the ship. Clifford rushed across to see Barnato in the water. He stripped off his top clothes and dived overboard. In a choppy sea he could not immediately reach Barnato. The ship's engines were stopped, but by the time a boat was launched Barnato was floating head downwards, drowned. It was three weeks before his forty-fifth birthday.

The body was taken back to Southampton and London for burial. Barnato's death caused more than merely a *frisson*. It was rumoured that the circumstances of his death were suspicious: had Joel been pulling at Barney's leg or pushing him? That was absurd, and in any case the verdict of suicide given by the coroner was just as disturbing. What knowledge had driven Barnato to his end? London that summer had seen a small bull market in 'Kaffirs'. Gold production figures had been encouraging, and the Jubilee euphoria helped. Barnato's death did not: there was a slump, especially and not surprisingly in shares connected with his name.

Respects were paid nevertheless. Two hundred carriages left Kate Joel's house near Marble Arch for the Jewish cemetery in Willesden. The Lord Mayor of London was among the mourners. So too was Alfred Beit. He had always despised Barnato but insisted on attending the funeral. In South Africa grudging tributes were paid. The Johannesburg Stock Exchange closed for a day, the Cape parliament adjourned. Rhodes was less sentimental. He was travelling to Rhodesia by train when the news came one night. His secretary decided not to wake him. When told the next morning Rhodes gave one of his famous displays of brutal indifference. What was Barnato to him? And did any of his entourage suppose that if one of them were to be killed he would lose a minute's sleep?

When death had seemed near once before, Barnato had chuckled at the thought of the fight to come over the chips. He would have enjoyed the reality of 1897. A bitter row ensued within De Beers as to whether his estate was entitled to any posthumous interest on account of his life governorship. In the end an *ex gratia* payment was made. Francis Baring-Gould begged for the vacant governorship, but Rhodes saw to it that David Harris was appointed instead.

Plenty of crocodile tears were shed for Barney Barnato. He had been a man more envied and hated than loved, and he had known it. He had born the resentments of colleagues and investors, and the fathomless contempt of such men as Merriman, to whom Barney always remained 'a rogue without even the redeeming quality of good manners'. Perhaps so, but there was more to it than that. Few had as much reason to hate Barnato as Louis Cohen, or so he thought. He could write bitterly that 'Barnato's name would have been drowned with him had it not been for the incredible harm he did in his life.' And yet when he came to pronounce his final verdict, Cohen remembered something else, going back all those years to Kimberley. Barnato was 'decidedly unscrupulous' – who could doubt that? – but 'had, nevertheless, a grain of gold in his character'. That was the fair last word.

14
People Whom We Despise

Beside the larger struggles being waged Barnato's death was no more than an episode, though a dramatic one. The mine-owners were threatened but undefeated. Rivalries and conflicts on the Rand grew more intense. There were several distinct parties in the struggle. The central group of magnates, the Corner House and Rhodes, were at odds with Kruger, but also with other mining groups. The Randlords' foes ranged on every side: the black workers, who were powerless; the poor whites or mass of Uitlanders who were more and more vociferously hostile to the 'capitalists'; some British politicians who thought that they were being used; and now a number of radical critics in England. The 'South African magnate' had already become a widely recognized figure in Europe during Kimberley's heyday. The Raid had brought these magnates unwelcome prominence in the public eye and had stimulated detractors who thought they saw what the Randlords were up to and set out to attack them with every weapon that came to hand, not hesitating to hit below the belt.

In the year of the conspiracy the magnates had been a notably ill-assorted and disunited group. They were still more so in the following year. The mine-owners who had stood outside the plot marked their displeasure with Rhodes and Beit in a practical way. In April 1896 there was a secession of twenty-two companies from the Chamber of Mines to form a rival Association of Mines led by Robinson, Albu and Goerz. They wanted to snub the Corner House and Rhodes, and they hoped to ingratiate themselves with Kruger and Pretoria. But though Kruger would accept carved lions from supplicant magnates (not to speak of more valuable *douceurs*) he was not swayed. J.B.R. and Barnato had no influence in Pretoria, Phillips thought, and if that smelled of sour grapes it was true enough.

The Corner House certainly had no influence through friendship but, being by now easily the largest mining group, it had other ways of making itself felt. Coup and filibuster had failed dismally. There were different pressures to be exerted. Kruger tried to stand free of the mines, but with every year he became more dependent on his revenues from them and more vulnerable to the mine-owners. They knew that, and so too did Chamberlain. Late in 1896 the Governor of Natal wrote to him: 'My reading of recent market operations is that certain people, headed by the house of Eckstein, have determined to try and get the reforms in another way which they failed to get by their muddle-headed revolution last year.' Chamberlain was always one to spot and borrow a good idea. He wrote to Phillips wondering what the effect would be of simply closing the mines and forcing the Transvaal government to make concessions. Phillips and his colleagues did not act on this suggestion. For one thing they knew how fragile were their relations with the Uitlander community whose aim, as one of them said, was to make money in peace and to avoid any further confrontation with President Kruger.

Still the old economic grievances were not settled. No answer came from Pretoria on the vexed *bewaarplaatsen* question. But the mines were much better placed to deal with their black labourers, the more so after Chamber and Association were reunited, and in April 1897 there was a 30 per cent cut in black wages as Goerz had suggested. The rate of desertion from the mines worried the companies who wanted 'a complete system of control over the natives' at the mines.

Even more did they want the blacks on the mines to be sober, whereas many of them spent much of their time drunk or incapably crapulous. In 1897 prohibition of a sort was brought in, but it was not seriously enforced. When Jan Smuts became the new Attorney General of the South African Republic he did his best to control the illicit liquor trade but a complete remedy in the mine-owners' interests was far off.

Within his own community Kruger was stronger than ever. He faced an election in 1898 and won it by a landslide.* The Uitlanders had a new political voice, the South African League, organized throughout the country, in the colonies as well as the Transvaal, to campaign for British supremacy. It appealed to the 16,225 men and women on the Reef whom the 1896 census reckoned came from the British Isles, with more to come: between 1895 and 1898, 8,600 people left the United Kingdom for South Africa. It was less attractive to the large number of Uitlanders who were not British, including 2,262 Germans and 3,335 Russian Jews, and of course not at all to the 'Asiatics' – 4,807 Indian traders – or the 42,533 blacks enumerated in the census, certainly an underestimate.

* Admittedly the electorate was derisorily small. He received 12,858 votes; his opponents, including the Corner House's friend Kotzé, fewer than 6,000 between them.

The politically conscious Uitlanders, and particularly the large mining houses, had an advantage of a different kind in the new High Commissioner, Sir Alfred Milner, deliberately chosen by Chamberlain as a capable, industrious and dogmatic man who would take a strong line against Kruger. He took up his appointment in May 1897. Milner sympathized warmly with the mine-owners and soon worked closely with them; believed that only a strong mining industry was the key to a new and presumably British South Africa; and had not been in Cape Town for nine months when he prophecied war to Chamberlain as the answer to the Transvaal question.

This coincided with the magnates' wishes, especially the hotheads Lionel Phillips and Percy FitzPatrick. Phillips had gone from Pretoria gaol to London and then back to Johannesburg, but despite the amnesty it was almost impossible for him to work there and he returned once more to the London office. He and Florence bought a house in Grosvenor Square. Onwards and upwards they ascended socially. Their son Harold was about to go to Eton; Florrie pulled every string she could to be presented at Court in Jubilee Year, but the 'Lying in State at Westminster' supervened. Phillips was a comparatively truthful and dignified witness, answering as best he could Labouchere's questions while all the while bitterly resenting 'the indecency of the attacks on me that had been published in a newspaper belonging to Mr Labouchere'. He fell ill under the strain of it and was unable to attend Barnato's funeral. He recovered to write an article in *The Nineteenth Century*, defending 'Judasburg' against accusations of cowardice and of deserting Jameson: 'The whole essence of the arrangement with Dr Jameson was that he should come to our assistance and not that we should go to his', which was true enough. Within the walls of the Wernher, Beit offices Phillips urged Beit on to more resolute action against the Transvaal government.

He egged on FitzPatrick in Johannesburg, though cautiously adding that in view of past events they should be careful how they went about things, especially anything 'which might be construed as conceived in spite'. For his part, FitzPatrick busily took part in the political affairs of the Rand. Complaining volubly to Beit about 'the shower of scandals which the Volksraad have poured out', he complained also about the lack of weight behind the one paper in Johannesburg which the Corner House controlled, the evening *Star*, comparing it unfavourably with Rhodes's *Cape Times* under the editorship of Edmund Garrett, he who had coined the name 'Judasburg'.

The press was one of the most important fields on which the

three-cornered contest between Randlords, Uitlanders and Kruger was played. 'The press was squared', wrote Belloc; 'How the Press was worked' was Hobson's title for a pamphlet. But in fact there was far from being merely a tame 'capitalist' press on the Rand. Kruger temporarily suppressed the *Star* in March 1897 and then tried to enforce press laws. The effort to control the *Standard and Diggers News* was prolonged but unavailing; at the end of 1897 the Corner House once more dropped its subsidy, and the *News* resumed its policy of promoting the interests of the white working class. The magnates looked around for some 'antidote to the poison administered daily' by the *News*. One answer was to resuscitate the *Star* under a 'first-class man from Europe'. There was a third paper in Johannesburg, the *Times*, owned by J.B. Robinson, but it failed in October 1898. Newspapers were not an unfailing licence to make money and on the Rand there were too many of them chasing too few readers in what was after all a small community compared with a European industrial city.

First-class men did not come cheap. In London the opinion of Harmsworth the newspaper magnate was sought: '£3,000 a year is insufficient for a first-class man . . . he gives two men connected with his papers £10,000 a year each'.* Then they found the man they wanted. W.F. Monypenny was an Ulsterman who worked for *The Times* in London, a journalist of high quality and a cut above any South African newspaperman (he was to be Disraeli's official biographer). His services were placed at the Corner House's disposal 'through influential quarters'. At that point the *Star* was an 'absolutely rotten' paper which no one wanted to read. It was not just partisan, it was tediously partisan, constantly polemizing against Pretoria in a way which made it easy for the *Standard and Diggers' News* in return to say that the mines were trying to run the country. 'As if we could help ourselves: we have the government, the Jews, the Liquor gang and many of the unemployed all against us.' Monypenny was the man to put the *Star* on its feet. But he did not have long; the fuse was burning which led to war.

By 1898 the atmosphere of Johannesburg was as charged once more as it had been in 1895. Morbid symptoms appeared, strange happenings took place, there was violence in the air. After Barney Barnato's death his heirs in business, the successful contestants for the chips, were his brother and his nephew Woolf. As Harry had settled in Nice

* Can this really have been true? It would be equivalent to more than £250,000 in today's money. Or was Harmsworth big-mouthing?

where he spent most of the year, it was Woolf Joel who took up the reins at 'Johnnies'. He was much depressed by Barney's death and having arranged his affairs in London he left for South Africa in a sombre frame of mind in January 1898. At his farewell dinner a guest failed to turn up, leaving thirteen at table. Like his uncle, Joel had become susceptible to omens. Like Barnato also he turned to the solace of religion. He made a large donation to his synagogue in St John's Wood, and took a collection of scrolls of the Law with him to present to South African synagogues.

Whatever peace of mind religion gave him was broken on arrival at Johannesburg by a series of threatening anonymous letters. The writer demanded £12,000, which he needed to 'avoid ruin or disgrace'. Unless Joel paid, he would be killed. 'Your death shall not be murder, but your own doing really, though I will willingly admit all blame for removing you to a better world, this or the other side of the river Styx, where Barnato may be glad to see you again.' The letter was signed 'Kismet'. When he received no satisfactory reply, Kismet threatened the life of Solly Joel as well as that of Woolf: Solly was so alarmed that he left for Cape Town, and Woolf agreed to meet the letter-writer in Barnato Park. 'Kismet' announced himself as Baron von Veltheim; a tall, well-dressed man with a strong German accent. He claimed to have known Barney Barnato and to have been owed money by him, but brought his demands down to a modest £2,500. Woolf Joel now made two disastrous mistakes. He believed that he could handle 'Veltheim' himself, without the attention of the police, which might anyway lead to unwelcome publicity; and he thought that the man could be bought off cheaply.

More notes came. On 14 March Veltheim accosted Harold Strange, a senior manager of 'Johnnies', and demanded to be taken to Woolf. In his office he threatened them both further: unless they paid him £200 immediately neither would leave alive. Woolf was sitting watchfully behind his desk; Strange was watching with as much attention. When the threats became more alarming, Strange tried to distract Veltheim's attention while he went for the gun he had put in his pocket. Veltheim was quicker on the draw. He fired, but missed, at Strange, who fired back, then three times at Woolf. When men burst into the office to overpower Veltheim, they found Woolf dead.

Veltheim was tried for murder. The prosecuting counsel should have been Smuts, but he had taken up his new post as State Attorney. Veltheim, it transpired, was really one Karl Kautze, born in Brunswick in 1854, a drifter and petty crook who had pulled various strokes in four continents. He had the confidence-trickster's nerve. In court he claimed that he had fired in self-defence. The jury was made up of Transvaal burghers who disliked mining magnates, who remembered Solly Joel's part in the conspiracy and Barnato's threats afterwards. They acquitted 'Veltheim' after three minutes to the judge's horror. Shortly afterwards Kruger deported Kautze from the Republic as a public menace.

It was an alarming episode. Few of his fellow financiers cared for Woolf Joel or his brothers. Louis Cohen nursed his own bitter hatred of all the Barnatos and Joels and, now ruined again by them as he thought, he poured out his resentment to Solly Joel: 'I am sure you who delight so much, as you have always told me, in "doing someone down", will revel in the chronicle of the methods by which you have so triumphantly filched my money and left me with nothing to thank you for except the memory of half my life wasted for you.' Woolf as well as Solly had done down many others in Johannesburg.

But that was scarcely reason for shooting him, and if a Rand magnate could be killed with impunity was it not, some must have thought, another good reason for seeking a new political arrangement? Uitlander passions were inflamed again at the end of the year when a Uitlander called Edgar was shot dead by a 'Zarp', and the policeman was likewise acquitted of murder. It was more grist to the mill of the South African League and Milner.

Thus in 1899 the tension was screwed tighter and tighter. The mines flourished. Output of gold had begun to accelerate again. In 1897 the Rand produced 3,034,678 ounces; in 1898, 4,295,608; and in 1899 production looked set to pass 5 million ounces.* But the problems of cost were the same as ever. Materials were too expensive, Kaffirs were too scarce. By August there were 96,704 'boys' working on the mines, but the Chamber of Mines reckoned that nearly 20,000 more were needed. As the political situation deteriorated, with these old grievances in mind, Friedrich Eckstein – who had himself recently been assaulted in public – complained about the weakness of the British government, and Samuel Evans of the Corner House wrote to London, 'I entirely share your view as to war being inevitable.'

Milner was in close contact with the Corner House and was in effect conspiring with them to bring about war to achieve the desired political ends. That was the view of historians later as of radical critics at the time – and not only radicals. Any acute observer could see what was happening. Salisbury deplored the corruption of the Pretoria

---

* In the event, with the mines closing at the outbreak of the war in October, production for the year was 4,008,325 ounces.

regime and Kruger's numerous provocations of the Uitlanders and
the British Empire; yet he hoped against hope that war would not
come. It was not only Kruger he blamed. With the conflict approaching, Salisbury gave his own weary, private verdict: Milner was
dragging England into war. And so 'we have to act upon a moral field
prepared for us by him and his jingoist supporters. And therefore I see
before me the necessity for considerable military effort – and all for
people whom we despise, and for territory which will bring us no
profit and no power to England.'

'We' who despised the Randlords was a far wider circle than the
Prime Minister and his friends. A reaction was brewing. Throughout
the 1870s and 1880s the South African mining magnates had become
celebrated; by the 1890s they were the object of widespread
resentment and dislike in England and Europe. The Jameson Raid
displayed their capacity for political intrigue; now critics accused
these ostentatious and domineering Rand capitalists of dragging
England into a war. Yet another woman visitor arrived on the Rand
in the year that war broke out. Violet Markham thought that the
mines were by no means a mixed blessing to South Africa. 'The
financial adventurer who has wandered to the Transvaal is not the
type of man who would demonstrate the higher side of European
civilization to the Boers.' Others agreed. Rhodes was adulated in the
jingoist press but loathed by anti-imperialist 'Little Englanders' and
humanitarians. Even in conservative circles his name aroused mixed
feelings. At the height of his fame he was blackballed from the
Travellers' Club and when Oxford proposed to confer an honorary
doctorate upon him many high-minded dons signed a round robin
deploring the proposal.

The Randlords were hated by opponents of Empire for their
political machinations, especially so after war broke out, dividing the
country and the Liberal Party. Little Englanders became 'pro-Boer',
idealizing the gallant burghers of the two republics and condemning
anyone who might be responsible for, and might gain from, the war.
Equally the Randlords were hated for their ill-gotten riches by
radicals and socialists. In an age when the great capitalist – Ruhr
industrialist of the 'Herr im Haus' school, French high financier,
American 'robber baron' – was everywhere the object of execration
among Left-thinking people, the Randlords were bound to attract
hostile attention.

This hostility was given a sharper cutting edge by the fact that so
many of the Rand magnates were Jews. By 'the Jews' Rouliot had
meant Albu, Goerz, Lewis, Marks and Lippert (it was still a tactless

way of describing them in a letter which would be read by Beit). To some outside observers and critics 'the Jews' might have meant all the Rand millionaires.

The end of the nineteenth century found Jews everywhere on the move. Millions of poor Jews poured in from eastern Europe and the *shtetl* to the slums of the lower East Side in New York or Whitechapel in London. Some had come to South Africa: the 3,335 Russian Jews in Johannesburg listed in the 1896 census was doubtless an underestimate. But there were also great families of rich Jews moving upwards: Bleichröder and Rathenau in Germany; Sassoon and Samuel in England; Guggenheim and Sachs in the United States; Rothschilds in four countries. Some accused these new millionaires of stimulating prejudice. Louis Cohen may have been what was later called a self-hating Jew, but he was not alone in ruefully reflecting that, 'A Jew has had always oppression and antipathy enough to combat, without the offence of the ignominy and hatred these parvenus engender.' The parvenus all experienced social resentment, but besides that they witnessed what had grown alongside their own ascent to wealth and power, a new kind of political anti-Semitism.

Everywhere in Europe the Jews had their enemies. In the 1890s France had seen its greatest outburst of Jew-hatred when Captain Alfred Dreyfus was falsely convicted of treason. In imperial Germany a self-styled anti-Semite party was growing (by the 1907 election it won sixteen seats in the Reichstag). The much larger militaristic Nationalist Party had adopted an official anti-Semitic policy in its Tivoli programme of 1892. In central Europe there was a sprouting of peasant or lower-middle-class clerical parties. They were 'reactionary' but also radical: the progenitors of fascism. The archetype was the Christian Social Party in Habsburg Austria, anti-Marxist but also anti-capital, anti-liberal and of course anti-Semitic. Its leader Karl Lueger was Mayor of Vienna, a city where anti-Jewish feeling had been stimulated by an influx of Jews and by the crash of 1873 and its repercussions. All of these political phenomena belonged loosely speaking to the 'Right'. But anti-Semitism was anything but a right-wing preserve. The German socialist Lassalle had called anti-Semitism 'the socialism of fools' and in no country at the turn of the century were radical parties free of this taint. Only the French socialists had been cured of any ambivalence towards the Jews by the Dreyfus affair.

In England especially the South African episode fuelled left-wing anti-Semitism. Labouchere's and Hobson's contempt for the magnates as swindlers and their dislike of them as Jews had always been intertwined. Their hostility grew as war approached, a war which both saw as a war for gold, the Randlords' war. In a sense these critics were right. The magnates had tried once more to get rid of Kruger. Between 1896 and 1899 they despaired of reforming the Transvaal in their own interests, especially after the suppression of the *Star* and the press law, Kruger's re-election and the threats to preserve the dynamite monopoly indefinitely. War was their last resource.

Neither Labouchere nor Hobson allowed radicalism to temper Jew-baiting. Labouchere to be sure was an unusual radical, a rich Etonian playboy turned journalist and parliamentarian. He was a man of great gifts whose political career was held back by his frivolity and irresponsibility. At the same time he owned and edited a weekly paper, *Truth*, which he founded in 1876, muck-raking in an elegant and sometimes brilliant way. One of the paper's regular satirical columns was 'Letters of Moses Levin of Whitechapel to Isaac Levin of Johannesburg'. Labouchere naturally saw conspiracies and he was not afraid to call them Jewish conspiracies.

Rivalling Labouchere in eloquence, J.A. Hobson, a journalist and an idiosyncratic economist of the Left whose originality was later saluted by Keynes, far excelled him in seriousness and intellectual weight. In 1899, with war imminent, he went to South Africa for the *Manchester Guardian*, greatest of English Liberal papers. His despatches formed the basis of his book, published in the following year, *The War in South Africa: Its Causes and Effects*. That book crystallized the ideas which Hobson turned two years later into *Imperialism*, which was in turn drawn on heavily by Lenin in his writings on the subject.

Hobson's theme was that a conspiracy had taken England into an ignoble war against the Boers, a people with whom the English should want no quarrel. The war had been engineered by a bought press – the 'Rhodesian press' in South Africa, controlled by Rhodes – and the weight of the capitalists' financial power. Hobson skilfully analysed, and dismissed, the ostensible arguments for the war which had broken out as he wrote. 'For what are we fighting?' he asks, and answers 'a small confederacy of international financiers working through a kept press'. The war was a disaster for everyone else in Britain and South Africa, 'but for the mine-owners it means a large increase in profits'. He damns the Randlords – FitzPatrick, Robinson, Hammond – as admitting out of their own mouths that 'good government' will bring better profit margins. He hammers away on the question of black labour, quoting Charles Rudd: 'there is a morbid sentimentality on the question of the natives'; and Albu: 'The native at the present time receives a wage which is far in excess of the exigencies of his existence.'

'For what are we fighting?' A simpler answer still is 'The Jews'. Writing in the *Manchester Guardian* in September 1899 he described Johannesburg as 'a weird mixture of civilization and savagedom': Uitlanders, Boers and 'Kaffirs, who are everywhere in White Man's Africa the hewers of wood and the drawers of water'. The town was British,

> but British with a difference which it takes some little time to understand. That difference is mostly due to the Jewish factor. If one takes the recent figures on the census, there appear to be less than seven thousand Jews in Johannesburg, but the experience of the streets rapidly exposes this fallacy of figures. The shop fronts and business houses, the market place, the saloons, the 'stoops' of the smart suburban houses are sufficient to convince one of the large presence of the chosen people. If any doubt remains, a walk outside the Exchange where in the streets, 'between the chains', the financial side of the gold business is transacted, will dispel it. So far as wealth and power and even numbers are concerned Johannesburg is essentially a Jewish town. Most of these Jews figure as British subjects, though many are in fact German and Russian Jews who have come to Africa after a brief sojourn in England. The rich, vigorous and energetic financial and commercial families are chiefly English Jews, not a few of whom, as elsewhere, have Anglicised their names in true parasitic fashion. I lay stress upon this fact because, though everyone knows the Jews are strong, their real strength here is much underestimated. Though figures are so misleading, it is worth while to mention that the directory of Johannesburg shows sixty-eight Cohens against twenty-one Joneses and fifty-three Browns. The Jews take little part in the Outlander agitation; they let others do that sort of work. But since half of the land and nine-tenths of the wealth of the Transvaal claimed for the Outlanders are chiefly theirs, they will be the chief gainers by a settlement advantageous to the Ouitlanders.

In passing Hobson managed to condemn 'the ignominious passion of Judenhetze', though he could scarcely be acquitted of it.*

---

* Years before, Hobson had written for his first paper, the *Derbyshire Advertiser*, about the poor Jewish immigrant in east London who, though 'an orderly citizen . . . is almost devoid of social morality. No compunction or consideration for his fellow-worker will keep him from underselling or over-reaching them . . . the superior calculating intellect, which is a national heritage, is used unsparingly to enable him to take advantage of every weakness, folly and vice of the society in which he lives.' And well before he had observed the bought Rhodesian press in South Africa, he had seen the London press 'falling more and more under the control of Jews'.

Hobson was a great radical journalist, the *Manchester Guardian* under C.P. Scott a great Liberal newspaper. If Hobson could write to Scott describing Johannesburg as full of 'the veriest scum of Europe . . . Jew power' it is not surprising what less fastidious publicists could say. The Irish Nationalist John Dillon told the House of Commons that most of Rhodes's associates were of 'the German Jew extraction'. Edward Carpenter, that remarkable libertarian socialist and champion of homosexual love, defended the Boers against the power of Johannesburg – 'a hell full of Jews' – and complained that British politicians were 'being led by the nose by the Jews'. The great labour leader and organizer of the London dockers John Burns told the Commons that the British army 'has become in Africa a janissary of the Jews', that wherever the South African question was examined, 'there is the financial Jew operating'. When he had gone to hear the Committee of Inquiry in 1897 he said 'I thought I had landed myself in a synagogue', and at the trial of the Raiders, 'I thought I had dropped into some place in Aldgate or Houndsditch'.

Although the war about to be fought was the Randlords' war, and although they were before long to have all that they wanted, there was a bitter legacy of resentment, prejudice and hatred.

15
War for the Rand

'Considerable military effort' Salisbury had foreseen; he could not have guessed how great it would be. Labouchere warned the House of Commons that the war might cost £100 million. He was laughed down. In the event the Anglo-Boer War – known then as the South African War or the Boer War, known still to Afrikaners as the Tweede Vreiheidsoorlog, the second war of freedom or independence, the first being the victory at Majuba in 1881 – lasted for two and a half increasingly frightful years and cost Great Britain several hundreds of millions.

In the final months before war broke out the two sides, Milner and Kruger, had taken irreconcilable positions. Kruger, a hedgehog not a fox, knew one big thing: to accede to the demands made by the British, or by the mining companies, meant the end of his independence. Milner for his part had decided that his friends the mine-owners, with whom he was in close touch, should have what they wanted, by peaceful means if possible, by war if necessary. War it was, beginning on 5 October.

In the next few months the South African Republic and its sister Dutch republic of the Orange Free State which had chivalrously taken its part might have won. But for all their skill as light cavalry (or rather as mounted riflemen) and as gunners the Boers lacked strategic sense and unity. If in those last months of 1899 they had driven straight through the Karroo for the Cape and across the Drakensberg passes for Durban they could have gained all South Africa before British reinforcements arrived. As it was they laid pointless siege to three British garrisons at Ladysmith, Mafeking and Kimberley.

The mine-owners welcomed the war and as far as they could helped to present it in a noble light. The best writer among them wrote a polemic to advocate the Uitlander case and took it to England where Wernher read and approved it. FitzPatrick's *The Transvaal from Within* was published in September just before the war began. In the mood of war hysteria it sold 10,000 copies before the end of October.

Others did different war work. Just before Kimberley was besieged Cecil Rhodes arrived there. After the calamity of 1895–6 Rhodes had returned to political life with astonishing speed and resilience. He emerged as leader of the newly-formed Progressive Party in the Cape, a coalition of jingoist, Dutch-hating, anti-Krugerite, moneyed imperialists and of Cape Liberals who hoped to introduce Coloured and other black citizens into the polity: a curious mixture at first sight but exactly prefiguring the Progressive Federal Party, the South African opposition eighty years later. In the 1897 election he almost led the party to victory. Once more the Dutch farmers of Barkly West stayed loyal to him. Instead the South African Party formed the government under W.P.Schreiner, and Rhodes retired north to dream again. He played little part in the events leading up to war – in contrast to the Corner House, he was not involved in the renewed Uitlander agitation, or in the South African League, or the press campaign in Johannesburg against Kruger.

He was hated as much as ever by his enemies, even more so now than before. Schreiner's sister Olive wrote furiously to Merriman: 'Rhodes is a *coward*' who could be frightened off, one accusation which was surely untrue. Rhodes disproved, if any man did, the comforting belief that bullies are always cowards. Olive Schreiner turned on Rhodes with a woman's vindictiveness the rage of one who had once admired him deeply. After the Mashona rebellion she denounced his adventures in a short novel, really a polemic, *Trooper Peter Halket of Mashonaland*. It was no great work of literature, but a heartfelt cry of rage at the cruelty of imperialism, at the wickedness of Rhodes's campaigns that had civilized Zambezia with Maxim-gun and noose. Two generations later copies of this book could still be found in homes in Rhodesia; few of them preserved intact the original frontispiece, an horrific photograph of the Chartered Company's oafish troopers standing round a tree that bears strange fruit, African bodies hanging with their bodies contorted and broken.

When Rhodes reached Kimberley he was glad to be there, even glad to be caught by the siege. Rhodes the misogynist was escaping from another woman. He had become entangled in an episode which cast much-needed comic relief on dark days in South African history. In May 1899 he had been in England, and sailed for the Cape on 1 June. On the ship he met Princess Catherine Radziwill who made a dead set at him. She was a bizarre figure, though no fraud, no 'von Veltheim'. Her name was real but after her husband had run off with another woman she had put herself beyond the pale of the court circles in which she had once moved by writing racy memoirs and had even fallen so low as to make her living as a freelance journalist. She had first met Rhodes in London in 1896 at dinner with Moberly Bell. Maybe she now wanted to entrap Rhodes and marry him, in which case she can scarcely have understood him. Maybe she hoped to compromise and blackmail him, another unlikely enterprise. She certainly made a splash in South African social and political circles. But more than that she continued to pester Rhodes at Groote Schuur. He would leave whenever she appeared, but then after one of her visits made the worrying discovery that some papers had disappeared from his office.

Apart from conveniently escaping from the Princess it is possible that Rhodes had gone to Kimberley precisely with the plan of focusing attention on the town and making it necessary for the British to lift the siege. He made a nuisance of himself to the British commander, criticizing the conduct of the defence, but threw in the resources of De Beer, manufacturing a large gun, 'Long Cecil', in the company's workshops.

The Boers soon lost the impetus of war. Their great moment came early in December with a string of victories, 'Black Week' for the British. Those early months showed up the British army in all its complacent incompetence. For a generation it had faced no stronger enemy than Afghan or Ashanti. Now the British soldier found an enemy whose musketry and field-craft were much better than his own. But the Boer command was riven – a perennial Afrikaner story – by internal differences. All three sieges were lifted, and Kimberley was relieved on 15 February 1900; massive reinforcements began to make their weight felt. Rhodes returned to Cape Town, Groote Schuur, and his importunate princess, from whom he once again escaped to Rhodesia.

The siege halted diamond-mining, and on the outbreak of war the whole mighty economy of the Rand had shuddered to a halt. Thousands of Uitlanders packed into trains for the Cape and Natal. A hundred thousand black workers fled also, trying to reach their homes on foot. They took no part in the white man's quarrel and rightly guessed that any black caught up in it would only suffer. Pro-Boer propagandists in England and the United States uttered dark threats of Johannesburg's sharing the fate of Moscow in 1812 but in the event the mines were left unharmed by the Transvaalers who even posted guards upon them against sabotage; they did not want to lose the sympathy of German and French investors.

Once the tide of war had been reversed the British pressed on into the Transvaal. Johannesburg was taken in May, Pretoria in June, and in September 1900 Kruger sadly left the Transvaal for European exile. In October Salisbury called his 'Khaki election' and won easily; like the army's victories on the veld it was hollow, and success was counted too soon. In a pamphlet against the war Olive Schreiner had told the British soldier that there were no laurels for him in South Africa and warned England that 'the hour of external success may be the hour of irrevocable failure'; a prescient warning for the new century. Even in the short term hopes were dashed.

The victors had hoped to start up the engines of the Rand straight away but it was not easy. Defeated in conventional battle, the Boers took to irregular war. Their columns rode on long commandos, hitting and running, disrupting any attempt to impose a settlement on the country, defying the British to catch them. The most daring commando of all was led by the lawyer turned general Jan Smuts, riding deep into the Cape Colony. From the taking of the Rand, it was to be an almost unimaginable two years before peace finally came, two years in which the British resorted to increasingly brutal means of counter-insurgent war, means that have become only too familiar in the century that was opening: blockhouses along the railway lines, and punitive measures against the enemy, the execution of Boers found wearing British uniform (though they had only taken them to replace their own rags) and other recalcitrants. To dry up the sea in which the fish swam, Boer women and children were rounded up and 'concentrated' in special camps where disease took a heavy toll. The British could not guess what they had spawned with their 'concentration camps', whose name was flung back at them in mockery by Hitler a generation later.

A few Rand mines started up in May 1901, and in December a war aim of a kind was achieved when the Johannesburg Stock Exchange reopened. By then several mining magnates had returned to the Rand. Those in Johannesburg in October 1899 had joined in the general flight to Cape Town. George Farrar contributed largely from ERPM funds to the raising of an irregular corps. He was commissioned and served as a major on the staff of the Cape Colonial division, seeing a good deal of the fighting. He was present at the siege of Wepener, where his brother Captain Percy Farrar was among those besieged; both brothers received the DSO.

Another who took an active part in the fighting was Abe Bailey. He served as an intelligence officer with the 11th Division and it was during the war that he met the young British war-correspondent Winston Churchill. Churchill was in the midst of one of his escapades, having been taken prisoner by the Boers and then escaped. Bailey talked to Churchill about what they would both do after the war: he wanted him to travel across the continent, from the Cape to Cairo. An important friendship was in the making.

The Corner House made large donations to the war effort, contributing to the upkeep of the Uitlanders' own unit, the Imperial Light Horse. Friedrich Eckstein watched the progress of the war, gloomily at first and then more cheerfully, from his office in Cape Town. The support which the mining houses gave was not ungrudging, or gratefully received. Beit became enmeshed in an interminable correspondence about the Hotchkiss quick-firing gun which he had presented to Brabant's Horse but which did not work in action. At the same time, when Ecksteins were forking out for £160,000 worth of railway engines and rolling stock, FitzPatrick complained, 'our firms have had to finance the richest government in the world. Is it not absolute rot? What would Disraeli, Palmerston and Pitt say to this?' What indeed. Fox and Gladstone might have said that the principal beneficiaries of a war could be expected to contribute to its expense.

All the same, FitzPatrick's relations with the British authorities were close. He became confidential adviser to Milner who was to administer the Transvaal after it was conquered. He house-hunted in Johannesburg for the great pro-consul. Barnato Park was one possibility, but in the event Milner made his official Johannesburg residence at Hennen Jenning's house, Sunnyside. Lionel Phillips had a good war in Homburg, Sicily and St Moritz as well as in London, raising funds for the war from those who had been 'killing Kruger with their mouths'. Well before the final peace he was able to congratulate Kitchener on the speed with which the mines were being reopened.

The Boer's guerrilla war was dogged but doomed. Peace came at last in May 1902, at a conference at Vereeniging on the banks of the Vaal. Sammy Marks, the old friend of Kruger, acted as mediator. Most of the country was conquered, most burghers had already become *hensoppers*, had recognized the inevitability of defeat. Now even the most irreconcilable *bittereinders* accepted that the struggle was over. Some of them smashed their rifles on the rocks outside the conference tents and vowed to go into exile rather than accept the British flag. In London the soldiers returned to a hero's welcome: the City Imperial Volunteers, a regiment raised by the Lord Mayor and recruited in the Stock Exchange, paraded down Piccadilly amid cheers and bunting; no building was more gaily got up than Wernher's London home, Bath House.

One man did not see the victory. By the beginning of 1902 Rhodes was very ill. The last thing he needed was to appear in a court case, but that was his fate. Catherine Radziwill's harassment had grown more and more bizarre. She had claimed to be his political agent, and had at least not discouraged rumours that she and Rhodes were to marry. Finally, in Rhodes's absence in England, she drew money on a promissory note with Rhodes's signature on it. It was Wilhelm Lippert's story again. Action was taken against her, and Rhodes was summoned back to Cape Town, where he gave evidence in court. She was exposed as a forger and, on the strength of an affidavit from Rhodes, charged with uttering a forged document. The journey to court in Cape Town was now too much for Rhodes. Instead he gave evidence before the magistrates at Groote Schuur. The Princess was present and tried to interject after Rhodes's evidence. She was silenced, and subsequently at the Supreme Court in Cape Town sentenced to two years' imprisonment. She had, it transpired, copied Rhodes's signature from a signed photograph.

This triumph also Rhodes did not live to see. He was moved from Groote Schuur to his little cottage by the sea at Muizenberg. Jameson, who had been through so much with his chief, was among those present at 6 a.m. on 26 March when Rhodes died.

'So much to be done, so little time.' Those were Rhodes's last words as reported by his friends and handed down to the hagiographers. He had made elaborate arrangements for what was to be done after his death. In all, his will had gone through six versions between the 'Confession' of 1877 and 1893. Their tenor was consistently extravagant and even preposterous.* The sixth will introduced Rhodes's scheme of scholarships.

A legacy of £100,000 was left to Oriel, his Oxford college. Then the terms of the Rhodes Scholarships were set out. There were to be sixty for the British colonies and two for every one of the United States. Scholars would be chosen to take up their places at Oxford on grounds of scholastic merit, but also on their qualities of manliness and moral character. No applicant was to be discriminated against 'on account of his race or religious opinions'. That was a gesture towards the Afrikaner. It was less helpful to others. Rhodes wished to soothe feelings between Briton and Boer rather than black and white, and for decades after non-whites in South Africa who thought to apply for Rhodes scholarships had a dusty answer.

Hated and adulated in his lifetime, Rhodes continued to be idolized after it. He was an astonishing phenomenon, deserving maybe

---

* In the third will Lord Rothschild was abjured to 'take Constitution Jesuits if obtainable and insert "English Empire" for "Roman Catholic religion" '.

Clarendon's verdict on Cromwell: 'he could never have done half that mischief without great parts of courage, industry and judgement'. But even in those racial attitudes he demonstrated an essential immaturity.

On 10 April Rhodes was buried at his chosen spot, View of the World, in the Matopos Hills of Matabeleland. Sixty years later his grave was visited by a perceptive observer. Evelyn Waugh reflected on the character of 'the Colossus'.

> There is a connexion between celibacy and 'vision' both at the lowest – Hitler – and the highest – the contemplative. Rhodes inhabited a world somewhere between. It is the childless who plan for posterity . . .
>
> There is an attractive side to Rhodes's character; his experimental farms; his taste in the houses he chose to live in. [But it is significant] that his scholarships were for Americans, colonials and Germans. The latin countries were excluded. For his obsessive imagination was essentially puerile. . . . He had a schoolboy's silly contempt for 'dagoes'; for the whole Mediterranean-Latin culture. . . . He saw in his fantastic visions of the future a world state of English, German and North Americans.

The dreams Rhodes dreamt were far from dying with him; but all the same, dreamers of his kind who plan for posterity are doomed to posthumous disappointment.

His old adversary survived Rhodes by little more than two years. Paul Kruger had a warm reception when he reached Europe. In Marseilles he was mobbed, in Holland and Germany he was greeted with enthusiasm even if the Kaiser had been told that it would be best if they did not meet. After the high points in German-Boer friendship, the Kaiser-Kommers and the Kaiser's telegram, that was a blow. Still, the simple truth which Kruger saw everywhere was that throughout Europe the Boer War had made England's name hated. He had bitter sorrows as well. His wife, whom he had left behind in Pretoria, died. And then came the news of capitulation. He had accepted that only those in the field could make that decision, but his exile's heart was crushed. He moved to the Riviera, then to Switzerland. On 9 July 1904 at Clarens by the shore of Lake Geneva he died, achieving one wish: 'Born under the English flag, I have no wish to die under it.' His body was taken back to the Transvaal and, with Milner's permission, buried next to his wife amidst scenes of ardent, repressed patriotism.

The British had won. The mine-owners had won. Victory was not quite sweet. It was not in Europe alone that the war had excited popular opposition. In England no war had ever been so fiercely opposed by such an important part of the population. There was the jingoist mood of war hysteria, the cheap press demonstrating its quality for the first time; but there was also an eloquent party of peace. In their pacifism the pro-Boers romanticized the Boers, just as the American peace movement was to romanticize the Vietcong. It was easy enough subsequently to show either faction the consequences of their enthusiasm, yet that would be to miss the point. As its opponents dimly saw, the question raised by the Vietnam war was not whether Communism had any moral purpose but whether the United States had. Their precursors had understood in the same way: it was not whether Afrikanerdom had any moral purpose, but whether the British Empire had.

The pro-Boers lost their immediate struggle but were not destroyed. One young politician, David Lloyd George, made his name in the campaign, and although the pro-Boers did not win the propaganda war, they had a signal success in persuading others about the Randlords' guilt. Lord Rothschild in London had warned Rhodes, 'Feeling in this country runs very high at the moment over everything connected with the war and there is considerable inclination, on both sides of the House, to lay the blame for what has happened on the shoulder of the capitalists and those interested in South African mining.'

Rothschild had in mind one parliamentary episode. In May 1901 Arthur Markham, the Member for Mansfield, made a bitter attack, in what *The Times* described as an 'almost incoherent speech', on the Corner House connexion. 'I charged Messrs Alfred Beit and Messrs Eckstein with being nothing more or less than thieves and swindlers.' What irked him was less the machinations leading up to the war than the machinations on and off the stock market. He subsequently repeated his allegations outside the privilege of the House of Commons, enabling Wernher, Beit & Co. to bring an action for defamation. Markham backed off and the case was settled out of court, but his attack was a symptom.

Even though the war was won and production had started up, Alfred Beit was not a happy man; probably he had not been happy since the fiasco of the Raid. Rhodes's death was a grave blow to him, removing an important psychological support. Back in South Africa in January 1903 Beit had a paralytic stroke near Salisbury. He returned to Europe, then he went home to England – not his birthplace but the country he had decided to make his home. He retired to his country house at Tewin Water in Hertfordshire which he only owned for three or four years, and there on 16 July 1906 he died. He was fifty-three; Rhodes had died at forty-eight; Kruger at seventy-eight.

Of all the magnates Beit had been the meekest and most self-effacing, yet he was perhaps the greatest. At his death and after, his friends and colleagues showered him with effusive tributes. There was 'something Christ-like about Beit', FitzPatrick said, conscious that it might seem extravagant praise: after all, the comparison is odd applied to a financier, however gentle and generous. Christ threw the money-changers out of the Temple, he did not issue them with new stock at par. In any case, Beit deserved a different compliment. He was not a saint but he was a genius, the master mind in making a success of the gold-mining industry on the Rand, as J.B. Taylor truly said.

Partly because the war had lasted so long, victory was bitter-sweet. The campaign against the Randlords had done its work. Seeds of resentment and hostility grew to bear fruit. Even the British soldiers who had fought and won the war had misgivings. In May 1900 the Witwatersrand had been taken by two brigades commanded by Sir Ian Hamilton. Thirty-four years later at the Annual Conference of the South African Veterans Association, Hamilton looked back to those far-off days, in jocosely reminiscent vein:

> Has it ever struck you about the stock-brokers who do turns on South Africans, and the Jews who lord it over Johannesburg, and the Governor of the Bank of England* – once a trusted young officer of Mounted Infantry, second to none as a lifter of Delarey's cattle – has it never struck you that this gold which they speculate belongs, by right of conquest, to the two infantry brigades of Ian Hamilton's column . . . we stormed Doornkop and with it we took half the gold in the world.

By then, Hamilton was echoing a theme that had played for a generation.

In his rage against the war, J.A. Hobson called it a 'Jew-imperialist conspiracy'. The idea that the war had been engineered by and fought for the benefit of the mine-owners gained wide currency and was taken up by a voice of satirical brilliance. Hilaire Belloc was twenty-nine when the Boer War broke out. He was a radical, but of a continental sort, a political Catholic in much the spirit of Karl

* Montagu (Lord) Norman.

Lueger, importing to England the passions of the anti-Dreyfusard cause. During the war which they bitterly opposed he and his friend G.K. Chesterton turned up the pressure in the dissenting press, especially the *Daily News*. After the war it was Belloc of all writers who caught the spirit of the age, the glittering, gaudy, sordid world that was Edwardian England. He caught it in his political novels, he caught it above all in his light verse. He entered parliament as Member for South Salford in the Liberal landslide of 1906 but was as much trouble to his own party as to the Tories. He had after all immortalized that election:

> The accursed power which stands on Privilege
> (And goes with Women, and Champagne, and Bridge)
> Broke – and Democracy resumed her reign:
> (Which goes with Bridge, and Women, and Champagne).

He despised and denounced *The Party System* in a book written with Cecil Chesterton, brother of G.K. The front benches were perpetually acting in collusion, they claimed. Why else had the 'Committee of No-Inquiry' been so ineffectual? Because the Liberal Campbell-Bannerman was 'as eager as any Tory could be to hush up anything that might discredit the Colonial Office'. That was a travesty of the facts but with just a tincture of truth.

Belloc never managed to infect English politics with his violence and prejudice. But he left a mark. All his fierce indignation against the plutocracy, against corruption, against an unjust war, against the Jews, welled up into one savage poem. Several years after the Boer War a Tory imperialist peer said in the House of Lords that the pro-Boers had 'confused soldiers and money-grubbers'. In his 'Verses to a Lord' Belloc in his bitterest satirical vein saluted the brave soldiers who had won the Boer War.

> You thought because we held, my lord,
> An ancient cause and strong,
> That therefore we maligned the sword:
> My lord you did us wrong.
>
> We also know the sacred height
> Upon Tugela side,
> Where those three hundred fought with Beit
> And fair young Wernher died.

The daybreak on the failing force,
The final sabres drawn:
Tall Goltman, silent on his horse,
Superb against the dawn.

The little mound where Eckstein stood
And gallant Albu fell,
And Oppenheim half blind with blood
Went fording through the rising flood –
My lord, we know them well.

The little empty homes forlorn,
The ruined synagogues that mourn
In Frankfurt and Berlin,
We knew them when the peace was torn –
We of a nobler lineage born –
And now by all the gods of scorn
We mean to rub them in.

Belloc could insult the Randlords; he could not damage them. Within a few years he and his cronies moved on to another Jewish scandal and Marconi replaced Beit as the centre of attention for the ignominous passion of Judenhetze.

In any case, this Jew-baiting confused the issue. The question was not what the mining magnates' origins were, or even how honest they were, but what power they had and how they used it. Others also took a sombre view. Richard Solomon was a lawyer and politician who had served in Schreiner's Cape government as Attorney-General. After the fall of that ministry in June 1900 Milner asked him to join the administration which would reconstruct the Transvaal. Before the peace was signed, in the last bitter stages of the struggle, Solomon wrote to Rose Innes: 'I don't like the way things are going, but I can't write. I would like to talk to you when I should talk more freely. I will only say this. I prophesy that the gold mines are going to rule South Africa.'

16
Usually in Park Lane

The war had been fought to make the Transvaal safe for the Corner House; now it was up to the partners to win the peace. Men and women, black and white, began to flock back to the Reef after Vereeniging. The white population of the Rand had been just over 50,000 in 1896; by 1904 it had reached 123,000. The mines opened up one by one. In 1902, 1.7 million ounces of gold were mined, as the mines got back into their stride. In 1906 the pre-war peak of production was passed.

Of all men Alfred Beit had seen the potential of the deep levels. He now saw the next step, the mining of low-grade ores. As mining operations had retreated further from the original outcrop the quality of ore treated had declined. But the answer to the question of what was payable could change also. Beit believed that as richer ores were exhausted it could yet be profitable to mine what had once seemed hopelessly unpayable grades. However, this must mean mining them intensively, in ever larger quantities; most of all, it meant controlling and reducing costs even more vigorously than before. The war, like all wars, brought inflation in its train and the cost of stores and materials for the mines rose. Pressure was greater than ever on the cost of labour, the 'full, cheap, regular supply of Kaffir and white labour' which Hobson had rhetorically claimed was the all-important object of the war.

After British victory the mine-owners had at last a pliable government on their side. They asked Milner as early as 1900 if his reconstruction administration would take responsibility for the recruiting of black labour after the war. Although Milner refused, he encouraged the setting up of the Witwatersrand Native Labour Association. But in 1902 and 1903 there was still a shortage of labour on the mines. The 'boys' had been scared away by the war, in which blacks had suffered cruelly. Although blacks were keen enough to work for cash, the arduous work of hand-drilling in the stopes, and the knowledge that wages fluctuated, often enough downwards, did not make the mines as enticing as they needed to be.

Milner had brought a team of young men out to get the Transvaal going again, 'A sort of kindergarten of Balliol men',* Merriman called them, and the name kindergarten stuck. They were there to provide all the things which had supposedly been absent under 'Krugerism': a modern bureaucracy, impartial justice, free trade, cheap goods, controlled labour. They were there to help the mines. They needed the mines and the mines needed them. Milner's closest association was naturally with the Corner House, which emerged in the post-war decade as not merely the largest but overwhelmingly the richest mining group.

Hermann Eckstein had died young, Taylor had retired, Georges Rouliot likewise retired at the beginning of the century. The two leading men in Johannesburg after the war were FitzPatrick and Samuel Evans, both made partners in Eckstein's in 1902. Wernher and Phillips were in London and Friedrich Eckstein followed them there. There were increasing tensions between London and the Rand. The senior men doubted the capability of FitzPatrick and Evans, and also deprecated their open involvement in politics. It would be too simple to say that FitzPatrick saw mining as a means to a political end, imperial glory and the apotheosis of British South Africa, while Wernher and Phillips saw Milner's administration as a means to rebuilding the mining industry; but there were strong differences of emphasis and enthusiasm.

In either case the great question of labour had to be resolved. Various expedients were toyed with. An attempt was made to recruit cheap miners in Italy, but they failed to turn up to meet their boat in Naples. A correspondent wrote to suggest that American blacks should be sent to South Africa: they could be useful 'notwithstanding the inherent laziness of the race' and the Americans would be glad to see them exported. And there was an idea of bringing workers from the West Indies. These were pipe-dreams.

One other dream was not fantastic and was in fact of high political importance for South Africa: white labour. Everyone knew that the modern economy of the Rand had been built upon a form of serf labour, even if they did not put it like that. South Africa's labour economy was quite unlike that of western Europe, or North America, or Australia. An elite of skilled white workers and artisans were supported by a mass of unskilled 'boys', hand-drilling and performing

---

* Most of them were in fact from the college where Milner had taught, New College.

other grinding manual work. The relationship between the two was wholly different from that between skilled and unskilled workers in Europe, wide as that difference was.

Could it be otherwise? Some men thought that it might be. They had various motives: working-class solidarity, a dislike of the mass of blacks living on the Rand, a wish that separate cultures should live separately, a sincere dislike of exploitation, and disquiet that the economy had become dependent on blacks. The creed of all-white labour had its apostle in F.H.P. Cresswell, a mining engineer who had come out to the Rand in the 1890s and become manager of the Village Main Reef mine, owned by the Corner House. Cresswell was convinced that the mines could be worked just as mines were in the northern hemisphere by 'civilized', that is, white labour. The much higher wages of whites against blacks would be compensated for by the much higher efficiency of mines run on the most up-to-date lines with the latest machinery. His goal was 'the freeing of the country from its dependence upon the Kaffir'.

'White labour' was anathema to the mine-owners. They never believed that white workers in smaller numbers could work as cost-effectively as blacks. White workers had always meant trouble and always would. FitzPatrick was appalled by the prospect of 'the "working man's paradise" and have this industry throttled by labour unions'. Unionized whites could demand pay and better working conditions as blacks could not. In all sorts of ways blacks had their advantages. No compensation needed to be paid if they were killed; their deaths were merely 'wastage'. Trying to prove his point, Cresswell conducted an experiment with white labour in his mines. South Africa would have been a very different country if he had won. He lost.

The wishes of the magnates conveniently chimed in with Milner's policy: 'Our welfare depends upon increasing the quantity of the white population, but not at the expense of its quality. We do not want a white proletariat in this country.' In view of the whites' position among the vastly more numerous blacks it was essential, Milner said, that the poorest whites should stand far above 'the poorest section of a purely white country'. There was a political point to this. All the antagonisms between Uitlander and Boer were to be resolved, as Milner saw it; the franchise question was to be answered at last. For that to come about, there must be an absolute and not merely a relative distinction between free men and helots (in a more precise meaning of the words than Milner had used in his despatches before the war which described the Uitlanders as 'helots').

Defeating white labour schemes was all very well. It did not answer the mine-owners' practical problem. The Transvaal Labour Commission reported at the end of 1903 that another 130,000 labourers were needed on the mines, far more than the WNLA could recruit. Another answer came to the mine-owners: Chinese labour. Coolies would be imported from China with the assistance of the Imperial Government in Peking. They would come under three-year indentures, to return when they had served their contracts; they would live apart form the rest of the community until the time of their return. As soon as the plan became known it was bitterly opposed, by Afrikaner nationalists now beginning to recover their confidence after defeat, and by white workers. George Farrar tried to reassure skilled workers. Whereas unskilled white labour on the Cresswell pattern would soon become skilled labour and compete against them, with Chinese their position would be absolutely secure.

Opposition was in any case unavailing. Chinese workers began to arrive in 1904. On one level the vast traffic was a brilliant success. Over three years, more than 60,000 Chinese were brought to the Transvaal, shipped 10,000 miles to a place they had never heard of, among people they could not understand. At the high point of the experiment in July 1906 17,513 whites were working on the mines at an average monthly pay of £26 15s, 102,420 blacks at 52s 3d, and 53,062 Chinese at 41s 6d.

Violence was rife on the Chinese compounds, where execution proved no deterrent to frequent murder, and so were gambling and opium smoking. Desertion was also common. But whereas a black worker who deserted from the mines could soon find friends to shelter him the Chinese deserter at loose on the veld was helpless. Desertion was one of the chief reasons why Transvaalers opposed the importing of Chinese. In January 1905 a new political party was founded, Het Volk, 'the people'. Its foundation was preceded by the emotional spasm of Kruger's funeral in December, and stimulated by disquiet over the Chinese labour question. Two of the most capable Boer generals, Louis Botha and Jan Smuts, were among Het Volk's leaders. This new party was no passing mood among the Afrikaners: already the defeated of 1902 were looking to settle the score.

If Chinese labour was an economic success it was a political mistake. The British Prime Minister Balfour and his Colonial Secretary Lyttelton had agreed to Milner's and thus the Corner House's arguments about the needs for Chinese labour without thinking far ahead of the political consequences. In fact the episode produced general but confused revulsion. Many people in England as well as

South Africa had a racial fear and dislike of the Chinese. Working-men everywhere saw 'coolie labour' as a threat to their own position, as indeed it was; the fear played an important part in the early American union movement. At the same time humane feelings were repelled by 'Chinese slavery' as it came to be called, and by the lurid stories of vice and savage punishment on the Chinese compounds. Balfour finally resigned in December 1905. In the course of the ensuing election of January 1906 'Chinese slavery' played an impor-tant part. The Liberals under Campbell-Bannerman won a landslide victory.

The Under-Secretary of State for the Colonies with responsibility for South Africa in the new Liberal government was a man who knew South Africa and the Randlords very well. Winston Churchill, the 'Blenheim rat', had entered parliament as a Conservative in the 1900 Khaki election, but then in 1904 left the Tories for the Liberal benches. His career had been watched with interest by old Joseph Chamberlain, another floor-crosser. Since the death of Rhodes, Chamberlain's own relations with the Rand magnates had been uneasy. He had warned them of the political danger of Chinese labour shortly before he resigned as Colonial Secretary in 1903. The follow-ing year Beit and other magnates had subscribed £25,000 towards a testimonial fund for Chamberlain. But Chamberlain was never one for misplaced gratitude. One of his last utterances before the stroke which ended his public life was to accuse ministers of colluding over the Chinese question with ' "magnates" who are not creditable acquaintances and who live in palaces, usually in Park Lane'.

On taking office, Churchill had plenty of advice from these 'magnates'. His friend Abe Bailey told him that Johannesburg was in an awful state financially and warned that there would be panic if Chinese labour was immediately ended. It was not immediately ended. Hilaire Belloc, now in parliament, naturally accused his own front bench of acting 'to propitiate the South African Jews'. Churchill had other troubles on the Rand. Chief of them was Robinson. J.B.R. had of course refused to join either the employers' cartel of the WNLA or the Chinese labour scheme. Instead he started his own recruiting agency for black labour. He complained bitterly that his agency was not being given equal treatment. Churchill was inclined to humour him. He recognized that J.B.R. had won an unintended political victory over Chinese labour, and he did not mind him running experiments with white labour to keep Cresswell quiet: after his dismissal from the mine he managed, Cresswell was becoming a formidable agitator. The new High Commissioner for South Africa was Lord Selborne, a high-principled man who had little enthusiasm for Robinson or indeed any of the magnates. He told Churchill in London that Robinson's agency was disreputable. Churchill said that he should not be so fastidious: 'If you choose to judge them by an unduly high standard, having regard to the general character of persons engaged in the gold-mining industry, nothing further will be done by us.'

In any case, J.B.R. was about to receive another reward. Responsible government for the Transvaal had been ordained, on a white-only franchise as promised in the Vereeniging treaty. In the elections of February 1907 Het Volk won a majority together with some English-speaking allies. The Randlord's own party, the Progressive, was defeated. Robinson had advocated responsible government and had supported Het Volk. On Botha's recommendation he was made a baronet, which he had been begging for.

Two elections, one in England and one in the Transvaal, had been overshadowed by the Rand gold-mines and Chinese labour. On the face of it each election had been won by the foes of the great mining houses. Yet the Corner House was not downhearted. For one thing, Percy FitzPatrick had won a seat and Cresswell had not. More than that, the Corner House had reason to hope for good relations with the new government. A year before the election Smuts had told Wernher that they had in many ways identical interests. And after the election, FitzPatrick and Smuts met in the Rand Club, shook hands, and over a drink made scathing conversation about their common foes. FitzPatrick's election was a convenient excuse for easing him out of the partnership, as Phillips, Eckstein and Wernher had long wanted to do. 'Fitz' was pensioned off on very generous terms to continue his political career.

Political difficulties were outweighed by the financial effects of Chinese labour. By 1906 the Transvaal became the world's leading gold producer; it was reckoned that profits of £5.4 million were directly attributable to the Chinese. They had solved the two labour problems, black and white. The mines could now get blacks on the terms they wanted, and they had broken attempts effectively to unionize white workers. As the Chinese began to go home, the mines found a new source of labour, Afrikaners. They were simple country boys, ignorant of industry or of unions and ready to accept lower wages. When in 1907 the Rand saw its first major strike of white miners (not its last), Phillips enthusiastically directed the bringing in of Afrikaners as scabs and strike-breakers. The new administrations rallied to the mine-owner's side. Phillips himself saw the irony of it:

'The British troops to keep English miners in order while the Dutch men are replacing them in the mines.'

At the same time Phillips was told sharply by his partners to keep out of overt politics, and in particular not to take a seat in the upper house of the new legislation. Instead he resumed the presidency of the Chamber of Mines, and made it his first task to sweeten relations between Johannesburg and Pretoria. And so, with these hiccups and interruptions, the gold-mines continued to flourish.

So too did the diamond-mines. Although the South African centre of financial gravity had shifted to Johannesburg in the 1890s, Kimberley remained important both to the economy of the country and to the personal fortunes of many of the magnates. In 1902 £4,949,508 worth of diamonds were mined in the Cape Colony, against £7,301,501 worth of gold in the Transvaal (gold production had only just restarted). By 1906 gold production had rocketed ahead: £24,606,336 against £9,596,643 in diamonds; in 1910 the figures were £31,973,123 for gold, £8,189,197 for diamonds. This last figure included the new diamond-mines in the Transvaal.

In the 1890s Thomas Cullinan had been a builder in Johannesburg. Like Percy FitzPatrick, he was Cape born of Irish descent and arrived on the Rand by way of Barberton. He built a good many of the most prominent buildings of early Johannesburg: the Chamber of Mines, the National Bank, the Robinson Bank, the second Rand Club building. But he had not lost the digger's itch. He prospected for diamonds in the Orange Free State, staked in part by J.B. Robinson. Diamonds were found there, not on the scale of Kimberley but proof that there were other diamond pipes in South Africa. Just before the Boer War diamonds were found in the Transvaal, near Van der Merwe, a station on the Johannesburg to Delagoa Bay railway line. Cullinan looked further north in the Magaliesberg mountains, on the farm Elandsfontein owned by an old Boer called Willem Petrus Prinsloo. There were diamonds present, as there had been on the banks of the Vaal in 1869. But where had they come from?

After the war, Cullinan bought the farm on behalf of the Premier Syndicate and began mining; his partners included Bernard Oppenheimer, eldest of the Oppenheimer brothers. Sammy Marks visited the diggings and was not impressed. But when Alfred Beit took George Farrar and the mining engineer brothers Hennen and Sidney Jennings to Elandsfontein, Sidney Jennings said that he thought a pipe had been found. By April 1903 true blue ground was found and the kimberlite had yielded nearly 100,000 carats. Sigismund Neumann in London was soon trying to get his Diamond Syndicate in on the act. A bargain was struck: within three months the Premier mine had opened a sales agency in London, headed by Neumann. By the end of 1904, 2,000 blacks were working on the mine. And on the afternoon of 26 January 1905 a miner found a large object on the side wall where he was digging. It was cut out and taken to Johannesburg, a stone of 3,106 metric carats, four inches long.

The Cullinan diamond was the largest diamond ever found then or since.* The *Transvaal Leader* suggested that the stone should be bought and presented to King Edward VII, and bought it was by the Transvaal government for £150,000. There was some difficulty over its presentation. The Tories were against acceptance and the King viewed the proposal 'with great disfavour'. But his misgivings were overcome and on 9 November 1907 it was presented to the King at Sandringham by Sir Richard Solomon representing the Transvaal in the presence of the Queens of Spain and Norway. Abraham Asscher, of a great Amsterdam diamond family, took it to Holland in his pocket. It was cleaved by his brother Joseph on 10 February 1908, and then cut into nine gems, which were incorporated in the Crown Jewels worn by two Kings and a Queen at three coronations since.

The romantic story of the Cullinan diamond could not have had its climax in a less propitious year. In 1907 the diamond market collapsed. There was a recession in the United States, where by now some 70 per cent of South African diamonds were sold. Demand plummeted and prices with it. De Beers and the Syndicate were strong enough to withstand a temporary slump, and with them Premier which was now part of the cartel. In Kimberley and Elandsfontein production was sharply curbed, to hold back a further drop in prices, and many miners in South Africa and many diamond cutters in Amsterdam were put out of work.

Although by the new century Julius Wernher's fortune was overwhelmingly founded on gold he still took a keen interest in diamonds. The year before the diamond slump his concern for the diamond market led him farcically astray. The possibility that gems could be artificially produced still fascinated men, and haunted them. If it could be done, it might spell disaster for the industry. In 1905 a Frenchman called Henri Lemoine let it be known that he had cracked the riddle. His discovery was touted by the French press, and the diamond magnates took fright. Lemoine met Wernher in London and arranged to show him his workshop in Paris. In the basement a great furnace

---

* In 1934 a 726 carat stone was found at the mine, which was cut into the Jonker diamond, and in 1954 a 426 carat stone, now the Niarchos diamond.

blasted and ground and extruded a flawless diamond. Wernher was convinced. With his generous backing, Lemoine said that he would set up a factory in the Pyrenees; the two men would safeguard the secret between them.

Wernher informed Jameson, now Cape Premier as well as a director of De Beers, who was still more alarmed: Kimberley was critical to the whole economy of the Cape. At Jameson's request, Francis Oats went to Paris. Oats was one of the oldest and wisest of Kimberley hands, a Cornish miner who had worked his way from mining engineer to director of De Beers. He watched the astonishing process and then examined the diamond which the furnace produced. It was too good to be true: a fine Kimberley stone. Lemoine was a skilful fraudsman but he met his match in Oats, who by his own sleight of hand proved that the 'furnace' was a nonsense. He told Wernher to open the formula which Lemoine had presented to him sealed; it too was a nonsense. Lemoine had gambled not unreasoningly that charges would not be brought as Wernher would be made to look such a fool. But Wernher swallowed his pride, reluctantly told the French police and provided the evidence with which Lemoine was convicted and sent to prison for six years.

Both South Africa and Europe guffawed. Louis Cohen was delighted at this trick played on the great Wernher. In Paris, Florence Phillips told her husband, 'They roar with laughter at Wernher's name.' *Le tout Paris* who had laughed included a young writer, Marcel Proust. As yet to eat the madeleine which would send him on his immense fictional journey, he was perfecting his style by parody or imitation of other writers. *Figaro* published a dazzling series of pieces by him recounting 'l'affaire Lemoine' in the styles of the French masters – Saint-Simon, Balzac, Michelet, Flaubert and Sainte-Beuve – the pastiches of his book *Pastiches et Mélanges*.

Wernher's pride may have been bruised but he remained one of the richest men in the world. As the reign of Edward VII progressed, the South African mining millionaires approached the zenith of their wealth and fame. By now a good two dozen of these fabulously rich men had installed themselves in England, living in London and the country in stupefying luxury. Percy FitzPatrick's address book lists them, a golden panorama of Edwardian plutocracy: Alfred Beit at 26 Park Lane and Tewin Water; his brother Otto at 49 Belgrave Square; Friedrich Eckstein at 15 Park Lane; George Farrar at 54 Old Bond Street and Chicheley Hall in Buckinghamshire; S.C. Goldmann, the financier and author of books on mining, at 24 Queen's Gate, 'Mikki' Michaelis at Tandridge Court in Surrey as well as Ben Alder in

Scotland; J.B. Taylor at Sherfield Manor in Hampshire as well as his sporting estate in Scotland.

When Taylor returned to South Africa on holiday after almost fifteen years he stayed with his old partner Phillips at the Villa Arcadia, the house where the Phillipses had moved when FitzPatrick took over Hohenheim, 'Phillips's Folly'. Not that Phillips had put folly, or extravagance, behind him. He too had acquired a residence in the English countryside, Tylney Hall in Hampshire. It was another white elephant: a vast house with a 2,500-acre estate. After Beit's death he complained to Eckstein that he had made financial sacrifices on the firm's behalf by returning to South Africa, but the fact was that he had spent enormous sums on his new house built by Herbert Baker, architect-in-chief to the Randlords. At the same time, Phillips was buying a great country property at Woodburn in the northern Transvaal. He repined when it seemed necessary to put Tylney Hall on the market but, as he knew really, his plight was of his own making.

Two others lived not quite in Park Lane, but round the corner: Sigismund Neumann at 146 Piccadilly and Wernher at Bath House. Besides, Neumann had a racing box at Newmarket and an estate in Inverness-shire, exciting Louis Cohen's derision: 'he is known as the "Mac Neumann" though he does not wear a kilt for fear of the wind'. Kilted or not Neumann frequently entertained the King from nearby Balmoral. He spent money generously, as a large donator to the Berkeley Synagogue in London and as a collector of Pre-Raphaelite paintings which he donated to the Johannesburg Art Gallery. His reward came in 1912 when he was made a baronet.

By then honours were quite showering on the Randlords. For all the electioneering over Chinese slavery, the Liberals who came to power in 1906 with Bridge and Women and Champagne were closely allied to this new plutocracy. Businessmen were thick on the government benches in the House of Commons. The party was kept afloat by donations from financiers and industrialists. In return, Campbell-Bannerman and Asquith were lavish with the King's favour. The Tories gave their own friends honours but made snobbish distinctions. Only two generations back Melbourne had deplored the granting of a baronetcy to a Jew and the Tories still restricted their honours to gentrified bankers and the like. But the stream of titles had begun after the Boer War. Sir Percy FitzPatrick's knighthood came in 1902, as did Sir George Farrar's, though only one baronetcy was created for a South African magnate by the Conservatives. Then the floodgates opened. Before the first great generation of mining magnates had passed, most of them could, like Sir Walter Elliot of Kellynch-hall, pick up the baronetage to find 'occupation for an idle hour, and consolation in a distressed age' in the form of their own names. Beit died before he could be honoured, but his brother became Sir Otto. Then there were Sir George Albu, Sir Sigismund Neumann, Sir George Farrar (advanced a rung from his knighthood), Sir Friedrich Eckstein, Sir Max Michaelis, Sir Joseph Robinson and Sir Lionel Phillips.

And of course there was Sir Julius Wernher. His was the first baronetcy, conferred by the Tories in 1905. It was fitting. With his quiet manners and ostensible lack of interest in politics, Wernher was the chief of all, a prince-financier. He held a kind of court in London both at Bath House and at his offices, 1 London Wall Building, but it was at Luton Hoo that he shone in all his splendour. He entertained magnificently there, the King among his guests.

An odder pair of guests were Sidney and Beatrice Webb who stayed with Wernher when they were writing their minority report for the Royal Commission on the Poor Law. Even that thick-skinned couple saw the irony of it. Beatrice said that she could not count the servants at Luton Hoo ; in fact the retinue included more than fifty gardeners and ten full-time electricians. With all the luxury, Wernher had private sorrows. He had tried to bring up his eldest son Derrick to be an English gentleman, as well as his heir in business but, as he told Phillips, Derrick fell into the hands of bookmakers and money-lenders. The friends who had led him into folly when he was still under age were the sons of the Speaker of the Commons, Lowther. 'I shall never trust him sufficiently to put him in the business and that is final', he wrote, adding sadly that he had only stayed on in the business for Derrick's sake.

He wrote this in 1910, a fateful year for South Africa. Since the end of the war in 1902 the question of what form the government of South Africa should take had been unresolved. The two Dutch republics had been conquered; Rhodes's ideal of a greater South Africa seemed near to achievement. Was it to be a loose federation or a closer union? Several of the mining magnates threw their weight behind closer union: Farrar, who had entered the legislative assembly of the Transvaal in 1907, Abe Bailey, who was the Progressives' whip in the Assembly, and FitzPatrick. After the Liberals came to power at Westminster and the Afrikaner coalition in the Transvaal, the Corner House tried to distance itself from open political activity. But the Union which was consummated in 1910 was satisfactory to the Corner House and to the rest of the mining industry.

Every section of white South African society compromised over Union; some sections made more or less sensible or far-sighted compromises than others. The Cape Liberals compromised over the question of equal rights. High-minded men in South Africa and outside hoped for the best, or were deceived. Hobson wrote to Smuts appealing for a franchise which would admit citizens of all races; Smuts's reply was at best disingenuous, saying that, although many Boers were reactionary on the question, their leaders were in advance of them. The Cape's special colour-blind franchise under which a few blacks and a good many Coloureds had the vote was entrenched in the constitution, for ever it was supposed. The Afrikaner nationalists of the Transvaal received the concessions they wanted in regard to the franchise, and the mining houses got a strong central government which was bound to be sympathetic to the industry, especially over the question of labour, black and white.

As one era ended, another began for South Africa politically; so also for the Corner House. In the year of the Union the detailed plans to wind up the old partnerships of Wernher, Beit & Co. and H. Eckstein & Co. were completed. The firms' holdings on the Rand passed to the control of Central Mining and the diamond interests to another private firm created for the purpose, L. Breitmeyer & Co. The former partners made new arrangements. Lionel Phillips's extravagance proved to have been such that in order to acquire a stake in Central Mining he had to borrow £90,000 from Wernher. At the same time, Phillips decided that after Union he would return to politics and he entered parliament in the first Union elections of September 1910, representing Yeoville on the Rand in the interest of Dr Jameson's Unionist Party.

The loan was one of the last transactions between Phillips and Wernher. Wernher had been gradually compelled by ill-health to withdraw from active business, but he could look on an astonishing achievement. By the time the partnerships were wound up the Corner House was the 'fifth province' of the Union, its total turnover larger than the budgets of two provinces of South Africa, the Free State, or Natal. It was the largest landowner in the country, effectively controlled De Beers, the South African National Bank, many of the country's coalmines and the largest newspaper group. It produced over 3 million ounces of gold annually and paid more than half the Rand's profits. This was Julius Wernher's realm and his inheritance when he died in London on 21 May 1912.

17
Settling the Score

One family conspicuously did not benefit from the shower of honours in Edward's reign or later. No Barnato or Joel ever received his sovereign's accolade. They were still unmistakably shady. What was it about them? The Isaacses and Joels had been poor Jews from the East End of London, but then Neumann, Albu and Phillips were Jews also. Barney Barnato's business methods did not stand close examination; nor always did the operations of the Corner House. The cousinage of Barnato and Joel seemed to do penance for the rest. They remained beyond the pale, too cocky, too flashy, too brazen, too impenitent. They had bad luck also.

By the turn of the century Barney Barnato and Woolf Joel were dead, and Harry Barnato was living in placid retirement on the Côte d'Azur. The remaining brothers, Jack Barnato Joel and Solly Joel, were active in business. The family gold-mining company 'Johnnies' still flourished, as did their diamond interests. Solly concentrated on diamonds and on high living. The head of the family was now in effect Jack, who worked hard at the business, though hampered by the fact that he never visited South Africa. He had never quite lived down that something in his past which too many people knew about, even if they did not know quite what it was. Peoples' pasts have a nasty way of catching up with them; so, dramatically, it was with Jack Joel.

He had inherited Barney Barnato's fondness for racing, and was seen in the Royal Enclosure at Ascot and in the Jockey Club Stand at Newmarket – though not of course a member of the Club itself; that would have been barely thinkable in an aristocratic age. In the early 1900s Joel got to know another racing man, Robert Sievier. Sievier had come a long way since he went out to South Africa in the 1870s. He had crossed the world, spent time in Australia, married and unmarried more than one wife, and had made and lost more than one fortune. He was a bumptious figure who came back from disaster with the bounce of Captain Grimes. At the very beginning of the century he was on one of his upswings: as well as owning a house in the country, he rode regularly to hounds, and he was a heavy and sometimes successful punter. In the autumn that the Boer War broke out he made a large killing by backing Proclamation in the November Handicap, recognizing as not everyone did the genius of the American jockey Tod Sloan.*

His luck continued in the New Year when he won £9,000 on Sir Geoffrey in the Lincoln, the start of a fantastic season's betting: between December 1899 and November 1901, £262,000 passed through Sievier's bank account, almost all of it lost and won on the turf. Then he bought a fine yearling filly, paying ten thousand guineas in competition with the trainer John Porter who was acting for the young Duke of Westminster. This handsome bay was named Sceptre; she was the best thing that ever happened to Sievier.

The year of King Edward's accession, 1901, saw Sceptre's two-year-old career begin with victory in the Woodcote at Epsom. But Bob Sievier had taken a severe pasting and knew that he would have trouble settling his bets. He made heavy weather of his troubles when he ran into Joel and then – for he was not a bashful or shy man – asked him for a loan. Jack Joel was now worth at least £4 million and was in a position to pay anyone else's debts of honour. He also had a reason for helping Sievier. He was just beginning a serious career as an owner and like his uncle Barney he was finding it hard to break into the charmed circle of top-class bloodstock. Here was a chance. He offered a loan to Sievier, with two conditions: 10 per cent interest, and first refusal on the sale of any of his horses. He had his eye of course on the brilliant two-year-old filly that the whole racing world was talking about.

On the strength of this agreement Sievier turned down a large offer for another of his horses, Lavengro. He was naturally astonished and appalled to receive a letter from Joel withdrawing the offer of a loan. When the two next met, Joel offered instead to buy Sceptre for £10,000. Now Sievier was not astonished, he was outraged: Joel was taking advantage of his knowledge of Sievier's financial predicament to snatch away his most treasured prize. Sievier told him what he thought of his offer. Jack Joel had grown used to getting his way and to brushing enemies aside, or buying them off. This time he had made a mistake. In Sievier he had found a man who had not only ability but an almost insanely vengeful temper. The feud began in earnest.

---

* It was Sloan who revolutionized flat race-riding style in England, leaning forward on the horse's withers with leathers right up; as well as enriching the language with rhyming slang for 'on my own'; 'on my Tod'.

Sievier told the whole world, or the part of it he knew, of this episode, and he bestowed the name of 'Promising Jack' upon his enemy. That was not all the joke. He named an indifferent two-year-old of his Promising Jack. When it was beaten at Windsor by a horse owned by Joel the winning owner received a telegram: 'You have won although you were second.'

On the face of things 1902, Coronation year, was a glorious one for Sievier. Sceptre won the Two Thousand Guineas, beating the colts, and then the One Thousand Guineas in the same week, both in record time. She was beaten in the Derby, starting at evens, but won the Oaks and the St Leger, the only horse ever to win four of the five English classics. These victories only increased Joel's chagrin at not having acquired the filly; but they did not restore Sievier's fortunes. He lost far more on the Derby than he won on the other races; he became embroiled in a disastrous slander action; and he was, unjustly as it happened, warned off the turf.

Now at the pitch of his misfortune, he thought of a way to rescue his affairs, and to settle the score. He would start a new paper. The *Winning Post* had forebears, especially the *Sporting Times* as it was formally called but which was known from the colour of its paper as 'the Pink 'un' in every club and mess in an Empire on which the sun never set. Like 'the Pink 'un', the *Winning Post* was to be a mixture of tittle-tattle, racing tips, and 'sauce'. For that matter it has had its successors, notably *Private Eye* which combines sexual gossip, jokes and the settling of private scores in much the same way.

The paper came out at a penny a copy and was bought for its well-informed racing notes, its cricket column by the great Archie MacLaren – captain of Lancashire and England, a friend and fellow-gambler of Sievier's – and its excruciating 'tabasco tales' or dirty jokes such as delighted the stock exchange and the smoking room. That and its profiles: the paper began a series called 'Celebrities in glass houses'. The first were innocuous enough. Then the celebrity chosen for 15 October 1904 was Mr S.B. Joel of Grosvenor Square. Sievier knew a good deal about Jack Joel but he needed the help of someone who knew more, and he soon found it. Louis Cohen was back in London eager to make a living by his pen, and he had his own scores to settle with the families of Isaacs and Joel. His most bitter resentments were against Barney Barnato and Solly Joel, but Jack Joel would do.

Later on Cohen wrote in the *Winning Post* under his own name, but the anonymous 'Celebrity' could only have been by him. Few indeed but Cohen could exactly remember the skeleton in Joel's closet; the

IDB charge and the hurried departure from Kimberley. Now it was displayed for *Winning Post* readers – and it was selling 50,000 copies an issue – to gloat over. Joel's lowly origins and slippery conduct were paraded. In addition to this exceptionally venomous attack came a succession of further insults against 'Joel, the notorious dealer in illicit diamonds'. Scarcely an issue passed without a sneering reference. Sievier in effect dared Joel to take legal action, but he took none.

Nevertheless he was not passive under these attacks. He became so enraged that he hired a former prize-fighter, Dan Murray, to beat up and cripple Sievier. It was an extremely foolish as well as brutal plan, since Bob Sievier had better underworld contacts than Joel. He got wind of the plot, bought off Murray, printed the whole story with incriminating evidence in the *Winning Post*: 'Joel has tempted others with his filthy lucre to do that which he was too cowardly to attempt in person, and thus proved himself worse than the ruffian he employed.'

Still no sound was heard from Joel; still the campaign of abuse was kept up. Whenever the paper contained some offensive reference to Joel, Sievier employed a sandwich-board man to walk up and down Grosvenor Square with a placard reading simply 'Joel'. Nor was Jack Joel the only member of his family abused. Solly Joel came in for some of the shots and not long before his death in 1908 Harry Barnato had to read how 'your third sister married Issy Nathan who kept the fried fish shop in Pie Lomad and who was known to multitudes as Issy Sloshy. His brother Sambo the Piccaninny drove an omnibus. These are your family connections and it gives me further delight to further record that you have a nephew at Eton College. Prodigious!'

The duel took several years to reach its climax. Rather than sue for libel, Joel told anyone who cared to listen that Sievier was trying to blackmail him. Through intermediaries Joel tried to silence Sievier who was once more nearly broke, but Sievier was a proud and, by his own lights, an honest man and he would not be cajoled or bought. And so Joel in his desperation thought of another extreme, and once more foolish, remedy. He tried to set a trap for Sievier which would appear to present him as a blackmailer and provide evidence for a criminal prosecution. Elaborate negotiations took place, mostly through the offices of Charles Mills, a bookie's runner who had almost made good as a successful commission agent, employed by owners and trainers to get their bets on at the best prices. He was not the man for Joel's plot. He told Sievier one story, Joel another, said that Sievier wanted £5,000 for his silence, took a cheque from Joel while watched secretly by a police officer, made a payment to Sievier.

in with his financial backing and, just as important, with the dashing publicity he provided driving Bentleys.

He was entitled not only to cash sums left to him by his father and by his uncle Woolf, but also to a proportion of the profits of Barnato Brothers from 1897 to his majority in 1916. The accounts produced by the Joels were vigorously contested by Babe Barnato, by his legal and financial advisers, and not least by his father-in-law. Babe had married Dorothy Falk, daughter of Herbert Valentine Falk, a Wall Street financier, who was so keen to see his son-in-law get his due that he gave up his seat on the New York Stock Exchange to help wage the battle in return for a proportion of the proceeds. It was another ten years before a settlement was reached; Solly Joel fought every inch of the way and was extremely dilatory in producing books for examination. In the end, Babe Barnato agreed to receive £900,000 and costs. The pressures of this unseemly suit did Solly no good; he died of a heart attack in May 1931. He was buried near Barney Barnato in Willesden. Jack Joel remained an efficient head of the family business for the rest of the difficult 1930s. He died in 1940, almost seventy years after his uncles had gone to the diamond-fields.

Other South African magnates besides the Joels adorned Sievier's *Winning Post*. Louis Cohen wrote a series of articles abusing not only the Barnatos and Joels but also J.B.R., or Sir Joseph Robinson, as he now was. As ever, Robinson stood quite apart from the rest of the mining industry, especially from the Corner House. He broke ranks over the recruitment of labour, he opposed the Progressives in politics. FitzPatrick, who detested him, accused J.B.R. of having lent Kruger large sums in the 1890s, which he may have done. In return, while urging working men to vote for Het Volk, Robinson added that the mining industry had once complained of the old dynamite monopoly. 'What is the position of the dynamite trade today?' he asked. 'The principal men of the Progressive Party' – that is the Corner House – 'have the Trade in their hands', which was true enough. The Corner House ignored Robinson as far as possible, but also persuaded the new Transvaal government in the form of Smuts to try to smooth him over – a thankless task. Obstreperous in politics, Robinson was also ferociously quarrelsome and litigious. He doubtless read and pondered the sneers at him in the *Winning Post* but he ignored them. Until, that is, Louis Cohen like many another hack collected his pieces to make a book. *Reminiscences of Kimberley* was published in 1911. It repeated most of Cohen's old chestnuts.

This time Robinson pounced. He sued for libel, employing Carson as his counsel. Cohen put up a half-hearted defence, producing a couple of dubious witnesses. He lost. Robinson won £1,100 damages and the book was withdrawn.

That was not enough for J.B.R. Cohen had tried to justify his attacks, and his witnesses had repeated the accusation of IDB against Robinson. Robinson first bankrupted Cohen, to whom David Harris wrote in commiseration, no doubt expressing the private sentiments of a good many in Kimberley and on the Rand. Then he instituted proceedings for perjury against him.

In the witness box Louis Cohen tried to be as jaunty as he could. He kept up the insults to the last. When asked if it was true that he bore malice against Robinson he replied: 'Certainly not. If I had any malice against that man I would have put in my book that I saw him flogged for the seduction of Elizabeth Rebecca Fergusson in the main street.' Asked what Robinson's reputation in Kimberley had been, Cohen replied that he had been 'that of a persistent and malicious libeller, a liar, a coward, and a seducer of men's daughters and wives. That was his reputation.' In answer to another question from counsel, Cohen continued, 'He was recognized as the foremost IDB.' Asked if he had anything else to add he replied, 'I think I have said quite enough.' He had indeed said too much. The prosecution easily showed that both of the witnesses whom Cohen had roped in had lied and that one of them had never set foot in South Africa. Cohen, the judge declared, had 'wickedly suborned that man Berger to tell deliberate lies'. He was sentenced to three years' imprisonment.

It was a triumph for Robinson. But he had his own bitter disappointments to come. At last he decided to retire. It was 1915, he was seventy-five. But he had no partners – he could never find anyone to work with him – and could not gently ease himself out as men at the Corner House had done. He had to sell his entire Rand interests. The buyer was Solly Joel, one of the few magnates with whom J.B.R. had kept up something like civil relations. Joel paid £4.5 million for the Randfontein and Langlaagte companies. Both Joel and in turn Robinson were in for unpleasant surprises. Joel's managers and engineers found that the mines had been very badly run and needed a great deal more capital expenditure to restore them to order. His financial auditors found something more alarming. Most of the financial groups conducted 'insider trading' and internal deals which enriched the partners or directors at the shareholders' expense. Robinson had indulged in these sharp practices on a vast scale. He had privately bought properties which he then sold at a huge profit to his own companies. A half share in the Waterval farm, for example, had been bought by J.B.R. for £60,000 and then sold to one of his companies within a month for £275,000.

When this and several other transactions came to light, Solly Joel brought an action against Robinson: as buyer he had been defrauded, and wanted to recover the profits which J.B.R. had surreptitiously made. After a protracted legal battle ending in the Supreme Court of South Africa, Joel won £462,000 damages from Robinson.

Worse was to come. No South African magnate had yet risen higher than a baronetcy on the scale of honours.* Rand 'lord' in fact was a misnomer. In 1922 Lloyd George decided to award Robinson a peerage. It was an amazing decision in view of Robinson's reputation and only a year after the 'secret profits' case. Lloyd George's skin had grown thick, his senses dulled, with the atmosphere of corruption in which he lived, not least through the sale of titles. Now a unique occurrence took place. Robinson's peerage had been gazetted, but on 22 June the House of Lords debated his elevation. A torrent of abuse was poured on Robinson's head. Not Cohen's tittle-tattle, but many another charge was brought. Rhodes was twenty years dead, but came back to haunt Robinson who had hated him so much. The Chairman of Consolidated Gold Fields was Lord Harris who led the attack. Lord Selborne reminded the peers of Robinson's support for Kruger before the Boer War and the Afrikaner Party after it. It was an unprecedented form of humiliation, with a humiliating outcome. A week later the Lord Chancellor read the Lords a letter from Robinson to Lloyd George:

> I have not, as you know, in any way sought the suggested honour . . . I am now an old man, to whom honours and dignities are no longer of much concern. I should be sorry if any honour conferred on me were the occasion of such ill-feeling as was manifested in the House of Lords. . . . I would wish, if I may without discourtesy to yourself and without impropriety, his Most Gracious Majesty's permission to decline the proposal.

The letter was humbug. It is hard to believe that he had not sought his peerage as actively as he had sought his baronetcy. He boasted subsequently that the proposed peerage had cost him nothing, which, knowing the methods of Lloyd George and his honours salesman Maundy Gregory, does not seem likely either.† At all events, J.B.R. never became 'Lord Verneuker'. Louis Cohen had a revenge of sorts.

---

* Though Abe Bailey was on Asquith's list for a large creation of peers during the Parliament Bill crisis of 1910, which was not in the event needed.

† Some candidates, however, were aware that Gregory occasionally welshed and, there being no redress, took precautions. Jimmy Buchanan, the whisky millionaire, is said to have dated his cheque to the Lloyd George Fund '1 January 19...' and signed it 'Woolavington'.

After this disastrous turn of events, Robinson decided to sell his house in Park Lane and his collection of paintings. Once again the old man's rapacity undid him. His paintings were remarkable – Rembrandt, Hals, Gainsborough, Constable – but he put such high reserves on them at Christie's that most were bought in. Back in South Africa he lived in morose retirement at Wynberg in the Cape; at the end he must have been like 'Citizen Kane': deaf, misanthropic, alone. Sometimes his family visited him. His daughter had married an Italian diplomatist and become the Countess Labia. One of her children was christened Lucio Mussolini – an unfortunate name but maybe an appropriate one for J.B.R.'s grandson. In his last years Sir Joseph commissioned a grotesquely laudatory 'biography' by the old hack Leo Weinthal, who had edited his Krugerite paper in Pretoria an age before. Its publication was greeted with derision.

Older than the other magnates Robinson had been first on the diamond-fields. He outlived almost all of them, dying on 3 October 1929, just short of his ninetieth birthday. The *Cape Times* waited a week until his will was published before delivering its verdict in a leader under the headlines '*Nil Nisi Malum*': 'The voice of his contemporaries is perforce silent about the evil which his long and unredeemed career compelled them, without known exception, to think of him. . . . They are under no compulsion to attempt the forlornly charitable task of saying anything good about him.' His will was 'scandalously repugnant', it 'stinks, too, against public decency' for one who 'owed the whole of his immense fortune to the chances of life in South Africa. He has not left a penny out of all his millions to the country which showered those immense gifts upon him', but then he was notorious for 'immunity against any impulse of generosity, public or private . . .' He should be a warning: 'those who in the future may acquire great wealth in this country will shudder lest their memories should come within possible risk of rivalling the loathsomeness of the thing that is the memory of Sir Joseph Robinson.' He had hated for most of his life, been hated in return, and hatred pursued him beyond the grave.

18
Hoggenheimer

Along with Rhodes, Beit, Barnato and Wernher, Robinson had dominated the first generation of Rand magnates, who lingered on over the first forty years of the new century. But the great names were fading and no longer actively in control of the mines. A new generation was on its way. There was room for fresh blood, for a financier of genius who might play the central part in the mining industry in the first half of the twentieth century which Alfred Beit had played in the previous century.

In 1886 Louis Oppenheimer had arrived in Kimberley to join his brother Bernard. They were two of the six sons of a cigar merchant, Edward Oppenheimer, and of his wife Nanette Hirschhorn who belonged to the small Jewish community of Friedburg near Frankfurt on Main. They belonged also to the larger network of kinship of German Jewry which had already played such an important part in the South African mining industry. It was to play its part in the careers of the Oppenheimers, cousins of the Dunkelsbuhlers and the Hirschhorns. The two brothers had followed their cousin Anton Dunkelsbuhler. 'Dunkels' was the representative for Mosenthal's before starting business on his own account, with offices in Hatton Garden, the London diamond centre, and Kimberley. Like most other Kimberley merchants he established an interest on the Rand and it was there that young Bernard Oppenheimer took up the firm's business.

Bernard was still concerned with diamonds: it was he who signed the agreement setting up the Diamond Syndicate* on Dunkels's behalf, and when he left Dunkels it was to manage Lewis & Marks's diamond interests. By then Louis had followed him into Dunkelsbuhlers, to Kimberley, to the Rand and back to London. Ernest Oppenheimer, born in 1880, was the fifth son. He benefited like so

---

* The purchase quotas agreed were: Wernher, Beit & Co., 23 per cent; Barnato Brothers, 20 per cent; Mosenthal & Co., 15 per cent; A. Dunkelsbuhler, 10 per cent; S. Neumann, 8 per cent; Joseph Brothers, 7 per cent; with four small companies sharing the remaining 17 per cent.

many other magnates-to-be from the excellent German education of the time and had the advantage of his family connexions.

But it was Ernest's own industry and flair which marked him out. Like Beit before him, he had an affinity for diamonds. Once the Boer War was over, in 1902, he inevitably went to Kimberley. From the beginning Ernest Oppenheimer showed that he was no ordinary businessman; he ruthlessly shouldered aside Leon Soutro, the firm's representative on the diamond-fields. His undisguised toughness began to make him enemies, even among people who began as friends, partners or kinsmen. He lodged with Fritz Hirschhorn, a cousin on his mother's side who was now Wernher, Beit's Kimberley representative, and met the great names of the diamond-fields: Alfred Beit, Solly Joel, Dr Jameson, David Harris. Back in London in 1906 Oppenheimer married Mary Pollak and used his dowry to set up in Kimberley. They had two sons, Harry Frederick in 1908 and Frank Leslie in 1910. He was soon an established man of the town and a councillor; in 1912 when Kimberley and Beaconsfield were merged he became first mayor of the new municipality, aged little more than thirty.

A generation before, Beit's first strength had been his natural genius for dealing with diamonds. So it was for Oppenheimer, but in changed conditions. Despite the dominance of De Beers and the Syndicate, several problems affected the diamond trade. There was a conflict of interest between primary seller and buyer, producer and merchant. The mines naturally wanted to dispose of their whole output, dealers naturally wanted to buy certain superior grades of stone. Knowing diamonds as intimately, physically as he did meant that Oppenheimer could see the problem from both sides and know what he was talking about. Besides, as a thrusting newcomer in the business he was in the thick of its difficulties. One was presented by Cullinan's Premier mine. There was also a fresh supply of alluvial diamonds which had been discovered in German South-West Africa. Then came the 1907 slump. The three matters were interwoven, and came down to the old conundrum, as it had been on the Griqualand fields: how to establish some form of effective monopoly over both production and sales so as to control prices. De Beers and the Syndicate wanted to bring the Premier within their orbit.

To the men who controlled Kimberley it was obvious that Cullinan should pay on their terms but Cullinan saw it differently. He resented the 'attacks constantly made upon us by these high priests of the diamond religion: the Taschi Lama of Jagers and the Dalai Lama of

Kimberley' (i.e. Harris and Oats): he said that he would not reduce production by a single carat.

Boasts would have carried more weight if Cullinan had in fact control of the Premier mine, but the acquisition of a controlling interest by the Transvaal government made things harder. In the event the Premier created its own, rival selling organization in London. The slump of 1907 led directly to a price war between the two, the very thing which the Kimberley magnates had been trying to avoid for a quarter of a century. De Beer's solution was to ride the storm; the Premier was a threat, but the overwhelmingly strong position of De Beers made it able to do so, and against so many similar threats has done so since. So it was with South-West Africa. During the scramble for Africa this desert land had been hacked out of the side of the continent with the greatest brutality towards its original inhabitants as the German Empire looked for its own place in the sun. The discovery of diamonds was an undeserved bonus for the Germans and a worry for De Beers.

Not everyone immediately appreciated the importance of the diamonds in German South-West Africa. Despite his role in the Lemoine affair, Francis Oats had lost his grip: he condemned the finds as of no significance, just as he had done with the Premier mine. In fact before long diamonds were coming out of 'South-West' in large quantities. These were sold separately, through the German Régie state monopoly.

De Beers' first answer was to buy shares in the producing company, to buy concessionary rights in South-West Africa from Philipson-Stow, who had previously acquired them, then to buy South-West diamonds direct, and to try to effect a working alliance between the Syndicate and the Régie. All these developments were watched by young Oppenheimer. He was a person to be reckoned with on the diamond-fields, though as yet nothing like a leading entrepreneur. Kimberley dominated diamonds, De Beers dominated Kimberley, and Oppenheimer was not in De Beers.

The years before the Great War were marked by social upheaval and national passions in South Africa. The strikes of 1907 on the Rand mines were followed by larger disturbances in 1911, 1913 and 1914. White miners' unions were more effectively organized now and more militant, though the mining houses still refused to recognize them. Looming in the background was the ever-thorny question of relations between white and black miners. Compared with the 1890s the mine-owners had one great advantage: a friendly government. It was Botha's very friendliness towards the mines that upset those who considered themselves to be Kruger's heirs. Smuts was in close touch with Lionel Phillips of the Corner House and agreed with him that white labour was impossible. With allies like these in power in Pretoria the mine-owners could be sure that no legislation would be passed enforcing over-rigorous safety standards. Accidents and disease continued to take a heavy toll of life, white but far more black.

In 1913 the government at first appeared to concede to the strikers, and the mine-owners also made concessions, though in truth they were 'not very important'. The words came from Phillips, a man with plenty on his mind. That December of 1913 he was shot at several times outside the Rand Club, surviving only by good luck. His attacker was a madman unconnected with the strike but it was a portent of nervy times. The following year there was a general mining strike. The government hit back at the unions and summarily deported nine labour leaders. Cresswell, now a labour leader, was imprisoned.

The strike placed a further strain on Botha's government which had already shown signs of fragility as more extreme Afrikaner nationalists like Hertzog kicked over the traces. In the eyes of unreconciled Afrikaners, the two great leaders of the Boer War, Botha and Smuts, were now traitors to their people. Botha put nationalist noses out of joint by the gift of the Cullinan diamond to King Edward, by his failure to support the Afrikaans language, and by his agreeing to unveil a statue of Rhodes on Table Mountain in 1912. To the nationalists, Botha had gone over to *rooinek* and Randlord, the English and the mining companies. He even made friends with Jameson, the man whose very name spelt enmity to the old Dutch republics and to the Afrikaner people. Botha's most dangerous foe was another famous Boer general, J.B.M. Hertzog. In 1912 Hertzog was still a member of Botha's cabinet but he began an internal opposition which became increasingly open; not the least of his charges was that the government was too friendly to 'foreign fortune-hunters': the Uitlanders again. The two men broke with each other and within little more than a year the South African National Party had split. Hertzog left to found the National Party, *tout court*: an idea waiting for its time to come.

As these struggles took place a new demon was born: 'Hoggenheimer'. He appeared first in the cartoons of respectable newspapers, then in more brutal Afrikaner nationalist propaganda and poor-white propaganda; the two were beginning to overlap as the miners and other white workers found objective common cause with Afrikanerdom. Hoggenheimer was a stock character, like John Bull or

Uncle Sam. More to the point are the silk-hatted 'boss' of old-fashioned Communist propaganda and the rapacious Jew of fascist propaganda, to both of whom he bore similarities. Hoggenheimer was the archetypal Randlord, the rich and greedy exploiter of working men who manipulated government for his own ends, and whose features in cartoons were needless to say grossly Semitic. He was not originally modelled on Ernest Oppenheimer* – the similarity of name was coincidental – but as it was Oppenheimer's fate to emerge as the supremely powerful mining magnate of the second generation it was his fate also to be identified with Hoggenheimer.

The South-West African problem was resolved when an agreement was reached between De Beers, the Syndicate and the Régie. For an alliance between commercial representatives of the British and German Empires the date of the agreement was supremely unpropitious: 30 July 1914. The shots fired at Sarajevo a month before were about to alter the whole course of world history.

For South Africa, the diamond industry, and 'Hoggenheimer', the war had important consequences. For millions it brought suffering, even if they were remote from the battlefields of Europe. In Kimberley 15,000 blacks were dismissed from their work at the mines, although they were kept under guard on the compounds for some time. During the war the ranks of the mining magnates were further thinned. The pattern of ownership had already been altered by mortality. In 1900 Adolf Goerz had died, most shadowy of all the first Randlords. His company, A. Goerz & Co., continued, but unlike so many other illustrious names in Kimberley and Johannesburg, Goerz had never shed his German nationality and the company kept its close links with the Deutsche Bank of Berlin and thus with the German investing class. In 1914 five of Goerz's eight directors were German subjects who lost their positions, and the ownership of the company was largely sequestered under the official seizure of 'enemy-held' property. Before the end of the war Goerz & Co. had become the Union Corporation with a different pattern of ownership, ripe for infiltration and takeover.

In 1911, when old Dunkels died, he left two sons, Ernest and Walter, but more importantly he left a company, Consolidated Mines Selection, which was to give Ernest Oppenheimer the power base he needed, denied as he was a toe-hold in De Beers. Two of the founding

---

* 'Max Hoggenheimer' was a character, a genial enough South African millionaire, in a musical comedy called *The Girl from Kays* by Owen Hall which came to South Africa from London in the same year as Oppenheimer, 1902; a Cape Town newspaper cartoonist saw the show and picked him up.

fathers of De Beers had died within a decade of each other. Frederic Philipson-Stow died in 1908, the year after his – yet another – baronetcy had been created. His heir Sir Eliot served with the 14th Hussars in the Boer War and again in the Great War. Long after his death old Stow's shadow was to fall across the ascent of Oppenheimer. In 1916 Charles Rudd died, Rhodes's very first partner. For some time he had been retired from the hurly-burly of the diamond business at the tranquillity of his house in Argyll. And in the same year another mine-owner turned Scottish country gentleman, Sir Sigismund Neumann, also died.

One other death was directly connected with the war. George Farrar's East Rand Proprietory Mines had been largely acquired by the Corner House. The gross inefficiency with which the mines were being run could no longer be concealed. Not only inefficiency, but also dishonesty: ERPM were falsifying the monthly return of gold production. All this made very nasty publicity as it was thoroughly aired by the Johannesburg papers. In the end Phillips needed a form of coup to ease Farrar out, and Farrar felt obliged to resign his parliamentary seat.

South Africa's immediate concern once she entered the war was with the two large German colonies, in East Africa – Tanganyika as it became – and South-West Africa, today Namibia. The decision to go to war on the British side and to invade South-West Africa precipitated the 'Rebellie', an insurrection by unreconciled Boers who did not wish to fight under the British flag. The rebels were soon put down. One of them, Jopie Fourie, was tried by court martial and shot.*

The campaign in East Africa lasted the whole length of the 1914–18 War; that in South-West was quicker. Sir Abe Bailey returned to the field, first in the suppression of the Rebellie and then in the South-West campaign as an assistant quartermaster-general. Sir George Farrar joined the colours once more, although he was fifty-five, to serve also as assistant quartermaster-general to the force under General Sir Duncan McKenzie. One night the military train in which Farrar was travelling crashed; he was gravely injured and died at Kubis in South-West Africa on 19 May 1915.

Other mining families took part in the war through their sons of the first generation. Like Solly and Woolf Joels' sons, Wernher's younger

---

* Rather less than seventy years later an Afrikaans drama-documentary of his life was broadcast on South African television. It showed Fourie in his cell the night before his execution, bitterly comparing his fate with the public resurrection of 'Sir Starr Jameson', as well he might.

son Harold served in the army; and Percy FitzPatrick's son Nugent
was killed in action.

Even for Ernest Oppenheimer the war brought physical danger.
While the conquest of South-West Africa and its diamond-fields was a
convenient prize of the war, there were also nasty moments early on,
during the spasms of anti-German feeling which ran through South
Africa as well as England. To be a German Jew by birth, even though
naturalized as a British subject since 1901, as Oppenheimer was, was
ironically to be regarded as a Prussian war-lord. There were riots in
Kimberley and nameplates bearing the Hunnish names of
Oppenheimer and Hirschhorn were torn down. Ernest Oppenheimer
was forced to resign as Mayor and left the diamond city vowing never
to return. For a time he left South Africa itself, working in Dunkelsbuhler's London office with his brother Louis. When he went back to
South Africa it was to Johannesburg, to represent the London-owned
Consolidated Mines Selection Company, Dunkelsbuhler's goldmining
company.

It was now that Oppenheimer conceived his great idea. The
original finance houses were taking new shape, with Robinson's sale
to Barnato, the winding-up of the Wernher, Beit partnership, and the
changing ownership of Goerz and of Dunkels's CMS. Most of these
concerns had been controlled from London. Might it not be the time
to launch a great new company domiciled in South Africa? This was
the genesis of what was to change the face of mining in the Transvaal.
The men closely concerned in it were Ernest Oppenheimer, with the
co-operation of his brother Louis, and W. L. Honnold, an American
mining engineer. Honnold interested Herbert Hoover, the American
financier and president-to-be, in the plan; the great American finance
house of J. Pierpoint Morgan & Co. joined in, and Smuts gave his
blessing. Thus on 25 September 1917 the Anglo American Corporation of South Africa was born.

From the beginning it was a solid company, but far from being an
all-important one. In 1917 Anglo produced 422,000 of the 8.7 million
fine ounces of gold mined on the Witwatersrand, less than 5 per cent.
To achieve the extraordinary pre-eminence it did within less than
thirty years, several conditions were necessary. One was simple
enough, the physical disappearance of the first generation of Randlords. The Corner House was the extreme case. As long as Beit, and
then Wernher, were alive, and the Corner House was a private firm,
its position was impregnable. Once they had died and the firm went
public it could be taken over from outside. In any case, all the mining
houses had shown a natural tendency towards corporate character.

The families of the original magnates retained financial interests in the companies which their forebears had started, but were less and less directly involved. They owned enormous fortunes, but as rentiers rather than active entrepreneurs. That was the case with Barney Barnato's children and the Joels', with Wernher's sons and Beit's nephew, and to a greater or lesser extent it applied to Albu, Neumann, Farrar and Bailey. New blood was needed.

Another condition was the presence of an entrepreneur who could appreciate the mining picture afresh. This was the role played by Ernest Oppenheimer himself in the first years after the war, which he only in fact just survived. In September 1918 he was on board a liner, the *Galway Castle*, on one of his journeys between South Africa and London. In one of the last U-boat forays of the war the ship was torpedoed off Land's End. More than a thousand drowned; Oppenheimer was one of only 150 people saved. Two years later he was knighted for patriotic services during the war. His brother Bernard had shown even more ardent attachment to the British Empire and was made a baronet. They were in the nick of time, as it was shortly forbidden for South Africans to accept British honours.

The first problem Sir Ernest wrestled with was diamonds, his first love. The diamond-fields of South-West Africa were amalgamated under the aegis of Anglo, but Oppenheimer did not carry any weight in the all-important diamond Syndicate. After the war there was a short slump. The Syndicate, with De Beers behind it, met the slackening of demand by dropping prices. This seemed to Oppenheimer the height of folly: what was the point of having production and sales effectively cartelized if the cartel could not hold prices? There was little he could do but protest. His position was clear. In founding Anglo he had stated his objectives: 'The hope that, besides gold, we might create, step by step, a leading position in the diamond world, thus concentrating in the Corporation's hands the position which the pioneers of the diamond industry (the late Cecil Rhodes, Wernher, Beit, etc.) formerly occupied.' And he had also said, 'I want to be a director of De Beers'. To bring this about meant overcoming very powerful opposition and the first sparring was acrimonious, as Oppenheimer tried to oversee a new producers' agreement. Naturally no one wanted to accept a lower quota; De Beers especially was worried about increasing production of diamonds outside South Africa. They wanted to retain their world-wide share of the market, which logically meant increasing their share of the South African market.

If he had formidable opponents against him, Oppenheimer himself was more formidable still. His manner was quiet, mild even, but beneath he was rock-hard, utterly determined and ruthless. When in 1925 quotas had been agreed but no agreement arrived at by the Syndicate, Oppenheimer made an offer to the South African govern-ment: he would buy out the country's entire production. The offer leaked to De Beers who reacted fiercely, offering their own terms to the government and expelling Anglo and Dunkelsbuhler's from the Syndicate. But Oppenheimer had other cards to play. He was on good terms still with Solly Joel, who himself had been toying with the idea of a single buying and selling company to replace the Syndicate: 'all the diamonds must be sold through one channel . . . a new syndicate shall be formed'. Now when the Syndicate offered to make peace Oppenheimer stood firm; it became a test of nerves. In the end, the Syndicate had to accept his terms, and in July 1925 a new Syndicate was created in which Barnato Brothers under Solly Joel and Anglo-cum-Dunkels held 45 per cent each, 'Johnnies' holding the balance. A new producers' agreement gave De Beers 51 per cent of capacity.

Even now Oppenheimer was not on the board of De Beers, which he resented. He had an ally there in Solly Joel, who had a low opinion of his fellow-directors and wanted an abler colleague. With his help, in July 1926, Oppenheimer finally forced his way on to the board. Now he was ready for the big play. He made a further alliance with the firm of Lewis & Marks, seeing that they had the largest quota for producing alluvial diamonds. Alluvials – diamonds found on the surface rather than mined in kimberlite pipes – were becoming a problem as more and more finds were made. In 1927 the government forbade further prospecting for alluvial diamonds in Namaqualand.

Lewis & Marks had also lost its original founders. In 1920 old Sammy Marks died in his eighties, followed by his cousin Isaac Lewis in 1927. These two had stood to one side of the stage and out of the limelight, playing no great part in public affairs, avoiding also the obloquy that other magnates had attracted (at least since their ownership of Hatherleys distillery in the 1890s had made them so unpopular with other mine-owners), not presuming to ascend into English society, not discarding their background. They died South Africans and Jews, with pious benefactions as well as their commer-cial legacies.

19
Revolt and Reaction

The new diamond-fields marked a climacteric in Oppenheimer's career. He insisted that it was not diamonds themselves which were the problem, however many there were, but rather the irrational exploitation of them: he was the spiritual heir of Beit, Rhodes and Barnato. When crisis came in 1927 it was caused not by a world slump but simply by over-production. Oppenheimer saw that resolute action by De Beers was essential. He bought out the newest large diamond-field, Alladin's Cave in Namaqualand, owned by Hans Merensky. And then he proposed that a great new amalgamation should be pushed through, a culmination of 1887–9, with himself as chairman of the all-powerful De Beers. He had by now won Rothschilds over to his side: they had not become what they were by failing to recognize financial genius when they saw it. Oppenheimer was quite capable of being devious, even unscrupulous, when he wanted. Only a little earlier he had told the company secretary of De Beers that the price of shares in the Merensky syndicate was too high, while buying 10,000 shares surreptitiously for himself.

There seems, in retrospect, an aura of inevitability about Oppenheimer's ascent to the top of De Beers just as about the amalgamation of forty years earlier. In reality nothing in life is inevitable; it was merely very likely that a man of Oppenheimer's flair and ambition would succeed. In December 1929 succeed he did. The fierce in-fighting over, Oppenheimer became chairman of De Beers. The following March he became chairman also of the enlarged and remodelled Syndicate, now the Diamond Corporation.

The leading men in the diamond business had seen Ernest Oppenheimer in all his toughness and determination. Several of them had tried to block his path, not only Henry Philipson-Stow but also men who had welcomed him as a young arrival in the diamond-fields: Sir David Harris (as he had now become) and his own cousin Hirschhorn. They had failed. Harris tried to win over the victor, but without success, and he was eased out of De Beers. A new dispensation was made for production and distribution by which Oppenheimer consolidated his own position further still: Anglo-American was granted 45 per cent of diamond sales under a new quota. As in 1914 the timing of this triumph was unpropitious. Oppenheimer's conquest of De Beers took place in the year of the Wall Street Crash which precipitated a decade-long slump for the industrial world and posed an acute problem for the diamond business. Maybe it was just as well that a masterful personality was now in command.

The determination seen in Oppenheimer's takeover of De Beers was in evidence on the Rand as well. Anglo had been founded in the first place as a gold-mining company, to exploit the little-developed Far East Rand (though the Barnato group had established interests there) and to acquire a position along the whole length of the Reef. This brought Anglo and Oppenheimer up against the old, central problem of gold-mining: how to control costs. As ever this problem had a large political dimension. In 1914 some had reckoned, though wrongly, that the Central Rand would be exhausted by 1930. It was understood that there were hundreds of millions of tons of ore on the Far East Rand, but mining it meant an enormous further capital outlay, and reducing labour costs. In the immediate post-war period costs were rising alarmingly. Working costs per ton mined were 19s 2d in 1917, 21s 7d in 1918, 22s 11d in 1919, and 25s 8d in 1920. Once again it seemed to the mine-owners that the solution lay in a direct attack on labour costs, by reducing black or white wages, or both, or by 'diluting' white labour with black. In December 1921 cost-cutting began on the coal mines with a 5s per day reduction in wages, but, despite the government's stamping on the unions in 1914, the labour movement was anything but quiescent. The coalminers went on strike and in January 1922 hundreds of them were dismissed. On the gold-mines, capitalists and workers took up their entrenched, irreconcilable positions. Lionel Phillips claimed that the wages paid to whites made mining untenable and one mining engineer reckoned that half of all white workers could be dispensed with, an enticing prospect for the mine-owners.

For the white miners there could be no compromise, no reduction in their numbers. In January the great Rand Revolt of 1922 broke out. What started as a strike soon turned into something more. The miners were heirs in part to old Boer traditions. During the 1914 strike they had adapted the commando into small, semi-military formations, the ultimate logic of the strikers' picket line. All along the Reef violence broke out. The miners were heirs to events further away also: the October Revolution and its aftermath. There was, the respectable classes thought, a 'foul conspiracy which seized on the strike as a means to Bolshevism . . . a calculated design to repeat on the Rand the unnatural outbreak of crime in Russia that horrified the entire world'. Phillips agreed. He despaired of the influx into South Africa of 'Communists, many of whom are Russian Jews and other violent elements from Eastern Europe'.

And so on that topsy-turvy stage the miners marched under the unforgettable, unforgotten slogan 'Workers of the world unite, and fight for a white South Africa'. In 1922, as in Kimberley fifty years before – and as fifty years later – the white miners had two enemies: the capitalist and the Kaffir, 'Hoggenheimer' the hook-nosed oppressor of working men, and the black man threatening to drag whites down to his station.

In the nature of things the black enemy was more vulnerable and easier to conquer than the boss. Brutal murder was dealt to blacks who had rashly stayed on the Reef and who were found by groups of miners. This was the ugliest part – though only a part – of the violence. Several hundred people were killed and Smuts was driven to use not only infantry but also tanks and aircraft to put down the rebels. At the end of the year, well after order had been restored, three of the miners' leaders were tried and condemned. On 16 November Taffy Long and his two comrades walked to the gallows at Pretoria Central Prison singing the 'Red Flag'. Jan Smuts refused to commute their sentences as he had refused to commute Jopie Fourie's in 1914. Hoggenheimer had won.

It was not so much a pyrrhic as an equivocal victory. Two years later Sir Ernest Oppenheimer entered national politics when he followed Barnato and Rudd as Member for Kimberley in the House of Assembly of the Union of South Africa. The year he was elected was 1924, when Hoggenheimer's two foes, Boer and white miner, began to enjoy their revenge. In the election of 1921, Smuts's South African Party, whose voters were mostly Afrikaners, had been allied with the mining magnates' old Unionist Party and had won. In 1924 the Labour Party under Cresswell made terms with the Nationalists under Hertzog. The two won enough seats to form the 'Pact' government. This government gave the striking miners of 1922 much of what they wanted. There was already both an informal and an official colour-bar in the mines. The Mines and Works Amendment Act of 1926 made the bar more vigorous yet, closely demarcating which superior jobs in the mines could be held by whites only.

During his fourteen years in parliament Oppenheimer spoke almost
exclusively on mining matters, like other 'Members for De Beers'
before him. He had in any case little time to devote to politics, as
Anglo's conquest of the Reef pressed remorselessly on. He had begun
with a physical knowledge of diamonds like Beit, and he also showed
Beit's profound grasp of gold-mining finance. Concentrating on the
Far East Rand, Anglo grew larger, absorbing one company after
another by means of share swaps in which shares were exchanged
between companies and financial groups. Oppenheimer was a master,
like the Corner House before him in the 1890s, in the art of raising
money by selling equity but retaining control. All the time costs were
watched with acute attention. As Oppenheimer reminded Anglo
shareholders in 1932, the whole Witwatersrand was 'really a low-
grade field'. That had been its story from the beginning.

   That year the South African government went off the gold standard
and the resulting rise in the gold price boosted the mining industry.
Other entrepreneurs were emerging on the Rand: this was the age of
A.S. Hersov and 'Slip' Menell, who together founded the 'Anglo-
Vaal', the Anglo-Transvaal Consolidated Investment Corporation.
But it was too late for any newcomers to challenge the pre-eminence
of Anglo-American, whose impetus was now unstoppable.
Oppenheimer's next two ventures were on the West Rand in alliance
with Rhodes's own company, Goldfields, and in the Orange Free
State where Anglo bought Abe Bailey's company, South African
Townships. The Free State mines were the new deep levels. Indeed,
the reef there lay so deep that it could not be properly exploited until
after 1945. At the same time Oppenheimer was following in Rhodes's
steps on the road to the north, where Anglo became a dominant
company on the North Rhodesian (now Zambian) copper belt. This
was another taste of the future, as Anglo stretched its tentacles deep
into the African continent.

   Relations between mining companies and labour, and politically
resurgent Afrikaners, were scarcely less strained on the diamond-fields
than on the gold-fields. There the question was not so much how
many white miners should be employed in proportion to black and at
what wages, but whether white miners could be employed at all in a
recession as severe as that of the 1930s. As ever, Kimberley's
economic problem was diametrically opposite to the Rand's: not
increasing production while reducing costs, but controlling output
while sustaining or raising the selling price. Ernest Oppenheimer had
a gift for the neat phrase. Why, he asked, should we sell a million for a
million pounds when we could sell half a million for two million

pounds? Now that he was master of De Beers he could apply that principle when the going was good. Until then he must weather the storms of depression by reducing output and retrenching. De Beers mines were placed on half-time working, and from 1932 onwards mines were closed one by one and their miners thrown out of work. This action was bitterly opposed by the government, but Oppenheimer had his way. The Diamond Corporation continued to purchase from outside producers as it was obliged to; a huge stockpile of stones, £15 millions' worth, was accumulated; the price was held.

In 1934 the monopoly system moved a step nearer perfection when Oppenheimer founded the Diamond Producers' Association. All producers in South Africa or South-West Africa were assigned quotas which would be purchased by the Association and which would then be sold by the Diamond Trading Company, a wholly-owned subsidiary of the Diamond Corporation, itself a member of the Diamond Producers' Association. Thanks to these devices the South African diamond industry emerged after the Second World War more powerful and lucrative than ever, with Sir Ernest Oppenheimer presiding over it as no one had done before.

He had broken his resolution never to set foot in Kimberley again, but most of his time was spent in Johannesburg. There he built the greatest of all Randlord palaces at Brenthurst on a slope facing the Parktown ridge where the magnates of the 1890s had lived. With all his commercial triumphs he knew private sorrow. His wife Mary died in 1933 and the next year Frank Oppenheimer died suddenly beside a swimming pool in Madeira, in his early twenties. Like his elder brother, Frank had been to school in England at Charterhouse and had then gone on to Cambridge – Harry had been at Christ Church, Oxford. Harry remained as heir to the newest but greatest of dynasties. Sir Ernest's brother Bernard had died in 1920 and had been succeeded in the baronetcy by his son Michael. Young Sir Michael lacked the family flair with money: he went bankrupt, went to South Africa and died prematurely in the same year as his Aunt Mary. He left behind him a widow, Caroline Ina, and in 1935 Sir Ernest married her.

The marriage was one solace to Oppenheimer; another was religion. In 1930 he became an Anglican. It has been suggested that he was prompted by political considerations, and it is true that South Africa seemed almost to be going Germany's way in the 1930s. The resurgent Nationalists openly toyed with anti-Semitic policies and tried to ban further Jewish immigration to South Africa. But the suggestion is obviously absurd: race, not faith, was the National

Socialist obsession, as Hitler showed by eagerly persecuting Christians of Jewish descent, including priests and nuns.

The decade of Oppenheimer's ascent to supreme power saw the first generation finally disappear. In 1930 two less famous bearers of famous names died, Sir Friedrich Eckstein and Sir Otto Beit, brothers of the men who had given their names to the Corner House firms. Eckstein had succeeded Wernher as chairman of the Central Mining and Investment Corporation, but had been forced out of his chair by the anti-German hysteria at the beginning of the Great War. He had subsequently abandoned South Africa and his house Warrington Hall at Doornfontein for Ottershaw Park in Surrey, a large country house which he had rebuilt. Like him, Otto Beit was made a baronet late in life. Beit shared his brother Alfred's love of painting and formed a tremendous collection, Murillos in particular, at Tewin Water and at 49 Belgrave Square, once the Duke of Richmond's house. Beit devoted much of his last twenty-five years to keeping his brother's name alive through the Beit Trust for charitable and public work; the year before he died, the Beit bridge across the Limpopo linking the Transvaal by road with Rhodesia was completed.

Then in 1931, along with Solly Joel, Percy FitzPatrick died. His last year had been full of vexation, disappointment and sadness. All FitzPatrick's sons predeceased him: Nugent in action in France in 1917, Oliver of typhoid in Mexico in 1927, and Alan in a shooting accident in 1928. FitzPatrick was a naturally buoyant, or at least aggressive, man, but it would have needed superhuman resilience to withstand these blows as well as the disappointment of his experimental citrus farm on the Sundays River in the East Cape. Poor FitzPatrick, so dynamic at the time of the Raid, died a heartbroken man.

He was followed in 1932 by Max Michaelis, who had retired from the Corner House thirty years before and devoted himself to collecting pictures. In 1935 one of the Corner House's old foes followed when Sir George Albu died in his late seventies. Although he had not possessed Beit's financial genius, he had been a remarkably gifted entrepreneur who had enjoyed the supreme quality of luck. The Meyer and Charlton mine had been defunct only a few years after the discovery of the Reef when he took it over. It was the foundation of the fortunes of his firm, G. & L. Albu, from 1895 General Mining and Finance, which had specialized in resuscitating neglected and apparently worked-out properties. Before the Meyer and Charlton was finally closed in 1932 it had produced 2,765,000 ounces of gold.

Few were now left who had known the Rand in the days of Kruger's republic. Lionel Phillips had retired from the business in 1924, and gone to live not at Tylney Hall, his house in Hampshire, but at Vergelegen in the Cape near Somerset West. Several of the magnates had followed Rhodes's good example and restored historic Cape Dutch houses. Phillips was a man with enemies but admirers also: John Buchan dedicated *Prester John* to him. His elder son Harold having died suddenly in 1925, Phillips was succeeded in the baronetcy by his younger son in 1936.

In 1940 the chapter was at last closed when the two remaining names of the Golden Age died, Jack Joel and Abe Bailey. Bailey had retired from active business. Given the comparatively small scale of the South African press, Bailey was a substantial press-lord, founding two of the chief Johannesburg papers, the *Rand Daily Mail* and the *Sunday Times*. He had left politics too when he lost his seat in the 1924 election. He was known as a man of strong opinions; he had an especial dislike of the large Indian community which had been growing in South Africa since the 1860s, mostly in Natal but also in Johannesburg, and would, if he could, have expelled them all. After his retirement he was a busy string-puller. Both his houses saw history made: in 1916 the meeting which resulted in Asquith's being succeeded by Lloyd George took place at 38 Bryanston Square;* and in 1933 Smith and Hertzog met to form their coalition government at Rust-en-Vrede, the house at Muizenberg built for Bailey by Sir Herbert Baker.

His greatest success as a manipulator came through his friendship with Winston Churchill. When Randolph Churchill had visited Southern Africa in the 1890s he had been put down for a Rand mines issue of 1,000 shares at par but complained that 'those damned Jews only mean to have me if I hang on' and sold his options at £3 a share. Those were the shares that went to nearly £50. Winston Churchill was less suspicious than his father. From at least 1909 he accepted Bailey's financial advice. In 1916 Bailey urged Lloyd George to have Churchill in his new government and four years later Bailey told Churchill, 'I guarantee you against any loss on all shares purchased on my advice.' The friendship was further cemented in 1932 by the marriage, short-lived though it proved, of Bailey's son John and Churchill's daughter Diana.

Bailey was not a man of broad cultural interests. The great love of his life was sport. He had captained a Transvaal cricket team in 1894 and financed a South African touring team that visited England in

---

* A mild irony, as Bailey nearly received a peerage from Asquith in 1910.

1904. But his obsession was above all with racing, in England and in South Africa where he built up a large stud. His first triumph in England was in 1909 when Dark Ronald won the Hunt Cup at Ascot and landed a large gamble. He sired Son-in-Law, who won the Goodwood Cup and the Cesarewitch and turned into a great sire himself. His son Foxlaw won the 1927 Gold Cup for Sir Abe. In 1936 Bailey had his one classic success when his filly Lovely Rose won the Oaks.

In the following summer Sir Abe came to England to see Golden Sovereign run in the Derby. It was a brave performance, as by now both of Bailey's legs had had to be amputated, one in England, one in South Africa – appropriately he said. Golden Sovereign did not justify the journey; two years later Bailey died.

When Rhodes died, Bailey had supposedly said, 'C.R.'s cloak has fallen on me.' To which Sammy Marks replied 'Yes, possibly, but I think it is a misfit.' That story was told by the last digger of all. Louis Cohen had been on the fields in 1872. He had made friends and enemies, had mocked, and been crushed. He had left prison after the Robinson case, revisited South Africa in 1921, and in 1924, back in England, published another book of memoirs, *Reminiscenses of Johannesburg and London.* In the 1930s the indefatigable hack was still turning it out; he was nearly eighty when he wrote the mildly salacious *Memoirs of Priscilla, Countess Whopper.* Fittingly enough, Louis Cohen outlived every one of his contemporaries. His life covered the story of the South African mines from Rhodes's ice-machine and Barnato's cigars to the emergence of Harry Oppenheimer as the crown prince of Anglo. In 1945 Cohen died, when he was just over ninety.

20
Legacies

Like his father, Harry Oppenheimer entered politics as a Member of the House of Assembly for Kimberley, 'Member for De Beers' in a long line. Like Sir Ernest also, he was elected by the voters of Kimberley in a fateful year for South Africa. Harry Oppenheimer had joined Anglo as heir apparent and began learning the mining business. In 1939 war was declared in Europe. Whether South Africa would take part was in the balance. On 4 September Hertzog put down a motion in favour of neutrality. Smuts won in the House, but his government was suddenly split asunder. Hertzog and his supporters opposed South Africa's entering the war and left the Government in order to make plans, and to wait. Smuts led the country into the war in which South Africa played a modest part, with some 200,000 South Africans serving in the forces. Only those who were white could fight; black and Coloured troops served as auxiliaries. Among the Nationalists opinion ranged from those who did not think South Africa should take part in England's war to those who actively supported Hitler and his New Order. A Nationalist rally in July 1940 called for a South African Republic and condemned Smuts's government and 'British–Jewish capitalist influence'.

The young Harry Oppenheimer served with the Fourth South African Armoured Corps until he returned home to look after Anglo's interests in the summer of 1942. During the war he met Bridget McCall, who was serving with the Women's Auxiliary Army Service. They were married in 1943 and had a daughter and a son, Mary and Nick. In 1948 he stood successfully as a candidate for Smuts's United Party; but Smuts himself and the party were defeated. The Nationalists won the election, and they have been in power ever since. Under the new Prime Minister, Dr D.F. Malan, the government introduced a policy of 'apartness' (*apartheid*), a rigorous and dogmatic system of racial separation and white supremacy, and implemented the policy ruthlessly. Jan Smuts himself, the last Boer general, lived to accuse the Nationalists of using 'the methods of fascism'.

The interests of the new government did not directly conflict with the interests of the mining companies, although the 'Nats' had a long record of hostility to big business. Hoggenheimer was their foe and they toyed with a form of socialism or at least statist *dirigisme*. Several industries were to be controlled by the state and to serve ulterior social purposes, especially to provide employment for poor whites, rather than to operate in a free market. For doctrinaire reasons the government wished fervently to keep to the use of migratory black labour, the workers returning to their homelands after each contract finished. The mining companies, on the other hand, were beginning to have second thoughts about black labour. Migrants had served the mines well enough in the past, but there were good practical reasons to use 'stabilized' labour, as was done on the copper belt in Northern Rhodesia, and the Orange Free State gold-fields could have initiated a new approach. When Ernest Oppenheimer opened a new shaft on the Welkom mine, he suggested that Anglo's policy ought to be 'to create, within a reasonable time, modern native villages which will attract natives from all over the Union, and from which the mines will ultimately draw a large proportion of their native labour requirement'.

This was just what the Nationalist government did not want. Its original aim, fantastic as it may seem and which was anyway overwhelmed by events, was a rigid and complete separation of races, the black helotry to be kept on their ancestral lands, where they would maintain their traditional cultures, rather than as would inexorably continue to happen, that they should settle in proximity to mines and towns and become urbanized and detribalized.

The price of gold remained at $35 an ounce, where Roosevelt had pegged it in 1934. The old gold standard had been suspended in 1914 and been swept away forever by the cataclysm. At $35, gold could be enormously lucrative if costs were controlled and production increased, a prospect which the new Free State fields seemed to offer. When the new fields took off on the stock market, there had been nothing like it since the Kaffir boom of the 1890s. Shares in one company, Ofsits, went from 44s 6d to 70s, in Western Holdings from 71s 6d to 118s 9d. Anglo was not the only company exploring in the Free State. Union Corporation, the Albu's old company, had established a mine in the Free State at St Helena. But it was Anglo who cornered the new fields, and in their chagrin the Union directors refused to have any dealings with them. For years to come Union was the only important mining group which did not have Oppenheimer's nominees on its board.

In the years left to him, Ernest Oppenheimer consolidated his supremacy in the mines, and in the post-war years Anglo continued the ferocious diversification which left it with a finger in almost every business pie in South Africa and many far beyond. He still went every day to the Anglo headquarters at 44 Main Street and would greet young employees with, 'What shall we do to make some money today, my boy?' He remained chairman of Anglo, De Beers, the Diamond Corporation and of a score of other companies, but after a heart attack in 1949, he began to withdraw from business to make time for good works and for his recreations. Though he had little education himself he spent large sums on educational charities. The old School of Mines at Johannesburg had developed into the distinguished University of Witwatersrand and Oppenheimer founded two chairs there as well as a faculty of engineering at Stellenbosch, a Commonwealth centre at Oxford and a laboratory at Cambridge. At home he accumulated a large collection of books and pictures, notably the work of Thomas Baines, the nineteenth-century South African explorer and topographical artist.

His son spent his forties preparing himself for his coming inheritance in the Anglo and De Beers empires. Politically his life was less satisfactory. Smuts died in 1950, the leadership of the United Party passed into incompetent hands and the Party began to disintegrate. Ernest Oppenheimer died in 1957 at the age of seventy-seven. He was paid the tributes which the very rich customarily receive. He was not, like Beit, called 'Christlike', but the Reverend Trevor Huddleston said that his compassion and faith had been 'astonishing in one with such vast material concerns'. Those concerns were surely his true memorial. In 1958 Anglo's share of South African gold production, 4.8 per cent forty years earlier, had risen to 4.5 million out of 16.5 million ounces, or 27.2 per cent. When the production of other companies in which Oppenheimer had a large share is taken into account, not to speak of his control of De Beers, there is no doubt that he was the heir to Beit and Wernher. Harry Oppenheimer stepped into his shoes as chairman of Anglo and De Beers,* and resigned his parliamentary seat to concentrate on business. However, as the United Party finally fell apart, Harry Oppenheimer's involvement in politics was by no means over.

The Oppenheimers had become not so much the greatest as the only Randlord family, if direct control of the mining industry were the test. But there were plenty of descendants of the first generation of

---

* Succeeding also to other seemingly hereditary positions such as Chancellor of the University of Cape Town.

mine-owners enjoying other spoils of one sort or another. Flashy and mistrusted as the family of Barnato and Joel had once been, with him they acquired respectability of a sort. The brothers Joel, Solly and Jack, endowed a Chair of Physics at the Middlesex Hospital Medical School in London but did not otherwise compete in the field of charitable donations. Solly Joel's eldest son, Stanhope, served with the RAF during the Second World War and settled in Bermuda after it. His brother Dudley became a Member of Parliament and then joined the RNVR when war broke out; he was drowned in 1941 trying to rescue some comrades. The only son of the murdered Woolf Joel was Geoffrey, who had a business career in Johannesburg interrupted by war service, during which he won the Military Cross. He became a director of 'Johnnies' and of De Beers, one of which was the creation of his great-uncle, Barney Barnato, who had been in on the dramatic amalgamation struggle which led to the creation of the other.

Barnato's eldest great-nephew was also the longest lived. Jim Joel, Jack's son, was chairman of 'Johnnies' until his retirement, which ended the family's direct involvement in the business which had made them rich. In September 1984 he celebrated his ninetieth birthday after a lifetime of seclusion. A succession of triumphs on the Turf had been crowned in 1957, when Royal Palace won the Derby. There were other descendants of Isaac Isaacs of Aldgate. Henry Barnato had a daughter, Lily, who inherited a million pounds from him. Barney's daughter Leah was twice married – to Alfred Haxton, a violinist, and to Carlyle Blackwell, a silent movie star in Hollywood – before her life slumped into alcoholic obscurity. Jack Barnato died after only two years of marriage, and his widow became Lady Plunket.

That left 'Babe' Barnato. He and all his cousins had met for one last time in 1939, at a party at Claridges to celebrate Fanny Barnato's eightieth birthday. During the war Babe served as a wing commander in charge of the defence of aircraft factories. Active flying was left to his remarkable daughter Diana. She had learned to fly before the war and married a pilot, Derek Walker. Diana Barnato Walker herself spent the war as a ferry pilot, flying aircraft from one airfield to another. After the war she became one of the first women to fly a supersonic aircraft at more than 1,000 mph. When she finally gave up the comparative safety of flying, it was to return to the fox-hunting field. Her brother Michael moved to the United States; he called his son Barney.

Two other families which found riches in the diamond-fields continue. John Bevil Rudd works for De Beers, which his great-grandfather helped to found. Another name from early Kimberley days was Sir Frederick Stow. After passing in reverse order through his third son, the title passed to the fifth baronet, Sir Christopher Philipson-Stow, who lives in Canada. Sir Cecil Newman, son of Sigismund, was educated at Eton and Balliol, and served in the First World War with the Norfolk Yeomanry. The full style of the baronetcy was Neumann of Newmarket, where Sir Sigmund enjoyed the racing. In 1936 Sir Cecil changed his name, not to MacNeumann as Louis Cohen had helpfully suggested, but simply to Newman: the Newmans of Neumarket as the racecourse joke went. His son became Sir Gerard Newman in succession to his father in 1955. He is a director of several companies in the metal industry, a farmer and former High Sheriff of Hertfordshire.

The line of Eckstein died out with Friedrich's son Bernard; FitzPatrick had no surviving sons; George Farrar no sons (though six daughters). J.B. Robinson's title passed to his son, also Sir Joseph, who from Eton and Trinity went in to the Union parliament and died childless in 1954, 114 years after his father J.B.R. was born. The third baronet was J.B.R.'s grandson, Sir Wilfred, who lives outside Cape Town. Albu was succeeded by his son and then by his grandson, who remains in South Africa. His other descendants spread far and wide. One daughter married the Anglican Bishop of Pretoria, two others married South Africans, and another married Captain Nigel Bengough; Lieutenant-Colonel Piers Bengough, the Queen's Representative at Ascot, is their son.

Three of the Randlords wrote books: Phillips, FitzPatrick and David Harris. One or two descendants have shown literary interests also. Sir Thomas Cullinan, as he became in 1900, had nine children and more than six dozen great-grandchildren. One of them, Patrick Cullinan, used part of his inheritance to set up as a publisher and to bring out a short-lived South African literary magazine called *The Bloody Horse*.

More striking is Jim Bailey. Sir Abe had a son by his first marriage who succeeded as second baronet. On his death the title passed to Abe's eldest son by his second marriage to Mary Westenra, daughter of Lord Rossmore. Sir Derrick Bailey has spent part of his life in South Africa and part in England as a race-horse owner, like his father, and a cricketer: he captained Gloucestershire in the 1940s. His younger brother, James, inherited a part of the fortune and for thirty years spent it in a way that would have disconcerted Sir Abe by publishing *Drum*, a magazine for black South Africans. The story came to a curious end in early 1984, when Bailey sold *Drum* and two other black publications, not to the large English-language newspaper groups – descendants of Rhodes's *Cape Times*, the Corner House's *Star* or Abe

262 The Randlords

Bailey's *Rand Daily Mail*, and still in effect Randlord-controlled through the Anglo empire – but to the Afrikaner company Nasionale Pers. *The Standard and Digger's New*'s revenge?

The Corner House no longer exists to control the press and to challenge governments, even if it has a direct spiritual heir in Anglo. But the men of the Corner House have their inheritors. Sir Lionel Phillips left a title, but thanks to his extravagance no vast fortune. His grandson, Sir Robin Phillips, lives in London. Alfred Beit had no children and his line ends with his nephew; Otto Beit, his brother, had four children: a son who died young, two daughters who married into English county families, and his heir Sir Alfred Beit who was eighty in 1983. The latter was a Member of Parliament for most of the 1930s and 1940s, but a silent one. Though well known in Society, he has for the most part led a life of blameless obscurity at Russborough, his great house in County Wicklow with a magnificent collection of paintings, playing a modest part in charitable work as President of the Wexford Festival Opera. He emerged from the shadows briefly in 1970, when the IRA kidnapped him and stole some of his paintings.

Although the Wernher family, too, have died out in the male line, theirs is the most dramatic story of all in terms of the enrichment of the English upper class by the Rand magnates. Sir Julius of Wernher, Beit & Co. married a Miss Mankiewicz and had three sons. The eldest became Sir Derrick, he whom Julius forbad to take over the business. He succeeded his father in 1912 and was in turn succeeded by his brother Harold in 1948. Sir Harold was the third and last baronet, a millionaire who lived like one: he maintained a house in Grosvenor Square well after most Mayfair houses had been abandoned and kept his father's magnificent country seat at Luton Hoo in Bedfordshire. He married Lady Zia, the daughter of a Grand-Duke of Russia.* They were friends of the Royal Family and outstandingly lucky racehorse owners. Sir Harold owned the incomparable stayer Brown Jack, who won at Ascot six years running; and he also won the King George VI and Queen Elizabeth stakes with Combat. Lady Zia's brilliant filly Meld won the 1000 Guineas, Oaks and St Leger in 1955, and in 1966 she won the Derby with Charlottown.

Their only son was killed with the 17th/21st Lancers in 1942, but both their daughters married Guardsmen. Myra Alice married Major David Butter of the Scots Guards and has five children. The elder sister, Georgina, married Lieutenant-Colonel Harold Phillips of the

---

* And a descendant of Pushkin, and thus of Peter the Great's Court blackamoor Hannibal.

Coldstream Guards. They had a son, who inherited Luton Hoo, and four daughters, no fewer than two of whom have married dukes, Abercorn and Westminster.

No man goes so far as he who knows not where he goes, said Cromwell. The penniless young man from Darmstadt who took the long sea voyage to the Cape in 1871 can scarcely have guessed what fate would bring him and his family in the next hundred years.

Some families disappeared, and not all followed this flawless pattern of assimilation. The elusive Goerz left behind a large company but no personal heirs. Sammy Marks had six children, his partner Isaac Lewis, two sons and a daughter. The eldest son, Henry, married an Irishwoman, settled at Combwell Priory in Kent and had two daughters, the elder of whom, Annabelle, married the 13th Viscount Masserene and Ferrard and has a son and a daughter. Isaac himself had bought an estate in Kent as well as a house in Piccadilly but spent his later years entirely in South Africa.

In any case, the brothers-in-law from the *shtetl,* who corresponded in Yiddish all their lives, never absorbed the English traditions of respectability and found the years after 1902 less congenial than the old Dutch Republic of the Transvaal and President Kruger, with whom Sammy Marks had got on so well. Both Marks and Lewis had been dutiful Jews, both made pious benefactions. Isaac Lewis contributed handsomely to the building of the synagogue at Vereeniging on the Vaal, where the treaty ending the Boer War had been signed and where the two partners had important business interests. Sammy Marks was even more munificent. He endowed chairs of Hebrew at the universities of Cape Town and Witwatersrand, started a Hebrew school in Pretoria, sponsored the Jewish Land Settlement Association of South Africa to encourage Jewish boys to become farmers, and made a large gift of money for the restoration of the synagogue in his home town of Szegind-Neustadt.

All of these legacies form a rippling reflection of the fate of the Jews. The synagogue in Lithuania was doubtless destroyed in the catastrophe which overwhelmed European Jewry. That catastrophe did not touch physically the South African Jewish community, small in number – some 120,000 – though not in importance. But the savage hostility which the Rand magnates had inspired was a chord in the crescendo of anti-Semitism which welled up in Europe in the nineteenth and continued into the twentieth centuries. Although the South African Jews survived, they felt the cold winds from Europe. The Afrikaner Nationalist movement veered towards open anti-Semitism: the Transvaal Nationalist Party barred Jews from membership for a time in the 1930s and 1940s.

The general election of 1948 might have seemed a disaster for South African Jewry. The defeated Smuts was a philo-Semite and a Zionist. Without Smuts the South African Jews were anxious. They still remember from 1948 a tremor of fear when the Nationalists came into office with their supposed Nazi sympathies. The anxiety was misplaced. The new Nationalist Prime Minister, D.F. Malan, had already privately assured Jewish leaders that they need not fear his government and, in fact, one of his first acts was to offer moral and financial support to Israel, which came into being as the Nationalists came to power. South Africa was one of the first countries to recognize Israel; South African Jews played a vital part in the first Arab-Israeli war, especially in the new Israeli air force; Dr Malan visited Israel when he was Prime Minister; and South African Jewry contributed more per head to the new State than any other Jewish community. From the beginning Zionism had taken root among the South African Jews: the founder of right-wing 'Revisionist' Zionism, Vladimir Jabotinsky, visited the country three times. And in any case, fairly or not the two countries, Israel and South Africa, were lumped together by their foes, the two 'pariah states' which perforce had large common interests in trade and military equipment.

In the early days of the Zionist movement, Herzl had appealed to a potential supporter. The effort ahead of him was great, he said, 'but it is not too great for Cecil Rhodes. . . . It is an effort which carries the colonial effort in it . . . what I want from you is not that you should contribute a few guineas to our fund or that you should lend them to me but that you should place the stamp of your authority on the Zionist scheme.' The two never met; Rhodes sent Herzl somewhat obvious advice by way of Stead: 'If he wants a tip from me I have only one thing to say, and that is, let him put money in his purse'.

If Zionism was not to be one of Rhodes's legacies, he had plenty of others. He had spent much of his life dreaming dreams of the future, plans which were to be his progeny. His schemes were matched and even surpassed by those of his friend and fellow-bachelor Alfred Beit: large bequests to South African universities, to Guy's Hospital in London and to Oxford, where the Beit Chair of Colonial History was founded in his memory. The greatest memorial planned by the two men − the Cape to Cairo railway − never came about. 'Beitbridge', taking the line on its first stage across the Limpopo from South Africa to Rhodesia, has become less of a link than a barrier. For all that, reminders of Cecil John Rhodes are everywhere. So much to be done and so little time. Like Beit, he made many educational bequests and even had a university named after him at Grahamstown in the East Cape, where the 1820 settlers had built their market town.

Much though he did in South Africa for cultural and educational purposes, it is Rhodes's own alma mater at Oxford that has benefited most from his largesse. Oriel, which received large sums from his will, had every reason to congratulate itself on its foresight in taking Rhodes on as an undergraduate. There are two chairs at Oxford named after Rhodes, one of American history and one, ironic though it may seem, of Race Relations. Rhodes never visited America, but his Rhodes Scholarships were to benefit Americans and Germans as well as 'colonials', according to his world view in which the Anglo-Saxon peoples could combine to conquer and rule. The scholarships were both generous and vainglorious, setting 'a model which has been followed in other countries, whose confidence in their "way of life" is so strong that they must only be known to be loved'.

As well as the scholarships, there is a physical memorial in Oxford: Rhodes House, standing amid its gardens to the north of Wadham. Like so many of the magnates' houses in Parktown, the rebuilt Groote Schuur and the vast Union Building overlooking Pretoria, it is the work of Herbert Baker. He was unkindly described by Kenneth Clark as 'a polite and thoughtful man with a positive genius for errors of design; in his public buildings every proportion, every cornice, every piece of fenestration was (and unfortunately still is) an object lesson in how not to do it'. There is certainly something odd about Rhodes House. It was built not merely as a memorial but almost as a mausoleum. This sepulchral quality struck some who have lived and worked there. Sir E.T. Williams worked at Rhodes House for many years after the war as a Rhodes trustee; after a convivial lunch with a friend at another college he would sigh, 'Ah, well, back to Lenin's tomb.'

The tomb itself is elsewhere. High up in the Matopos hills is the ancient burial place of the Matabele Kings, World's View, which Rhodes came upon while walking after his parley with Lobengula. The name is apt: by some quirk of topography the hilltop seems to present an endless panorama, reaching far, far away. Near the summit is a squat, ugly memorial 'To Brave Men', Major Allan Wilson and his troops, who died as a man during the Matabele war of 1893. To one side is the grave of Rhodes's faithful, disastrous friend Jameson. And at the crest is the grave of Rhodes himself, a slab lying amid irregularly placed boulders, from under which the lizards scurry in the evening sunlight.

Epilogue
The Gold-Reef City 1987

The true memorial to Rhodes is not in the Matopos or at Grahams-town or Oxford. Nor is Beit's memorial his Beitbridge, nor Barnato's and Robinson's their graves. Their true legacy and that of all the Randlords is the Reef, the great arc of mines along the Witwatersrand, and at its centre Johannesburg. What would have happened to Southern Africa without the mines is an imponderable of history. We can only know what happened. There came together three elements: the incomparable riches which nature had left beneath the soil of the Transvaal, European financial expertise and lust for wealth, and cheap black labour. Together they transformed the country utterly.

Together they made the landscape of the Rand, which is not beautiful in conventional terms but has a haunting quality of its own. Its great man-made fells are described by the leading South African novelist of her generation, as the Karoo once was by Olive Schreiner: 'A horizon of strange hills. Some of them were made of soft white sand, like the sand of the desert or the sea, piled up colossal castles . . . others, cream, white, buff-coloured and yellow and worn into rippling corrugations by the wind, built up horizontal ridges.' And at the crest of the ridge is Igoli, the city of gold.

It is more extraordinary still today than ninety years ago. The heart of town is a mass of huge, undistinguished skyscrapers thrust up as if fed from roots deep below in the golden soil. In the day Market Street is as busy as ever it was, thronged with shoppers and office-workers. At night it is no longer so: in the early evening Marshallstown and all of central Johannesburg are turned off as if by a light-switch. The nightlife of 1895 with bars, hotels and brothels is quite gone. The centre of the city is purely a place for making money by day, rather than spending it by night.

To find the heirs of the men and women who made the golden city you must go further out. Today, still, fifty miles to the north the Boers rule in Pretoria, Paul Kruger's direct descendants, the Afrikaner Nationalists who came to power in 1948 to take their revenge for the

Jameson Raid and the defeat of 1902. Since then they have hung on to power with a tenacity, brutality and *slim* agility which have quite confounded their enemies. The whole world turned against South Africa, genuinely loathing its system of racial oppression, but also finding there a convenient target and focus for ritual anathematization. The Communist countries, the United Nations, the bourgeois democracies of Europe and North America, independent Africa, all have scorned and abused South Africa, but to little effect. The ring of white-ruled states forming a cordon round the Afrikaner heartland crumbled in the 1970s, but far from the destruction of South Africa being imminent, as the ninetieth anniversary of the Raid approached, the South Africans, fortified still by golden riches, browbeat their black neighbours into timidity.

Inside the country a mixture of force and guile has held the line. In the 1950s the Afrikaner ideologues, led by a professor of sociology, Hendrik Vervoerd, tried to devise a system which once and for all would entrench white supremacy and codify it with nice precision. White supremacy had existed for decades, centuries even in South Africa; it was if anything strengthened under Smuts; under Vervoerd it was deified. Only the whites would constitute the political nation, rigidly set apart from Coloureds and Indians and, above all, blacks. The blacks were to be expelled to their tribal 'homelands' or Bantustans, which were to be granted a measure of political autonomy, and from whence black workers would come only as temporary sojourners to the white lands, above all to the Rand.

It did not work. Economic pressure and the fierce, magnetic pull of gold drew a great mass of Africans to the towns of the Reef. There these heirs to the 'Kaffirs' or 'boys' who had dug the first mines settled permanently. There they enjoyed the same solaces of rotgut and music as the compound workers had known in the 1890s. There were migratory workers also, still coming from Mozambique and Lesotho to live in gloomy barracks and then go home with what money they had saved. But by the 1980s they were far outnumbered by permanent black inhabitants on the Reef. The white population of Johannesburg has grown to 500,000, while the vast township of Soweto to the south-west is home to more than three million blacks.

Conditions in the mines have improved from the early days. The level of fatalities remains high, but black workers are no longer simply 'wastage' if they are killed. Still, nothing can change the essentially back-breaking nature of work in the mines. Thousands of feet under ground the black miners huddle to work along the drives of the mines, given their orders in Fanakolo, a made-up *lingua franca*. In the stopes,

slanting upwards, a few feet high, they crouch at the rock face drilling
holes for the blasting charges. No one who has ever seen miners at
work will disagree with Orwell that:

> It is even humiliating to watch . . . it raises a doubt about your own
> status as an 'intellectual' and a superior person generally, for it is
> brought home to you, at least while you are watching, that it is only
> because miners sweat their guts out that superior persons can
> remain superior. You and I and the editor of the *Times Lit. Supp.*
> and the Nancy poets and the Archbishop of Canterbury and
> Comrade X, author of *Marxism for Infants* – all of us really owe the
> comparative decency of our lives to poor drudges underground.

But then, life for blacks on the Reef, even the mine workers, is not
mere drudgery and oppression. They hate white rule and from time
to time rise up in nameless rebellion against it. In 1960 after the
massacre at Sharpeville, in 1975 when Soweto exploded, and again in
1985–86 when the townships were alight with police gunfire and
with 'necklaces', the black South Africans seemed to be shaking the
country to its foundations. But their resistance was qualified and
ambiguous. Soweto is a grim compound, and it is the most prosperous black city in Africa. Its inhabitants want to be rid of white
supremacy, but they want to go on enjoying the benefits of the
industrial society which the Randlords created. This must be remembered in order to understand why the Afrikaner Nationalist regime
approached its fortieth anniversary in office, despite nearly forty
years' of predictions by its enemies of its imminent overthrow. The
blacks, too, have compromised with gold.

So, too, have the heirs to the Uitlanders of 1895, the English-
speaking whites of the Reef. They still chafe under 'Krugerism' – say
'Verwoerdism', or 'Bothaism' – but they have in practice accepted it.
The poorer whites, heirs to Aylward and his diggers, have suppressed
their dislike – maybe never very real – of the Boers, but have
maintained all their old hostility to capitalist and Kaffir. Their distilled voice is found in the Mine Workers' Union, still railing at
Hoggenheimer. The Anglo-American Corporation has made halting
attempts to mitigate the colour-bar and employ blacks in skilled jobs;
in the pages of the Union's paper, *Die Mynwerker*, Anglo (AAC) is
attacked as the 'Advancement of Africans Corporation'.

In one sense the Africans have advanced, ironically, despite the
efforts of the Mine Workers' Union. If Aylward's heir, Arrie Paulus of
the MWU, is a power in the land so, for the first time, are the heirs to
those who led the first doomed black workers' strike at Kimberley a
century earlier. Recognizing that classical or pure apartheid was no

longer plausible, the South African government introduced, in 1984, a new constitutional system in which Coloureds and Indians have for the first time some semblance of national political power. Black Africans were excluded. But even as the new constitution came into operation, the impossibility of ignoring the majority of the population of the country was recognized. In 1985 the new President, P.W. Botha, spoke of what would have been unspeakable not many years before, the need to grant some political rights to 'urban blacks', those once supposedly temporary sojourners in Soweto and the other townships of the Reef. Just as important, for the first time black trade unions had been recognized, and began to strike.

All of these shifts and turns by the Nationalist government were made easier by the power which it fundamentally enjoyed. And that power was founded on gold. In the late 1960s the price of gold was at last set free. It took off like a rocket, touching $800 an ounce in 1979. Then it fell back, but even at something under $300 it remains – together with all the other minerals, not only diamonds and coal but also the copper and manganese and vanadium and uranium which a capricious providence has strewn beneath the soil of South Africa – the prop and mainstay of the country.

That is one of the many strange and quite unpredictable consequences of the Randlords. As they sailed to the Cape in the 1870s, they might have guessed, and certainly hoped, that some of them would make great fortunes. They could not have guessed at the coming importance of the new South Africa, their South Africa.

They could not have guessed at other political repercussions. The Randlords were heroes to a few and villains to many. Rhodes was idolized in his lifetime and after; today the influence he longed for has everywhere receded. Empire has gone, Rhodesia itself is no more. The adulation which he had once received turned sour; for Spengler, Rhodes had been 'the first precursor of a western type of Caesar', but those words had an unhappy ring after the reigns of Mussolini and Hitler.

It was beyond guessing that the mining magnates' fame would lead to the last time in English history when anti-Semitism had a tincture of political respectability; or that these magnates would have such profound, if indirect, consequences on the Left. From his observation of the Rand in 1899, J.A. Hobson drew the conclusions which he set out in *Imperialism*. These were, in large part, a fallacious extrapolation from or generalization of the particular, and he has distorted perceptions of Africa to this day. He was not wrong about South Africa, but the generalized materialist explanation of imperialism is wrong precisely because the Rand is the only part of Africa south of the Sahara really worth owning. That did not stop Lenin appropriating Hobson's theories wholesale, and it is not too much to say that distant memories of the Transvaal have conditioned Communist thinking to this day.

That would be a perplexing thought for the Randlords' heirs, should they ponder it. They have other concerns. They sit today in the leafy northern suburbs of Johannesburg, Houghton, Rosebank and Lippert's Saxonwold or, as in 1895, in the bar of the Rand Club, a prosperous but nervous and isolated community. At its head is Harry Oppenheimer, in his seventy-ninth year in 1987 and one of the richest men in the world. Heir to Rhodes and Beit, FitzPatrick and Phillips and their Progressive party, throughout the 1960s and 1970s he almost single-handed paid for the new Progressive party, the 'Progs' or Progressive Federal Party, which opposed the ruling Nationalists with a programme of mild reform. He could afford to: his Anglo empire controls a large part of the world's gold output, and a great deal else besides. Almost a century after the amalgamation battle, De Beers and its cartel, the Central Selling Organization, still control the world diamond market. Recent episodes, when Zaire or Australian producers tried to loosen its tentacles, have only shown how all-powerful De Beers remains.

Today the tensions of the 1890s still contort the Reef. The Uitlanders railed against the Boers, as another—perhaps the last—all-white election took place in early 1987. Where there had been exoduses of Uitlanders to the south in 1895 and 1899, there was a greater exodus, as more and more English-speaking whites fled the country. Many remained, some to keep the flag of white liberalism flying. And yet there was something factitious about these passions. What had enabled the white liberals of South Africa to be liberal was the wealth of the mines, which was at the same time what had given white rule its extraordinary endurance.

Sometimes in Igoli you can feel the earth move. Far below, the drive of a worked-out and disused mine has collapsed, sending up an artificial tremor. There are other tremors besides shaking South Africa: internal political dissension, violent forays from outside, riot in the townships, strikes on the mines, American trade boycotts, the uncertain price of gold. At times the Boers and the Randlords have seemed to be losing their nerve, but never for long. This land, so poor and empty and far away in 1869, remains enormously rich and powerful. And whoever inherits its future inherits a world the Randlords made and may look back on a strange story of the power of gold to shape men's destinies.

Biographical Notes

Albu, Sir George (1857–1935) b. Berlin, name metathetic form of 'Blau'; to South Africa 1876, diamond merchant in Kimberley 1876–89; to Rand and formed partnership of G. & L. Albu with younger brother **Leopold** (1860–1938); resuscitated Meyer & Charlton mine, formed General Mining and Finance Corporation; outside Reform Plot, seceded from Chamber with Robinson 1896; baronet 1912. Johannesburg home, Northward, built by Herbert Baker; active in Jewish philanthropy; eldest daughter married Bishop of Pretoria.

Arnot, David (1821–94) Lawyer and agent of Griqua chief Waterboer, succeeded in keeping diamond-fields out of hands of Dutch republics, obtained 1800 square miles of land; amateur botanist.

Bailey, Sir Abe (1864–1940) b. Cradock, E. Cape, educated in England; to Barberton 1886, Rand 1887 as stockbroker and independent speculator; Reformer, imprisoned after Raid; intelligence officer in Boer War, captured by Boers; Cape Parliament, 'British' and Progressive in politics but 'Boer to the backbone' on black question and wished to expel all Indians from South Africa; founded South African Townships with Julius Jeppe; KCMG 1911; served in South-West African campaign, Union Parliament 1915–24; baronet 1919. Cricketer and racehorse owner; friend of Churchill; might have been made peer by Asquith.

Barkly, Sir Henry (1815–98) MP 1845–1848, Governor of British Guiana 1848, Jamaica 1853, Cape 1870; supervised dispute over Griqualand West, then digger's revolt; retired 1877.

Barnato, Barney (1852–97) b. Barnett Isaacs, Aldgate, east London, son of publican, grandson of rabbi; to Kimberley 1873 as koppie-walloper (diamond-dealer), boxer and amateur actor; formed Barnato Bros 1874 with brother **Harry** (1850–1908), acquired claims from Barnato Diamond Mining Co 1881; 'defeated' in amalgamation battle, but Life Governor of De Beers 1888; Cape Parliament 1889–97; to Rand, formed Johannesburg Consolidated Investment Corporation, 'Johnnies', 1889, floated great quantity of stock, damaged by crash of 1895; outside Reform Plot; presented marble lions to Kruger in conciliation; died 'by drowning while temporarily insane' after jumping from liner; left £1 million; Harry left £5.8 million.

Beit, Alfred (1853–1906) b. Hamburg, to Kimberley 1873; brilliant diamond expert, partner in Jules Porges & Co. with Wernher 1880; played vital but invisible role in take-over battle, Life Governor of De Beers; founded firms of Wernher, Beit & Co. in London, H. Eckstein & Co. in Johannesburg, 1889; grasped importance of deep-level mines, floated Rand Mines 1893; compromised by Raid, resigned from board of Chartered Company. Shy and repressed, 'Little Alfred', as FitzPatrick said, was the outstanding South African financier; left £2 million to charity, rest of fortune to brother **Sir Otto** (1865–1930), who worked for Corner House in Johannesburg, as stockbroker in London, devoted last years to Beit Trust; baronet 1924.

Breitmeyer, Ludwig (1853–1930) Employee of Porges, partner in Wernher, Beit 1895–1910; acquired Corner House diamond interests on dissolution of partnership 1911; founded L. Breitmeyer & Co.

Cohen, Louis (1854–1945) b. Liverpool, Jewish father, Irish mother, to Kimberley 1872; early partner of Barnato; light journalist and diamond merchant; to Rand as stockbroker, made and lost several small fortunes, friend and then enemy of Barnatos and Joels; wrote scurrilous, entertaining memoirs, first in newspapers, then as books: *Reminiscences of Kimberley*, *Reminiscences of Johannesburg and London*; destroyed by Robinson in libel action 1911; imprisoned for perjury 1913; last years in England.

Chamberlain, Joseph (1836–1919) b. London, manufacturer and radical municipal politician in Birmingham, MP 1876; broke with Liberals over Irish Home Rule; Colonial Secretary 1895–1903; denied complicity in Jameson Raid – in it 'up to his neck' said Rhodes. Broke with Tories over tariff reform 1903; 'He lives in the present and changes his opinions with incredible ease' (Paul Cambon); wanted to marry Beatrice Potter (Webb).

Cresswell, Frederick Hugh Page (1866–1948) b. Gibraltar; mining engineer, to South Africa 1893; served in Boer War; manager of Village Main Reef mine post-war; proponent of 'white labour'; dismissed by Corner House; Union Parliament 1910 as labour leader; imprisoned 1913; served as Lieutenant-Colonel in Great War; Minister of Labour under Hertzog 1924–5, 1929–33.

Cullinan, Sir Thomas (1862–1936) b. Elandspost, E. Cape, Irish descent; builder, to Barberton 1884, to Rand, began own brick and tile works at Olifantsfontein; built second Rand Club; prospected for diamonds; raised Cullinan's Horse in Boer War; bought Prinsloo farm, site of Premier mine, for £52,000 1902; Cullinan Diamond found 1905; Union Parliament 1910–15.

Currey, John Blades (1829–1904) b. London, to South Africa 1850, to diamond-fields; secretary to Griqualand West Governor 1872; suppressed gambling, favoured Coloured diggers; retired 1875 after diggers' revolt; friend of Merriman and Rhodes.

Dormer, Francis Joseph (1854–1928) b. Leicester, to South Africa 1875 as journalist; fought in Ninth Frontier War 1877–78; bought *Cape Argus* 1881, backed by Rhodes; bought *Star* in Johannesburg 1888; founded *Rhodesia Herald* and *Bulawayo Chronicle*; resigned when told by Rhodes and Beit of conspiracy 1895; attacked British policy over Boer War; in business for last decades.

Dunkelsbuhler, Anton ('Dunkels') (1850–1911) To Kimberley from Germany as diamond merchant; built up large business South Africa and London; founded Consolidated Mines Selection, forebear of Anglo-American.

Eckstein, Hermann (1847–93) b. Hohenheim, near Stuttgart, son of Lutheran pastor, to South Africa 1882; manager of Phoenix Diamond Mining Company at Dutoitspan; to Rand and founded H. Eckstein & Co (Corner House) 1887; to London 1892 shortly before death. Younger brother, **Sir Friedrich**, joined Corner House late 1880s; in Johannesburg until 1901; partner in Wernher, Beit 1902–10; chairman of successor Central Mining Investment Corporation 1912–14; retired from mining business in anti-German scare of 1914; baronet 1929.

Evans, Samuel (1859–1935) b. North Wales; civil servant in Egypt 1883; to South Africa; to Johannesburg to join Corner House, important gold-mining executive post war, pioneer of hygiene and safety methods in mines.

Farrar, Sir George (1859–1915) b. Cambridgeshire, to South Africa representing family engineering firm 1879; partner of Carl Hanau on Rand; chairman East Rand Proprietory Mines; sentenced to death 1896 after Raid; served in Boer War; knighthood 1902, baronetcy 1911; proponent of Chinese labour post-war; led Progressive Party in Transvaal, founded Unionist Party, Union Parliament 1910–11; resigned when ERPM in difficulties; died on active service after railway accident. Champion athlete in youth; founder of Johannesburg Turf Club; member of White's and Boodle's in London.

Ferreira, Ignatius Philip (1840–1921) b. Grahamstown, E. Cape; to Kimberley, Barberton; on Rand when gold discovered 1886; gave name to 'Ferreira's town' quarter of Johannesburg; owned mine, but bought out by Corner House.

FitzPatrick, Sir James Percy (1862–1931) b. Kingwilliamstown, E. Cape, Irish family, educated Downside; to Barberton as transport rider, origin of *Jock of the Bushveld* (1907); worked for Corner House; entered 1895 conspiracy, wrote polemical *The Transvaal from Within* 1899; partner in H. Eckstein & Co. 1902–07; left for politics, Union Parliament 1911–20; started unsuccessful model citrus farm in E. Cape; famous killer of snakes.

Goerz, Adolf (1857–1900) To South Africa as representative of Deutsche Bank; founded company 1889, which became A. Goerz & Co 1897, later Union Corporation.

Goldman(n), Charles Sydney (1868–1958) b. Cape, financier and author of books on gold-mining; war correspondent in Boer War; MP in UK for Penryn and Falmouth 1910–18; married daughter of 1st Viscount Peel.

Hammond, John Hays (1855–1936) b. San Francisco, mining engineer; met Barnato 1893, to work for him on Rand, then worked for Rhodes; pioneer deep-level mining; Reformer, condemned to death; to United States 1899; anti-Boer propagandist; professor of mechanical engineering at Yale; represented President Taft at funeral of George V 1911.

Hanau, Carl (1855–1930) b. Friedburg, Germany; to South Africa; company promoter and director; representative of Barnato Bros, partner in S. Neumann & Co., close partner of Farrar's at creation of ERPM.

Harris, Dr Frederick Rutherfoord (1856–1920) Qualified as doctor; from England to South Africa 1882; became friend of Jameson, and Rhodes's shady man of affairs or confidential agent; to see Lobengula with Jameson 1889; first chairman of Chartered Company; Cape Parliament 1894–1900; in England July–November 1895 as part of conspiracy, tried to implicate Chamberlain in plot; ignominious performance at 'Commission of No-Inquiry', returned to England; Tory MP 1900.

Harris, Sir David (1852–1942) b. Canonbury, N. London, cousin of Barnato, to Kimberley 1871, 'koppie-walloper'; served in 'Ninth Frontier War', Bechuana Rebellion of 1896–7; director of De Beers; in Kimberley throughout siege; succeeded Barnato as Member for Kimberley in Cape Parliament; then Union Parliament, 1911–29; KCMG 1911; wed in first Jewish marriage on diamond-fields; trustee of Kimberley synagogue.

Harris, George Robert Canning, 4th Baron (1851–1932) Minor statesman and major cricketer – captained Eton, Oxford, Kent and England; Governor of Bombay 1890–95; chairman Consolidated Goldfields of South Africa 1899–1929; denounced Robinson in House of Lords debate 1922.

Hirschhorn, Friederich Heinrich ('Fritz') (1867–1947) b. Frankfurt, to Kimberley, worked for Wernher, Beit; helped form Diamond Syndicate; director De Beers 1905; fell out with cousin Ernest Oppenheimer, opposing his election to chairmanship of De Beers; resigned 1930.

Hobson, John Atkinson (1858–1941) b. Derby, UK, journalist and economist; to South Africa 1899 for *Manchester Guardian*, wrote *The War in South Africa* (1900); vehemently pro-Boer; denounced 'Jew-imperialist conspiracy'; *Imperialism* (1902) greatly influenced Lenin.

Imroth, Gustav (1862–1946) b. Friedburg, Germany; to Kimberley 1880; worked for Barnato, helped found 'Johnnies'; London 1896; Johannesburg 1909; managing director 'Johnnies' 1911–20; retired as advisor to Barnato Bros and Solly Joel; supported athletics in Transvaal; married to Florence Hirschhorn and thus related to Ernest Oppenheimer.

Jameson, Sir Leander Starr (1853–1917) b. Edinburgh, qualified as doctor, to Kimberley 1878; well known as 'the Doctor' ('Dr Jim' invention of English press); became associate of Rhodes; to see Lobengula 1899 and 1890, treated his gout successfully; administrator of Matabeleland and Mashonaland 1889–94; led Raid 1895–6; sent to England and imprisoned, but released; Cape Parliament 1900–10; leader of Progressive Party 1903; Prime Minister

1904–10; Union Parliament 1910–12; baronet 1911; director Chartered Company, president 1913; *homme à femmes* but died unmarried.

Jennings, Hennen (1854–1920) b. Washington, USA, mining engineer to Venezuela; to Johannesburg for Corner House 1889, applied MacArthur–Forrest process to deep ores; to London 1898.

Joel, Solomon Barnato ('Solly') (1865–1931) b. London, son of Barney Barnato's sister; to Kimberley, worked for uncle; to Rand; imprisoned as Reformer 1896; helped organize diamond syndicate, became largest shareholder in De Beers, invested in Far East Rand; large racehorse owner, collector of pictures, backer of theatrical ventures. Elder brothers: **Jack Barnato Joel** (1862–1940) to Kimberley, but left there under cloud; head of 'Johnnies' in England; and **Woolf Joel** (1863–98) to Kimberley, then Rand 1888–93; director of De Beers 1889–96; took over Barnato Bros after uncle's death; murdered 1898 by 'von Weltheim'.

Kruger, Stephann Johannes Paulus (1825–1904) b. Cape, went north on Great Trek as boy; opposed British acquisition of Transvaal; elected President of South African Republic 1883 (and 1888, 1893, 1898); resisted mining magnates of Johannesburg; after outbreak of Boer War 1899, into exile 1900; died Switzerland; believed Earth was flat; asked after visit to England what impressed him: 'The size of the sheep'.

Labouchere, Henry (1831–1912) Rich Etonian radical, MP for Northampton 1880–1905, witty rather than effectual parliamentarian; owner-editor of *Truth*; foe of Rhodes and other Randlords or 'astute Hebrew financiers'.

Lanyon, Sir William Owen (1842–87) b. Belfast, N. Ireland, professional soldier, with Wolseley on Ashanti campaign 1874; administrator Griqualand West 1875; administrator Transvaal 1879; KCMG 1880, dismissed after retrocession of Transvaal 1881.

Lewis, Isaac (1849–1927) b. Szegind-Neustadt, Lithuania; to Kimberley 1870 via England; partnership with Marks, loses in power struggle; to Transvaal, mine-owner but more interested in concessions.

Lippert, Edward (1853–1925) b. Mecklenburg, Germany, cousin of Beit; to South Africa, walked to Barberton; won confidence of Kruger and gained dynamite concession, worth £360,000 in five years; wife, Marie, gifted writer and artist; fervent pro-Boer; returned to Germany and died in poverty.

Lobengula (1830–94) Paramount chief of Matabele 1870; tolerant of Europeans, even though those whom he met were 'mostly scoundrels' (Evelyn Waugh); induced to make mineral concession to Rhodes's representatives 1888; blamed concession on Lotje who was executed with all wives and household; won early clash with Rhodes's troops of Chartered Company but lost Matabele war of 1893 and died mysteriously.

Marks, Samuel (1843–1920) b. Szegind-Neustadt, Lithuania; to South Africa 1868 via England, dealer on diamond-fields; squeezed out, to Barberton, Pretoria; mining interests in Transvaal in partnership with brother-in-law Lewis, but Lewis & Marks's also concerned in concessions – Hatherley distillery – and land speculation; gave statue of Kruger to Republic, thought Boer War 'preposterous blunder'; supported Het Volk post-war, in Senate after Union; founded chair of Hebrew at Witwatersrand.

Merriman, John Xavier (1841–1926) b. Somerset, England, to South Africa as child, land surveyor; Cape Parliament 1869; Kimberley as diamond dealer, but never successful in business; in various Cape governments between 1875 and 1893; broke with Rhodes; unsuccessful again on Rand; wished to conciliate Boers 1900–01; Prime Minister (the last) of Cape 1908–10; Union Parliament 1910–24. Successful fruit and wine farmer near Stellenbosch.

Michaelis, Sir Max (1860–1932) b. Eisfeld, Saxony, to South Africa 1875, founding partner in Wernher, Beit; retired 1901; collection of art and benefactor of museums; KCMG 1924.

Milner, Sir Alfred, 1st Viscount (1854–1925) b. Hesse, journalist, civil servant; High Commissioner for South Africa 1897–1905, oversaw beginning of Boer War; Governor of conquered Dutch colonies 1901–15; intimate of several Randlords; Rhodes Trustee.

Mosenthal, Adolph (1812–82) b. Hesse, to South Africa 1840 with brother **Joseph** (1813–71); formed family trading firm; founder of Cape Town Jewish congregation. Son **Harry** (1850–1915) director of Rand Mines.

Nellmapius, Alois Hugo (1847–93) b. Budapest, to South Africa 1873, claim-holder at Pilgrim's Rest; friend of Kruger, leading concessionaire, especially Hatherley distillery, whose liquor known as 'Mapius'.

Neumann, Sir Sigismund (1857–1916) b. Fürth, Bavaria, to South Africa 1870, Kimberley, Barberton, Rand and formed own company; with well-chosen expert advice became leading mine financier; coal-mine owner post-war; entertained lavishly in London and Scotland.

Oats, Francis (1848–1917) b. Cornwall, mining engineer, to Kimberley 1875; manager of Victoria Co 1887; director of De Beers 1890; mistook value of diamond finds at Premier Mine and in South Africa, but exposed Lemoine fraud.

Oppenheimer, Sir Ernest (1880–1957) b. Frankfurt, to London then in diamond trade, then Kimberley, represented Dunkels; Mayor of Kimberley 1912–15; formed Anglo-American Corp. 1917, conquered diamond trade; chairman of De Beers 1929; Union Parliament 1924–48; greatest twentieth-century mining magnate; philanthropist, knight 1920. Son **Harry ('HFO')** (1908–) educated Charterhouse and Christ Church, Oxford; succeeded as chairman of Anglo 1957–82 and De Beers; MP for Kimberley 1948–58.

Philipson-Stow, Sir Frederick (1849–1908) b. Mowbray, Cape, to diamond-fields as lawyer; became large claim-holder in De Beers mine, and associate of Rhodes; Life Governor De Beers 1888, but fell out with Rhodes and resigned 1898; baronet 1907. Son **Robert Frederick** (1878–1949) director De Beers 1914, and tried to resist Oppenheimer's conquest.

Phillips, Sir Lionel (1855–1936) b. London, to Kimberley, journalist and mine manager, made and lost fortune; to Rand 1889 for Corner House with whom he remained; sentenced to death after Raid 1896; Union Parliament 1910–15; baronet 1912; able businessman but personally spendthrift, building 'Phillips's Folly', Hohenheim, in Parktown, and buying large house in England.

Porges, Jules (Yehuda, Julius) (1838–1921) b. Prague, migrated to Paris and became leading diamond dealer; to Kimberley 1875; employed Wernher and Beit as representatives on diamond-fields; on his retirement 1889 J. Porges & Co., interest transferred to Wernher, Beit & Co.

Reyersbach, Louis Julius (1859–1927) Wernher, Beit's representative in Kimberley 1894–1900; partner in H. Eckstein & Co. 1902; formed own firm in London.

Rhodes, Cecil John (1853–1902) b. Bishops Stortford, to South Africa 1870; built up diamond business while also attending Oriel Coll., Oxford; formed De Beers Mining Co. 1880, won amalgamation struggle; Life Governor De Beers Consolidated 1888; launched Goldfields of South Africa 1887, but rarely visited Rand; British South Africa Company 1889; Cape Parliament 1881; Prime Minister 1890; resigned 1896 after implicated in Raid, but rebuilt public career; at siege of Kimberley 1899–1900. Brother **Francis (Frank) William** (1851–1905) served in Royals; became CJR's associate; sentenced to death with Reformers 1896; *Times* correspondent in Sudan 1898; intelligence officer Boer War; 'Frank might just as well be somebody's butler' (C.J. Rhodes).

Robinson, Sir Hercules, Lord Rosmead (1824–97) High Commissioner for South Africa 1880–89, 1895–6, when was embroiled by Rhodes in Raid conspiracy.

Robinson, Sir Joseph Benjamin (1840–1929) b. Cradock, Cape, to diamond-fields, dealer, then large-scale owner, but lost in amalgamation struggle; early to Rand, sold and bought large properties, made large fortune on West Rand; spent time increasingly in England; friendly with Kruger before Boer War and to Afrikaner Party after; baronet 1908, but 'blackballed' for peerage 1922; universally distrusted and disliked.

Rothschild, Lord (Nathan Meyer) (1840–1915) b. London, educated Cambridge, joined family firm of N.M. Rothschild & Sons; MP 1865–85, when elevated to House of Lords as first professing Jew; closely associated with Rhodes and Beit in De Beers amalgamation battle and in Rand mining ventures.

Rouliot, Georges (1861–1917) Joined Corner House 1893; partner H. Eckstein & Co 1894–1901; chairman Chamber of Mines 1897–1901.

Rudd, Charles Dunnell (1844–1916) To South Africa before diamonds found; to fields; early partner of Rhodes; formed original De Beers Co; joint managing director Goldfields 1887; obtained 'Rudd concession' from Lobengula; extricated Goldfields from politics after Raid.

Sauer, Hans (1857–1939) b. Smithfield, Orange Free State, qualified as doctor, Edinburgh, to practice in Kimberley 1882; 'great smallpox war' with Jameson and other doctors 1884; to Rand, London, Rhodesia; imprisoned after Raid, though not part of conspiracy; married FitzPatrick's sister; wrote memoirs *Ex Africa*.

Shaw, Flora (1852–1929) Colonial correspondent of *The Times*; admirer of Rhodes; implicated in Raid conspiracy; married Frederick, Lord Lugard, pro-consul, 1902; invented name 'Nigeria'; DBE 1918.

Smuts, Jan (1870–1950) b. N.W. Cape, met Rhodes 1888: 'keep an eye on that young fellow Smuts'; to Cambridge, called to Bar; shocked by Jameson Raid; State Secretary of Transvaal 1898; wrote Boer pamphlet *A Century of Wrong*; commando general in Boer War; minister in Botha's Government 1910; Prime Minister 1919–24 suppressed strikes of 1907 and 1914, Afrikaner 'Rebellie' 1914, Rand revolt 1922; fought Germans in East Africa 1916–18; Prime Minister 1939–48; philosopher of 'holism'.

Southey, Sir Richard (1808–1901) To South Africa in settler family 1820, civil servant; Colonial Secretary of Cape Colony 1864–72; concerned in Griqualand West adjudication; retired 1872, but recalled as Lieutenant-Governor Griqualand West 1872–75, retired 1878; ardent Freemason.

Sievier, Robert (1860–1939) b. London, to South Africa 1876; fought in Kaffir wars; to Australia; returned to England; racehorse owner and punter; friend and then bitter enemy of Solly Joel; started *Winning Post* to persecute Joel.

Taylor, James Benjamin (1860–1944) b. Cape Town, to Kimberley 1871; early partner of Beit; worked with brother **William Peter** (1859–1944); to Barberton as stockbroker; to Rand; helped found Corner House; retired 1894.

Weinthal, Leo (1865–1930) b. Graaft-Reinet, son of German-Jewish wool merchant; photographer, to Pretoria 1887; then journalist on pro-Kruger papers owned by J.B. Robinson; latterly British imperialist; wrote *The Story of the Cape to Cairo Railway* and hack life of Robinson.

Wernher, Sir Julius (1850–1912) b. Darmstadt, to South Africa 1871 for Porges, whom he persuaded to buy heavily in mines; built up French Company partnership with Beit; 'understanding' with Rhodes before amalgamation struggle; organized diamond syndicate; founded Wernher, Beit & Co. to exploit Rand, director from London; left huge fortune and estates; great-grandfather of two duchesses.

Williams, Garner Fred (1842–1922) b. Michigan, USA, mining engineer, to South Africa 1884, Pilgrim's Rest; dismissed Rand discoveries; to Kimberley as general manager of De Beers 1887; reorganized and developed mines; involved in Raid; wrote *The Diamond Mines of South Africa*; succeeded by son as general manager.

Bibliography

The **unpublished** sources for the history of the South African mines are patchy and unsatisfactory. Most of the mining houses and most mine-owners of the first generation left few, if any, papers: none for Barnato and the Joels, Robinson, Goerz, nor Albu, except for a box of General Mining and Finance Corporation papers in the Johannesburg Public Library. Some, but far from all, of Rhodes's papers are either at Rhodes House, Oxford, where they have been excellently catalogued by Mrs Williams, or at the Cory Library at Rhodes University, Grahamstown. Others have been destroyed, inadvertently as in the fire at Groote Schuur, or possibly wilfully. There are papers held by De Beers at Kimberley which I was not able to see.

Indeed, far too many documents have, it would seem, been destroyed deliberately at some point or other, and far too many of those that survive are inaccessible to the student – a sign of touchiness on the subject to this day. The Chamber of Mines in Johannesburg holds a large archival collection which may not be consulted; Mr D.A. Etheredge saw it in 1948, while writing his dissertation 'The Early History of the Chamber of Mines 1887–97' (Witwatersrand, 1949). There is some talk of an official history of the Chamber, which would be better than nothing.

There are miscellaneous collections touching on the mines at the Jagger Library of the University of Cape Town, the Cullen Library of the University of Witwatersrand (catalogued in a 'Guide to the Archives and Papers', 3rd edition 1975, with 'Cumulative Supplement 1975–79'), the National English Literary Museum and Documentation Centre at Grahamstown, which holds FitzPatrick's papers, and the South African Library, which holds Merriman's. There is a 'Guide to the South African Manuscript Collections in the South African Library' in the Grey Bibliography Series, as well as the official archives in Cape Town and Pretoria. Besides these archives' individual inventories, there is a 'List of Archivilia in South African Archive Depots'; and J.D. Pearson *A Guide to Manuscripts and Documents in the British Isles relating to Africa* (1971) – place of publication in London unless otherwise named.

Given this exiguity of sources, it is fortunate that the most important of the original mining houses should have left its papers intact. The archives of H. Eckstein & Co. are held in Johannesburg by the Corner House's successor company, Barlow Rand, with an 'Inventory' by M. Fraser, a rare and scarce document but with copies available in England at Rhodes House Library and the Business Archives Council in London. While expressing again my gratitude at having been allowed to consult this treasure trove, I must also express the hope that it will be made available to all students on equal terms, as is apparently not the case at the moment.

The classic **bibliography** is 'Mendelssohn' in its latest edition, *A South African Bibliography to the Year 1925: Being a Revision and Continuation of Sidney Mendelssohn's South African Bibliography* (1910) (4 vols, Cape Town and London, 1979). There are also R. Musiker *South African Bibliography* (1970) (henceforth the place of publication is London unless otherwise specified) with 'Supplement 1970–76' (Johannesburg, 1977); J.F. Muller *et al.*, *South African History and Historians: a Bibliography* (Pretoria, 1979); *The Scoma Directory of Libraries and Special Collections on Africa* compiled by Robert Collinson, 3rd edition revised by John Roe (1973); and H.F. Conover *Africa South of the Sahara: A Selected Annotated List of Writings* (Washington, 1963).

A number of more **specialized bibliographies** have been published – several by the librarianship schools of South African universities. From Cape Town there is N.M. Southey 'Kimberley and the Diamond Fields of Griqualand West 1896–1900' (1946); (and see also J.H. Schepers *Die Ontdekking van Diamantse in Suid Africa: 'n bibliografie* [Johannesburg, 1949]); E. Freedberg 'The Jewish Contribution to the Development of South Africa in the Nineteenth Century' (1959); B.H. Watts 'Gold-Mining in the Transvaal' (1964); D.M. Sinclair 'The Orange Free State Goldfields' (1967); A. Moggridge 'The Jameson Raid' (1970) (and see also M.G. Holli 'Joseph Chamberlain and the Jameson Raid: a bibliographical survey' in *Journal of British Studies* III [1964]), and M.J. Stern 'South African Jewish Biography 1900–1966' (1972). From Johannesburg, T.R. Shaw 'The Growth of Johannesburg from 1886–1929: a list of articles in serial publications' (1963); B. Hughes 'Personal Reminiscences of Early Johannesburg in Printed Books 1884–1895' (1966); J.S. Winter 'First-hand accounts of Johannesburg in English-language Periodicals 1886–1895' (1967).

The **newspapers** of the period are detailed in the *Union List of South African Newspapers* (Grey Bibliography No 3) (Cape Town, 1949). D.H. Varley gives *A Short History of the Newspaper Press in South Africa 1652–1952* (Cape Town, 1952).

There are numerous **general histories**. F.R. Cana *South Africa from the Great Trek to the Union* (1909) is still worth reading. M. Cole *South Africa* (1961) is partly an historical but principally a geographical survey including useful geological information. E.A. Walker *A History of Southern Africa* (3rd edn, 1957) is the book against which subsequent historians have reacted. C.W. de Kiewiet *A History of South Africa: Social and Economic* (Oxford, 1942) was a brilliant survey which changed the course of South African historiography and among other things gave the gold-mines their true importance. The same author wrote another masterpiece, *The Imperial Factor in South Africa* (Cambridge, 1937). The reinterpretation of Walker produced M. Wilson and L. Thompson *The Oxford History of South Africa* (2 vols, Oxford, 1969) and there is also T.R.H. Davenport *South Africa* (1977). As yet the anti-liberal, Marxist or *marxisant* school of history has not produced a full-dress interpretation of South African history. An interesting discussion of the conflict between the two schools, liberal and radical, is H.M. Wright *The Burden of the Present* (1977).

One last historical school should not be forgotten: F.A. von Jaarsveld *The Afrikaner Interpretation of South African History* (Cape Town, 1944); a sideways look at the same theme is W. de Klerk *The Puritans in Africa* (1975).

Of **primary sources**, other than documents and newspapers, the most important are the Annual Reports of the Chamber of Mines from 1890. Confusingly, it was the Witwatersrand Chamber until 1897, then the Chamber of Mines of the South African Republic from the ninth to the eleventh Reports, then the Transvaal Chamber for the twelfth Report, thereafter the Chamber of Mines of South Africa. The Reports are indispensable, especially in view of the unavailability of the Chamber's records. Other contemporary works on mining are discussed below.

The most useful recent addition to our knowledge of the subject comes through published **correspondence.** A.H. Duminy and W.R. Guest edited *FitzPatrick: Selected Papers 1888–1906* (Johannesburg, 1976), with notes and biographical index. Better still is a book edited by M. Fraser and A. Jeeves, *All That Glittered: The Selected Correspondence of Lionel Phillips 1890–1924* (Oxford, 1977), with all sorts of information. Best of all is *Selections from the Correspondence of J.X. Merriman 1870–1924* (4 vols, 1960–69), beautifully edited by P. Lewsen.

Memoirs are notoriously unreliable unless they are written at the time, sometimes even then. Contemporary descriptions may be of the first importance, as with Anthony Trollope *South Africa* (1878), reprinted in a helpful edition by J.H. Davidson (Cape Town, 1973); or Lady Florence Dixie *In the Land of Misfortune* (1882). J.W. Mathews *Incwadi Yami* (1887), F. Boyle *To the Cape for Diamonds* (1873), G. Beet *The Grand Old Days on the Diamond Fields* (Cape Town, n.d.), and J. Angove *In the Early Days* (Kimberley and Johannesburg, 1910), complete the early story of the diamond-fields.

Barnato's foe, R.W. Murray, wrote his *South African Reminiscences* (Cape Town, 1894). The Black Flag rebel, A. Aylward, took to the pen with *The Transvaal of Today* (1881). A decade later another part-time journalist, Lord Randolph Churchill, published *Men, Mines and Animals in South Africa* (1892). There are W.C. Scully *Reminiscences of a South African Pioneer* and *Further Reminiscences of a South African Pioneer* (both 1913), H.R. Abercrombie *The Secret History of South Africa of Sixty-Five Years in the Transvaal* (Johannesburg, 1951), G.S. Fort *Chance or Design? A Pioneer Looks Back* (1942), C. Jeppe *The Kaleidoscopic Transvaal* (1906), F. Johnson *Great Days* (1940), H. Graumann *Rand, Riches and South Africa* (1936), and E. E. Kennedy *Waiting for the Boom: A Narrative of Nine Months Spent in Johannesburg* (1890).

These bring us to the memoirs of the Randlords proper, or of those close to them. L. Phillips *Some Transvaal Problems* (1905) is political rather than reminiscential, as is J.P. FitzPatrick *The Transvaal from Within* (1899), a polemic of historical importance which received an answer of sorts in the melancholy P. Kruger *Memoirs* (1902). Also in answer to FitzPatrick are E.B. Rose *The Truth about the Transvaal* (1902) and C.W. van der Hoogt *The Truth About the Boers* (1900). Phillips also wrote *Some Reminiscences* (1924), less interesting than Mrs Lionel Phillips *Some South African Recollections* (1899). Florence Phillips's life is told in T. Gutsche *No Ordinary Woman* (Cape Town,

1966). FitzPatrick published *Some South African Memories* (1932). D. Harris *Pioneer, Soldier and Politician* (1931) is notable even in this field for its inaccuracy. Two brothers wrote: W.P. Taylor *African Treasure: Sixty Years Among Diamonds and Gold* (1932), and J.B. Taylor *A Pioneer Looks Back* (1939). H. Sauer *Ex Africa* (1937) and J.H. Hammond *Autobiography* (2 vols, New York, 1935) are both interesting rather than objective (and both, incidentally, handsomely printed). A later addition to the field is S. Joel *Ace of Diamonds* (1958). That leaves a favourite, L. Cohen *Reminiscences of Kimberley* (1911) and *Reminiscences of Johannesburg* (1924). Cohen prefaces *Kimberley* with a line from *I Pagliacci* : 'And the story he tells you is true'. A more apt epigraph from opera might be '*Se non è vero è ben trovato*', but if his books are rarely to be trusted, they are at least very funny.

Biography reveals the huge discrepancy of information between different personages. There is nothing on Wernher, Albu, Goerz or Neumann. On the greatest genius of the first generation of magnates there is only G.S. Fort *Alfred Beit* (1932), which is of little value. On the founder of the Corner House in Johannesburg there are only the few pages of M. Fraser's 'Hermann Ludwig Eckstein 1847–1893' in *Africana Notes and News* 22 (3) (1976), FitzPatrick, on the other hand, has two lives: J.P.R. Wallis *Fitz* (1955), the view from the Rand Club window, and A.P. Cartwright *The First South African* (1961). Another who deserves a decent biography is J.B.R.; he has only the laughable L. Weinthal *Memories, Mines and Millions* (1929). H. Raymond *B.I. Barnato* (1897) has the advantage of interviews with Barnato. R. Lewisohn *Barney Barnato* (1937) is worthless; S. Jackson *The Great Barnato* is useful, but would be much more so if it had references. N. Helms has written *Thomas Major Cullinan* (1974). P. Lewsen *John X. Merriman* (New Haven and London, 1982) is an important recent work.

This paucity of information makes other **reference** books more important still. Several magnates appear in the (British) *Dictionary of National Biography*. Though variable, it is often better than the *Dictionary of South African Biography* (4 vols so far published), which is, of course, indespensible, but infuriating. Its four volumes are not in alphabetical order as a series (or even on one occasion within a single volume). Many people who should have been included are omitted; the best essays – e.g. on Rhodes – are much better than the worst. E. Rosenthal *Alphabetical index to biographical notices in 'South Africa' 1892–1929* (Johannesburg, 1963) is useful.

If other magnates have been too little written about, Rhodes approaches the league of Napoleon and Wagner as the object of biographical overkill. That can be seen from the 'Bibliography of Cecil John Rhodes' in Edinburgh University Library. The official *Life of Rhodes* by L. Michell (1910) was preceded by H. Hensman *Cecil Rhodes: A Study of a Career* (1902), 'Vindex' *Cecil Rhodes: His Political Life and Speeches* (1900), and the curious W.T. Stead *The Last Will and Testament of C.J. Rhodes* (1902). It was succeeded by P. Jourdan *Cecil Rhodes* (1910), *Cecil Rhodes: His Private Life by His Private Secretary* (1911), which does not live up to its title; G. Le Seuer *Cecil Rhodes* (1913); B. Williams *Cecil Rhodes* (1921), which is worth reading; and J.G. McDonald *Rhodes: A Life* (1927). W. Plomer *Cecil Rhodes* (1933), S.G. Millin *Rhodes* (1933) and A. Maurois

Cecil Rhodes (1933) all have literary, rather than historical, interest. J.G. Lockhart *Cecil Rhodes* (1933) adumbrated J.G. Lockhart and C.M. Woodhouse *Rhodes* (1963), which ought to be the definitive life, but is not. J. Flint *Cecil Rhodes* (1976) has the merit of brevity, and other merits besides. On different tacks are H. Baker *Cecil Rhodes by his Architect* (Oxford, 1934), and B. Roberts *Cecil Rhodes and the Princess* (1969) on the Galitzine affair. One article should be mentioned: J. Butler 'Cecil Rhodes' in the *International Journal of African Historical Studies*, x, 2 (1977).

After this exhausting, if not exhaustive, list it must be said that a satisfactory Life of Rhodes is still to seek. Why? Is it because, as hinted in chapter 9, and to borrow Gertrude Stein's words in another context, 'there is no there there'? The looming gap between his deeds and his unfathomable personality remains.

On Rhodes's friends and associates there are G.T. Hutchinson *Frank Rhodes: A Memoir* (n.p., 1908) and I. Colvin *Life of Jameson* (2 vols, 1922), intelligent and well-written Tory imperialism. Minor characters are dealt with in H. Kaye *The Tycoon and the President* (Johannesburg, 1978) on Nellmapius and E.M. Bell *Flora Shaw* (1947) on the dangerous woman journalist. A biographical article, J.M. Smalberger 'Alfred Aylward' in *South African Hist. Journal*, 7 (1975), illuminates the Black Flag revolt. In contrast to Rhodes's friends, one of his chief critics has two biographies: A.L. Thorold *Henry Labouchere* (1913) and H. Pearson *Labby* (1936), as well as R.J. Hind *Henry Labouchere and the Empire 1880–1905* (1972). Another foe of the Randlords, Sievier, has his story told in J. Welcome *Neck or Nothing* (1970).

The last generation is covered in T. Gregory *Sir Ernest Oppenheimer and the Economic Development of South Africa* (1962), which is packed with information but unreadable, as well as A. Hocking *Oppenheimer and Son* (Johannesburg, 1973), another exercise in public relations, and E. Jessup *Ernest Oppenheimer: a Study in Power* (1979), another serious book with no references. Finally two multi-biographies: S. Mendelssohn *Jewish Pioneers of South Africa* (1912); and P.H. Emden *Randlords* (1935) – it is ungenerous to say so of a book whose subject and title I have stolen, but it is not very good.

Histories of the **mining houses** lead first to A.P. Cartwright, who has made a speciality of popular books on South African mining: *Gold Paved the Way* (1967) is on Goldfields; *The Corner House* (1958) describes the firm's early years, with the subsequent history in *Golden Age* (1968); *Valley of Gold* (1962) tells the East Transvaal story; *The Gold Mines* (1962) is a general history of the Randlords. These are readable and lively accounts, making good use of unpublished material, and telling the magnates' story from their own point of view. Similar are the official histories, using some unpublished sources, *The Gold Fields 1887–1937* (1937) (it is said that a new history is in hand for the centenary), *The Story of 'Johnnies'* (Johannesburg, 1957); and there is a cyclostyled history of Goerz's Union Corporation by I. Greig (1947). A couple of other popular histories are worth noting: W. Macdonald *The Romance of the Golden Rand* (1933) and D. Jacobssen *Fifty Golden Years of the Rand* (1936).

The early story of the **diamond-fields** is told in M. Robertson *Diamond Fever:*

South African Diamond History 1886–9 from primary sources, (Cape Town, 1974). O. Doughty *Early Diamond Days* (1963) is less useful. B. Roberts *The Diamond Magnates* (1972) makes excellent use of newspapers, as does the same author's *Kimberley: Turbulent City*, (Cape Town, 1976). There are several contemporary books of a more or less technical sort written to encourage prospectors or investors: C.A. Payton *The Diamond Diggings of South Africa* (1872), E.J. Dunn *Note on the Diamond Fields* (Cape Town, 1872), H. Mitchell *The Diamonds and Gold of South Africa* (1889), Chaper *Note sur la région diamentifière de l'Afrique australe* (Paris, 1880), W.E. Boutan *Le diamant* (Paris, 1886), L. de Launay *Les diamants du Cap* (Paris, 1897), F. Abraham *Die Diamant-Gesellschaften Sudafrikas* (Berlin, 1887). G.F. Williams's classic *The Diamond Mines of South Africa* appeared in numerous British and American editions, originally in one volume (New York, 1902), then two vols (London, 1906). P.A. Wagner *The Diamond Fields of Southern Africa* (1914) has been reprinted Cape Town, 1971). H.A. Chilvers *The Story of De Beers* (1939) has surprisingly valuable, because indiscreet, information. S. Ransome *The Engineer in South Africa* (1903) is hair-raising on the subject of labour.

Several learned articles must be mentioned: J.B. Sutton 'The Diggers' Revolt in Griqualand West' (1875) in the *International Journal of African Historical Studies*, 12 (1) (1979), J.M. Smalberger 'IDB and the Mining Compound System in the 1880s, in *South African Journal of Economics* (Dec. 1974), J.M. Smalberger 'The Role of the Diamond-Mining Industry in the Development of the Pass-law system in South Africa' in *International Journal of African Historical Studies*, 9 (3) (1976). Two sprightly romps bring the story up to date: T. Green *The World of Diamonds* (1981) and E.J. Epstein *The Diamond Invention* (1982).

The **gold-fields** produce far more books. Their prehistory is found in D. Ayuso *Viajes de Mauch e Baines* (Madrid, 1877); then come E.L. and J. Gray *A History of the Discovery of the Witwatersrand Goldfield* (Johannesburg, 1940) and J.B. Taylor and W.R. Morrison *Recollections of the Discovery of Gold on the Witwatersrand* (Cape Town, 1936). W.P. Morison *The Gold Rushes* (2nd edn, 1968) has useful information, as has J.H. Curle *The Gold Mines of the World* (1902). W. Bleloch *The New South Africa* (1901) explains the Gold Law and the *bewaarplaatsen* question. An oddity is a novel set in early Johannesburg, *The Gentleman Digger* by Anne, Comtesse de Bremont (1899).

Some of the **technical** books on the Rand mines are in reality little more than punters' guides for the Kaffir market. That is particularly true of the works of the Randlordling C.S. Goldmann: *Financial, Statistical and General History of the Gold and Other Companies of the Witwatersrand* (1892), *South African Mines* (1896) and *Goldmann's Atlas of the Witwatersrand and Other Goldfields* (1899), which is obviously useful. E. Glanville *South African Goldfields* (1888) and T. Reunert *Gold and Diamonds in South Africa* (1893) belong in this category, as does L. Kessler *The Gold Mines of the Witwatersrand and the Determination of their Value* (1904).

In contrast, books by **mining engineers** are of the highest value. F.H. Hatch and J.A. Chalmers *The Goldmines of the Rand* (1895) with its statistical information and its many fascinatingly beautiful maps is the book which stimulated Professor Blainey and the ensuing controversy (see below). Hatch and G.S. Corstophine produced *The Geology of South Africa* (1905). G.A. Denny *The Deep-Level Mines of the Rand* (1902) is very useful, as is R.R. Mabson *The Statistics of the Mines of the Transvaal* (1901), L.V. Praagh (ed.) *The Transvaal and Its Mines* (1906) less so.

The keen Continental interest in the mines is demonstrated by H. Dupont *Les mines d'or de l'Afrique du sud* (Paris, 1890) and F. Abraham's several books, including *Dreissig Jahre südafrikanischer Bergwerks Industrie* (Berlin, 1898) and the invaluable *Methodische Wertbeurteilung der Witwatersrand Goldbergbau-unternehmungen* (Berlin, 1901); one of Abraham's books was translated as *The New Era of the Goldmining Industry in the Witwatersrand* (1894).

Subsequent books include S.J. Truscott *The Witwatersrand Goldfields* (1902), E. Sturzenegger *Rand Gold Mines* (1931), the indefatigable H. Sauer *The Witwatersrand Goldfields* (1933), O. Letcher *The Gold Mines of Southern Africa* (1936) and C.B. Jeppe *Gold Mining on the Witwatersrand* (2 vols, Johannesburg, 1946).

More **general** books on the Rand may begin with popular works: E. Rosenthal *Gold! Gold! Gold!* (Johannesburg, 1970), J.R. Crisp *The Outlanders* (1964), J.R. Shorter *The Johannesburg Saga* (Johannesburg, 1970), and J. Gray *Payable Gold* (Johannesburg, 1937). From the Boer side there is J.L.P. Erasmus *Die Rand en sy Goud* (Pretoria, 1949) and E.L.P. Stals *Afrikaners in die Goudstad* (Pretoria, 1978). There is, of course, next to nothing from the black point of view, but the poorest of the Reef are no longer ignored by history (even if they have been in this book). C. van Onselen *New Nineveh*, *New Tyre* (2 vols, 1982) are studies in the social and economic history of the Rand which give Marxist social history a good name. They include the particularly important essay on 'Randlords and Rotgut'.

Rather than go through the book bibliographically chapter by chapter, I shall deal with various subjects rather at random. R.V. Kubicek *Economic Imperialism in Theory and Practice: South African Gold Mining 1886–1914* (Durham, N.C., 1979) is an important survey of this controversial field, which has in its turn been criticized as well as praised. G. Lanning and M. Mueller *African Undermined* (1979) is an interesting left-wing polemic which touches on South Africa. J.S. Galbraith *Crown and Charter: the Early Years of the British South African Company* (Berkeley, 1975) speaks for itself as do *The Story of Johannesburg Stock Exchange* (Johannesburg, 1948), L. Hermann *A History of the Jews in South Africa* (Cape Town, 1935), and G. Saron and L. Holtz *The Jews in South Africa* (Cape Town, 1955).

On the **Raid**, H.M. Hole *The Jameson Raid* (1930) was superceded by J. van der Poel *The Jameson Raid* (Cape Town and Oxford, 1951), a landmark; E. Pakenham *Jameson's Raid* (1960) attempted to refute van der Poel, but with only partial success, as the new edition (as E. Longford, 1982) appeared to concede. E. Garrett and E.J. Edwards told *The Story of a South African Crisis* (1897). J.S. Marais *The Fall of Kruger's Republic* (1961) is important. There are also Mermeix *Le Transvaal et le Chartered* (Paris, 1897), E.M. Drus 'The Question of Imperial Complicity in the Jameson Raid', *E.H.R.*, October 1953, and E.H. Drus 'A Report on the Papers of Joseph Chamberlain relating to the Jameson Raid', *Bulletin of the Institute of Historical Research*, xxv, and R. Blake 'The Jameson Raid and "The Missing Telegrams"' in H. Lloyd-Jones, V. Pearl and B. Worden (eds) *History and Imagination* (1981).

Approaching the **Boer War**, J.A. Hobson *The War in South Africa: Its Causes and Effects* (1900) is important. The same author's *The Psychology of Jingoism* (1901) and 'Capitalism and Imperialism in South Africa' in *Contemporary Review*, LXXVII (1900) are also worth reading. Of many histories of the war, there are R. Kruger *Goodbye Dolly Gray* (1959) and the later and longer T. Pakenham *The Boer War* (1979), which gives due weight to the part the Randlords played. P. Warwick (ed) *The South African War* (1980) has an important essay on 'The Gold Mining Industry' by P. Richardson and J.J. Van-Helten.

P. Richardson *Chinese Mine Labour in the Transvaal* (1982) is definitive if austere. Here may be mentioned two other studies of labour: S. van den Horst *Native Labour in South Africa* (1942) and F.A. Johnson *Class Race and Gold* (1976). N. Herd *1922: The Revolt on the Rand* (Johannesburg, 1965) is all that I can find on an important subject. The Randlords' own architect is treated sympathetically in two books by D. Greig *The Domestic Work of Sir Herbert Baker in the Transvaal* (1958) and *Herbert Baker and the Houses he Designed in South Africa* (1968).

Several **articles** in learned journals are valuable. G. Blainey 'Lost Causes of the Jameson Raid', *Economic History Review*, series 2, II, 18 (1965) began an intensely interesting controversy. It has been followed by R.V. Kubicek 'The Randlords in 1895: a reassessment', *Journal of British Studies*, xi (1972), the same author's 'Finance Capital and South African Gold-mining 1886–1914', *Journal of Imperial and Commonwealth History* 3(3) (1975), I.R. Phimister 'Rhodes Rhodesia and the Rand', *Journal of South African Studies*, I, I (Oct. 1974), and N. Etherington 'Theories of Imperialism in Southern Africa Revisited' in *African Affairs*, 81 (July, 1982). On the earlier period there is N. Etherington 'Labour Supply and the Genesis of South African Confederation in the 1870s' in *Journal of African History*, 20(2) (1979). Then there are J.J. Van-Helten 'German Capital, the Netherlands Railway Company and the Political Economy of the Transvaal 1886–1900', *Journal of African History*, 19(3) (1978) and the same author's 'Empire and High Finance: South Africa and the International Gold Standard 1890–1914' in *Journal of African History*, 23(4) (1982); A.A. Mawby 'Capital, Government and Politics in the Transvaal 1900–07', *History Journal*, 17(2) (1974); D. Denoon ' "Capitalist Influence" and the Transvaal Government during the Crown Colony Period 1900–09' in *History Journal*, II, 2 (1968), 'Capital and Capitalists in the Transvaal in the 1890s and 1900s' in *History Journal*, 23, 1 (1980), and 'South Africa in a Comparative Study of Industrialization' in *Journal of Development Studies* 7(3) (1971); A. Atmore and S. Marks 'The Imperial Factor in the Nineteenth Century: Towards a Reassessment' in *Journal of Imperial and Commonwealth History*, III (Oct. 1974), and S. Marks and S. Trapido 'Lord Milner and the South African State' in *History Workshop*, 8, (1979). Last, but far from least, G.M. Frederickson *White Supremacy: A Comparative Study in American and South African History* (Oxford, 1981), a brilliant book, illuminating the subject on all sides.

References

Direct quotations and some other passages in the text are identified here by page number and catch-phrase of their last few words. References are provided selectively and common-sensically; obvious quotations or those whose source is implied are not further explained. The place of publication is London unless otherwise stated.

These abbreviations have been used:

CJR   The papers of Cecil John Rhodes (MSS Afr.227–228) in Rhodes House Library, Oxford

DSAB   *Dictionary of South African Biography*

Duminy *FitzPatrick, South African Politician, Selected Papers* 1888–1906, ed. A.H. Duminy and W.R. Guest (Johannesburg, 1976)

Fraser *All that Glittered: Selected Correspondence of Lionel Phillips 1890–1924*, ed. M. Fraser and A. Jeeves (Cape Town, 1977)

HE   The archives of H. Eckstein & Co., in the Barlow Rand Archives, Johannesburg

Lewsen *Selections from the Correspondence of J.X. Merriman 1870–1924*, ed. P. Lewsen (4 vols, Cape Town 1960–69)

NEL   National English Literary Museum and Documentation Centre, Grahamstown

SAL   South African Library, Cape Town

PREFACE

p. xvi   *of academic history'* *Times Lit. Supp.* 16 Jan. 1981 p. 58

p. xvi   *inconvenience and distraction'* S. Johnson, *A Dictionary of the English Language*, Preface

p. xviii   *also about today.'* C.W. de Kiewiet, *The Imperial Factor in South Africa* (Cambridge, 1937) p. 1

p. xviii   *change of euphemism'* E. Waugh, *A Tourist in Africa* (1960) p. 140

p. xix   *useful measuring sticks* G. Blainey, *The Rush that Never Ended* (Melbourne, 1963) p. vi

PROLOGUE: THE GOLD-REEF CITY 1895

p. 2   *Sodom and Gomorrah'* quot. T. O'Brien, *Milner* (1917) p. 153

p. 3   *state of construction'* quot. T. Gutsche, *No Ordinary Woman* (Cape Town, 1966) p. 62

p. 3   *by unscrupulous dealers'* quot. C. van Onselen, *New Nineveh, New Tyre* (Johannesburg, 1982) vol. 2 p. 62

p. 4   *in the gold-reef city'* A. Austin, 'Jameson's Ride' *The Times* 12 Jan. 1896

p. 4   *misery and vice . . . in a bar'* *Standard and Diggers' News* 24 July 1893

p. 4   *still in Johannesburg* Olive Schreiner to J.X. Merriman 17 Mar. 1899, Lewsen
        vol. III, pp 42–3

p. 4   *state of civilisation'* E. Moberley Bell, *Flora Shaw* (1947) p. 107

p. 5   *from the Almighty'* *The Rand Club 1887–1957* (Johannesburg, 1957) p. 41

p. 5   *but don't count'* J.A. Hobson, *The War in South Africa* (1900) p. 61

p. 6   *and formidable people'* R.S. Churchill, *Winston S. Churchill* Part II 1900–1914
        Companion Volume (1969) p. 665

p. 10   *declined to adopt us'* Duminy pp 20–21

1: THE LAND

p. 14   *cotton-spinner's lady'* F. Boyle, *To the Cape for Diamonds* (1873) p. 229

p. 16   *the gold standard'* Wernher Beit & Co. to H. Eckstein & Co., 4 Aug. 1893 quot.
        J.J. Van-Helten, 'Mining and Imperialism' in *Journal of Southern African
        Studies* 6, 2 (1980)

p. 16   *policy to pursue.'* H. Sauer, *Ex Africa* (1937) p. 4

p. 17   *South African mining.'* C.W. de Kiewiet, *A History of South Africa, Social and
        Economic* (Oxford, 1942) p. 96

p. 20   *prosperity or dominion'* quot. K.N. Bell and W.P. Morell (eds), *Select Documents on
        British Colonial Policy 1830–1860* (1928) p. 488

p. 20   *not worth owning'* A.J.P. Taylor, *Germany's First Bid for Colonies* (1938) p. 2

p. 21   *of an eruption'* Bell and Morell, *op. cit.* p. 501

p. 23   *sullen and dispossessed* R. Kipling, 'The Voortrekker'

p. 24   *properly sustaining them'* Bell and Morell, *op. cit.* p. 502

p. 25   *en borst spelt'* M. Robertson, *Diamond Fever* (Cape Town, 1974) p. 74

p. 26   *been discovered there.'* *Geological Magazine* Jan. 1869

2: IN THE EARLY DAYS

p. 27   *finger-like leaves'* O. Schreiner, *The Story of an African Farm* Ch. 1

p. 27   *that never came* ibid. Ch. 2

p. 29   *called "Toispan"'* Robertson, *op. cit.* p. 219

p. 29   *ahead of us again'* ibid. p. 225

p. 31   *surprised his critics* L. Weinthal, *Memories Mines and Millions* (1929) p. 76

p. 31   *no proof of love* L. Cohen, *Reminiscences of Kimberley* (1911) pp 92–3

p. 31   *wives and daughters'* report of Robinson's libel action against Cohen in Cohen
        papers, Cullen Library, Univ. of the Witwatersrand

p. 37   *on the diggings.* G.F. Williams *The Diamond Mines of South Africa* (1902) p. 172

p. 38   *utterly forbid it.'* *Diamond News* 31 Jan. 1872

p. 39   *shovelling and sieving'* J.W. Matthews, *Incwadi Yami* (1887) p. 99

p. 39   *well-trimmed dark beard'* A. Rudd, *Charles Dunnell Rudd* (privately printed, n.d.
        [1981]) p. 7

p. 40   *not wisely but too well* Cohen, *op. cit.* pp. 74–5

3: THREE MEN

p. 41   *traits of my race.'* D. Harris, *Pioneer, Soldier and Politician* (Johannesburg, 1931)
        p. 6

p. 41   *in the early 'eighties' idem.*

p. 42   *hole in the heart.* See J.E. Shee, 'The Ill-Health and Mortal Sickness of Cecil John Rhodes' in *Central African Journal of Medicine* II.

p. 43   *like anthills . . . , fifty buckets.* B. Williams, *Cecil Rhodes* (1921) pp 27–8

p. 43   *he cared nothing.'* Cohen, *op. cit.* p. 47

p. 44   *an absurd price?* J.B. Lockhart and C. M. Woodhouse, *Rhodes* (1963) p. 55

p. 46   *Signor Barnato! . . . for good stones.'* *Diamond News* 26 Nov. 1872 quot. Roberts, *op. cit.* p. 29

p. 47   *bitterness they caused me'* H. Raymond, *B.I. Barnato: A Memoir* (1897) p. 32

p. 48   *Beits of Hamburg'* G.S. Fort, *Alfred Beit* (1932) p. 103

p. 49   *energy and enterprise. idem.*

p. 49   *fortune to be made* Lockhart and Woodhouse, *op. cit.* p. 74

p. 50   *I'll hit you again".'* Raymond, *op. cit.* p. 166.

p. 50   *better join hands.'* Lockhart and Woodhouse, *op. cit.* p. 78

p. 51   *most loyal of Austrians* A.J.P. Taylor, *The Habsburg Monarchy* (new edn. 1948) p. 17

p. 52   *and bloody Faith* P.B. Shelley, 'Feelings of a Republican on the Fall of a Bonaparte'

p. 52   *unsolved puzzle'* M. Hess, *Rome and Jerusalem* (New York, 1918) p. 71

p. 53   *called "Judenhetz"'* [in orig. *'yüden hetze'*, *sic*]' Cohen, *op. cit.* p. 64.

p. 53   *heavy credulous Boers.'* 'W.E.T.', *The Adventures of Solomon Davis* (1879) p. 96

p. 54   *connected with IDB'.* J.R. Couper, *Mixed Humanity* (1892) p. 77

p. 54   *themselves so successfully.* Cohen, *op. cit.* p. 143

p. 54   *heads when caught'* Boyle, *op. cit.* p. 183

p. 54   *disgusted at the laziness'* *Diamond Field* 25 Jan. 1872 quot. in J.M. Smalberger, 'The Role of the Diamond-Mining Industry in the Development of the Pass-Law System in South Africa' in *International Journal of African Historical Studies* 9, 3 (1976) p. 422

p. 55   *'Regulations monstrous'* quot. *ibid.* p. 425

p. 55   *ask and they receive'* *Diamond Field* 8 Aug. 1872 quot. *ibid.* p. 434

p. 55   *attempting to deprive them'* Southey to Barkly 18 Aug. 1873 quot. in M. Macmillan, *Sir Henry Barkly* (Cape Town, 1970) p. 209

p. 56   *ringing in my ears.'* Cohen, *op. cit.* p. 81

p. 56   *doing everybody.'* *ibid.* p. 63

4 : NO SPOT MORE ODIOUS

p. 57   *make them more cautious'* Merriman to C.S. Manuel 5 May 1873, Merriman MSS, SAL

p. 58   *close your pockets'* Cohen, *op. cit.* p. 155

p. 58   *the native son.* Oral tradition

p. 58   *most desirable . . . better known . . . amorous pair.'* Cohen, *op. cit.* pp 41–2

p. 59   *to hold claims.'* G. Paton to Merriman 20 Sept. 1874, Lewsen, vol. 1 p. 9

p. 59   *assembling in arms'* A. Wilmot, *The Life and Times of Sir Richard Southey* (1904) p. 278

p. 59   *ruin for the natives'* *Diamond Field* 28 Nov. 1874 quot. in Roberts, *op. cit.* p. 44

p. 60   *in contented idleness'* Cape Colony Department of Crown Lands and Public Works, 'Report on Immigration and Labour Supply' (1875)

p. 60   *reign of the capitalist'* 23 June 1874, Merriman MSS, SAL

p. 62   *the few remnants'* quot. in Roberts, *op. cit.* p. 56

p. 63   *Dutch flavour about it'* A. Trollope, *South Africa* (1st edn 1875) ed. J.H. Davidson (Cape Town, 1973) pp 4–5

p. 63   *reaping machines' ibid.* pp 246–7

p. 63   *detestable place' ibid.* p. 9

p. 63   *mean to the eye . . . rather than air' ibid.* p. 358

p. 64   *human agency . . . work down there . . . delight to the eye . . . on the other set . . . soon be heard.' ibid.* pp 360–1

p. 65   *property of one firm.' ibid.* p. 360

pp 66–7 Source for Map of Kimberley mine: G.F. Williams, *The Diamond Mines of South Africa* (1902) pp 276–7

p. 68   *in South Africa'* Lord Rossmore, *Things I Can Tell* (1912) p. 227

p. 68   *the identical antique pipe'* L. Cohen, *Reminiscences of Johannesburg and London* (1924)

p. 68   *or even America'* Sauer, *op. cit.* p. 75

p. 69   *a life giver.' Winning Post* 2 May 1908 quot. in B. Roberts, *Kimberley, Turbulent City* (Cape Town, 1976) p. 165

p. 70   *reputable Protestant family' Dict. of Nat. Biog.*

p. 72   *land of Poland'* Cohen, *Kimberley* p. 152

p. 72   *objectionable form'* quot. in Roberts, *Kimberley* p. 168

p. 72   *Massett's visit* see *Cape Times* 8 Apr. 1880 quot. *ibid.* p. 170

p. 73   *as pestilential as possible'* reporter in Smalberger papers. 13.1.6., Jagger Library, Univ. of Cape Town

p. 73   *ill-dressed and ugly . . . many mines.'* A. Trollope, *An Old Man's Love* (1884) pp 113–14

p. 74   *should be unknown* J.B. Currey MSS pp 226–7, SAL

5: THE RIDGE OF GOLD

p. 78   *British population here.'* Wolsley to Hicks Beach 13 Nov. 1897

p. 80   *was the originator'* H. Kaye, *The Tycoon and the President* (Johannesburg, 1978) p. 29

p. 85   *signs of fatigue.'* J.B. Taylor, *A Pioneer Looks Back* (1939) p. 98

p. 86   *left them behind ibid.* p. 103

p. 86   *might be auriferous* Struben collection, A115, Cullen Library, Univ. of Witwatersrand

p. 87   *the whole Rand' idem.*

p. 90   *stupidity of Rudd* Sauer, *op. cit.* p. 117

p. 91   *not worth hell room." ' ibid.* p. 129

p. 91   *a small minority'* see p. 78

6: BUBBLE AND AMALGAMATION

p. 93   *right well he acted'* Cohen, *Kimberley* p. 78

p. 93   *that Bubble year. Winning Post* 10 Apr. 1909 quot. Roberts, *Diamond Magnates* p. 117

p. 93   *keepers of cows'* J.G. McDonald, *Rhodes: A Life* (1941) pp 1–2

p. 94   *to accomplish this.'* Roberts, *Diamond Magnates* p. 112

p. 95   *450 or 400 feet* quot. in H.A. Chilvers, *The Story of De Beers* (1939) p. 65n.

p. 95   *the great thieves'* A. Trollope, *South Africa* pp 365–6

p. 95   *prove his guilt.* R. Churchill, *Men, Mines and Animals in South Africa* (1892) p. 45

p. 96   *the honest shareholder'* Merriman to Currey 27 June 1885, Lewsen, vol. 1 p. 90

p. 96   *is very great . . . only absolute remedy'* quot. in J.M. Smalberger 'Illicit Diamond Buying and the Mining Compound System' in *South African Journal of Economics* Dec. 1974

p. 96   *For some ten days'* S. Ransome, *The Engineer in South Africa* (1903) p. 66

p. 97   *strictly reasonable lines . . . this humiliating process.' ibid.* p. 67

p. 97   *eyes of the natives.'* quot. in Roberts, *Kimberley* p. 216

p. 97   *position of our Kaffirs.' Daily Independent* 16 July 1883

p. 97   *a common Kaffir' Diamond Field Advertiser* 15 Oct. 1883

p. 98   *like an epidemic.'* Rhodes to Merriman 8 Apr. 1883, Lewsen, vol. I p. 122

p. 98   *When I passed* Sauer, *op. cit.* p. 81

p. 99   *not intelligently worked.'* Posno to Merriman 4 Jan. 1884, Merriman MSS, SAL

p. 100 *best position strategically'* Chilvers, *op. cit.* p. 53

p. 101 *Satan and J.B. Robinson.' Winning Post* 12 Sept. 1908 quot. Roberts, *Diamond Magnates* p. 72

p. 101 *chosen tribe . . . religious matters' Diamond Field Advertiser* 7 Nov. 1882 quot. *ibid.* pp 142–3

p. 101 *religious animosity' Independent* 8 Feb. 1884 quot. *ibid.* p. 143

p. 101 *barely 'scaped from judgement'* R. Kipling, 'Gehazi'

p. 104 *or assist themselves.'* F. Dixie, *In the Land of Misfortune* (1882) p. 182

p. 105 *dear old Kimberley' Diamond Field Advertiser* 8 July 1887, quot. Roberts, *Diamond Magnates* p. 188

p. 107 *how on earth' ibid.* p. 195

p. 107 *get the shares' ibid.* p. 200

p. 108 *such power as this* 20 Aug. 1858 Raymond, *Barnato* p. 44

p. 109 *monastery of labour* Moberley Bell, *op. cit.* p. 105

p. 109 *in the shade'* Matthews *op. cit.* pp 221–2

p. 110 *represent us in Parliament'* Raymond, *op. cit.* pp 79–80

p. 110 *employers in Parliament' ibid.* pp 79–80

7 : THE CORNER HOUSE

p. 111 *Witwaters Randt.'* 25 May 1887, HE 124

p. 112 *Molten opals.'* Cohen, *Johannesburg* p. 20

p. 112 *aristocracy of the land' ibid.* p. 25

p. 114 *and lost in hours* Ann Merriman to J.X. Merriman 9 Dec. 1888, Lewsen, vol. I p. 277

p. 114 *Gibraltar of South Africa'* S. Jackson, *The Great Barnato* (1970) p. 99

p. 114 *over the chips'* Merriman to Ann Merriman 23 Feb. 1889, Lewsen vol. I p. 283

p. 115 *a beautiful row . . . in each others arms.'* Lowry to Currey 3 Dec. 1890, CJR, Rh.C 10/1

p. 116 *names absolutely stink.'* Merriman to Ann Merriman, 22 Apr. 1890 Merriman MSS, SAL

p. 116 *Nous verrons!'* Charles Mills to Merriman, 7 Feb. 1889, *ibid.*

p. 116 *there is no change.'* quot. A.P. Cartwright, *The Corner House* (Cape Town, 1958) p. 93

p. 116n *Princes are thieves'* Merriman to Ann Merriman 23 Feb. 1889, Lewsen vol. I p. 283

p. 119 *a similar experience.'* Gutsche, *No Ordinary Woman* p. 62

p. 121 *and secrecy'* F.H. Hatch and J.A. Chalmers, *The Gold Mines of the Rand* (London and New York, 1895) p. 89

p. 122   *concur with me.'* Phillips to Wernher, Beit 3 Jan. 1890, Fraser p. 34
p. 122   *exception is nonsense.'* quot. Cartwright, *Corner House* p. 100

8: CONCESSIONAIRES AND WIREPULLERS

pp   the principle sources for these pages are D.A. Etheredge, 'The Early History
125–33   of the Chamber of Mines 1887–1897' MA dissertation, Wits. Univ. 1949,
         and A.P. Cartwright, *The Dynamite Company* (Cape Town, 1964)
p. 125   *doubts about it'* Phillips to Eckstein, 22 July 1892, Fraser p. 57
p. 126   *by its fumes'* Witwatersrand Chamber of Mines, First Annual Report p. 11
p. 130   *and East Coast'* Phillips to Wernher, Beit 8 Nov. 1890, Fraser p. 44
p. 130   *reduction of native wages'* Chamber, First Annual Report p. 10
p. 130   *heavy additional expense' ibid.* p. 9
p. 131   *influx of natives' ibid.* p. 10
p. 131   *paid far too much.'* Phillips to Beit 13 Aug. 1982, Fraser p. 59
p. 132   *am picking up.'* quot. Cartwright, *Corner House* p. 138
p. 133   *will slip by me'* quot. *ibid.* p. 119
p. 135   *Johannesburg wire-pullers . . . Rand mines . . . Mr Rhodes . . . Mr Rudd acts' Financial*
         News 7 Mar. 1892
p. 135   *some other country* [quot. in] *. . . combination . . . worth acquiring deplorably low ebb'*
         unidentified press cutting November 1893, HE 59

9: THE YOUNG BURGHER

p. 138   *long, withdrawing roar* M. Arnold, 'Dover Beach'
p. 139   *tropic stars* J. Ruskin, *Lectures in Art Delivered before the University of Oxford*
         (Oxford, 1870) pp 27–30
p. 139n   *gin and beer.'* J. Morley, *Life of Gladstone* VI, p. 14
p. 140   *Pall Mall Gazette* C. Newbury, 'Out of the Pit: The Capital Accumulation of
         Cecil Rhodes' in *Journal of Contemporary History* I, 27
p. 140–2   *to my country . . . Anglo-Saxon influence . . . with those men take it . . . our schools . . .*
         was needed . . . such an object interest of humanity. quot. J. Flint, *Cecil Rhodes*
         (1976) pp 248–52
p. 145   *race in South Africa' ibid.* p. 53
p. 145   *of serving office . . . in South Africa' ibid.* p. 54
p. 146   *£1500 for a half'* Rudd, *op. cit.* p. 11
p. 147   *concession lately obtained'* quot. R.V. Kubicek, *Economic Imperialism in Theory and*
         Practice (Durham, N.C., 1979), p. 91
p. 148   *the yearly meeting'* quot. Kubicek, *op. cit.* p. 92
p. 148   *a gentle stimulus'* Flint, *op. cit.* p. 168
p. 149   *Company a Fenian* Merriman to Currey 28 Sept. 1887, Lewsen, vol. I p. 268
p. 150   *speculative element* CJR, 17 Sept. 1890 C 7
p. 150   *the 'ground floor'* Davies to Frank Rhodes 23 Nov. 1894, GFSA papers
         M6.16.079, Cory Library, Rhodes Univ.
p. 151   *oust the Boer'* Phillips to Beit 16 June 1894, Fraser p. 78
p. 151   *Kruger's government.'* Beit to Phillips 6 Dec. 1894, HE 61

10: ALL THE CONSPIRATORS

p. 152   *Dutch and Boer'* J.H. Hammond, *The Truth about the Jameson Raid* (Boston, 1918)
          p. 20
p. 153   *care a fig'* quot. J. van der Poel, *The Jameson Raid* (1951) p. 7
p. 153   *language as foreign'* Raymond, *Barnato* p. 192
p. 154   *quite prepared.* H. Belloc, 'Lord Lundy'
p. 159   *profitless support'* 1 July 1895 CJR C10/2
p. 159   *penalties in your time.'* Barnato to Rhodes 7 Apr. 1894 CJR C7
p. 159   *ghettos of Warsaw'* L. Weinthal, *Memories, Mines and Millions* p. 122
p. 159   *trail than benefits . . . rivers of tears'* ibid. p. 112
p. 160   for story of Hatherly distillery see C. van Onselen, 'Ranlords and Rotgut' in
          New Nineveh, New Tyre (2 vols, Johannesburg, 1892) vol. 1 p. 44
p. 160   *out of our pockets . . . right side'* Phillips to Beit 12 Aug. 1894, Fraser pp 81–2
p. 161   *ulterior motive'* Phillips to Beit 26 Mar. 1894, HE 3
p. 161   *make great fortunes'* Weinthal, *Memories* p. 123

11: BEAUTIFUL, BOUNTIFUL BARNEY

p. 167   *them tightly pressed'* E. Longford, *Jameson's Raid* (London, 1982) p. 147
p. 167   *the stock markets'* J.A. Hobson, *The Evolution of Modern Capitalism* (1928), p. 269
p. 169   *because of its rarity.'* *South Africa* 11 May 1895
p. 169   *Robinson may rot!* Jackson, *op. cit.* p. 184
p. 171   *philanthropy of Barney . . . supposed to be'* Fitzpatrick to Taylor 26 Oct. 1895,
          NEL, Q2=A/LB
p. 171   *most successful enterprises.'* Jackson, *op. cit.* p. 185
p. 171   *nor need to fear'* ibid. p. 185
p. 171   *the Stock Exchange.'* *The Times* 8 Nov. 1895

12: THE BAFFLED BAND

p. 173   *personality of Rhodes . . . all but impossible'* Flint, *op. cit.* p. 118
p. 173   *President J.B. Robinson'* ibid. p. 188
p. 173   *is this correct? . . . British flag.'* ibid. p. 189
p. 174   *prompted this appeal.* quot. Longford, *op. cit.* p. 31
p. 175   *in your favour.'* quot. ibid. p. 177
p. 176   *or two at least'* quot. Longford, *op. cit.* p. 181
p. 177   *immediate flotation.'* ibid. p. 184
p. 177   *floated next Saturday.'* idem.
p. 177   *a reformed republic.'* ibid. p. 186
p. 180   *a mere item'* quot. Cohen, *Johannesburg* p. 258

13: A GRAIN OF GOLD

p. 181   *Anglo-Boer conflict'* quot. T. Pakenham, *The Boer War* (1979), p. 9
p. 181   *the Jameson Raid'* quot. ibid. p. 5
p. 181   *all our misfortunes'* Lewsen vol. IV, p. 292
p. 182   *belong to England'* quot. Longford, *op. cit.* p. 193
p. 182   *it will ruin me'* quot. ibid. p. 199
p. 183   *may be hanged'* W.S. Blunt *My Diaries* (1919) p. 260

p. 183	*Kaffir lawyer . . . no gentleman . . . no judge'* Jackson, *op. cit.* p. 120
p. 184	*they have not.'* H. Belloc, 'The Modern Traveller' vi
p. 184	*Matabele tomorrow.'* Flint, *op. cit.* p. 204
p. 185	*ordinary Kaffir labourer'* 1 Jan. 1897, HE 62
p. 185	*at any price* George Rouliot to Wernher, 18 Jan. 1897, HE 174
p. 185	*were not surprising* 23 Jan. 1897, HE 62
p. 186	*blood of African natives' Truth* 10 Jan. 1896
p. 186	*divided the profits'* H. Pearson, *Labby* (1936) pp 279–80
p. 186	*and packed the jury.'* Longford *op. cit.*
p. 187	*extent been manufactured'* Newton, *Lord Lansdowne* (1939) p. 141
p. 188	*up to the neck?* Longford, *op. cit.* p. 188
p. 189	*interest was at stake.* See R. Blake, 'The Jameson Raid and the "Missing Telegrams" ' in *History and Imagination* ed. H. Lloyd-Jones *et al.* (1981) p. 236
p. 189	*man of honour* R.C.K. Ensor, *England 1870–1914* (Oxford, 1936) p. 234
p. 190	*abominable Raid'* *op. cit. DSAB*
p. 192	*Barnato Brothers lives.'* Jackson, *op. cit.* p. 200
p. 192	*cause of everything.' ibid.* p. 199
p. 193	*his money that goes.'* Rouliot to Wernher, 4 Jan. 1897, HE 174
p. 195	*of good manners'* Merriman to Currey 30 May 1891, Lewsen vol. II p. 216n
p. 195	*he did in his life.'* L. Cohen, *Kimberley* p. 217
p. 195	*in his character' ibid.* p. 225

14: PEOPLE WHOM WE DESPISE

p. 196	*influence in Pretoria* Rouliot to Wernher, 18 Jan. 1897, HE 174
p. 197	*revolution last year.'* quot. J.S. Marais, *The Fall of Kruger's Republic* (Oxford, 1961) p. 137
p. 197	*over the natives'* Chamber of Mines, Tenth Annual Report, p. 5
p. 198	*to Mr Labouchere'* quot. Gutsche, *op. cit.* p. 137
p. 198	*should go to his'* quot. *ibid.* p. 138
p. 198	*conceived in spite'* Phillips to FitzPatrick, 12 June 1897, Fraser p. 105
p. 198	*poison administered daily'* 4 Feb. 1899, HE 66
p. 198	*man from Europe'* 29 Nov. 1897, NEL:Q2=A/LB XVII
p. 199	*£10,000 a year each'* 17 Dec. 1895, HE 65
p. 199	*influential quarters'* 21 Jan. 1899, HE 66
p. 200	*ruin or disgrace . . . see you again.'* Jackson, *op. cit.* p. 236
p. 201	*wasted for you.'* L. Cohen to S.B. Joel 18 May 1899 Cohen Papers, Cullen Library, Univ. of Witwatersrand
p. 201	*war being inevitable'* S. Evans, 21 Sept. 1899, HE 22
p. 202	*no power to England.'* Newton, *op. cit.* p. 157
p. 202	*to the Boers.'* quot. C. Van Onselen, *op. cit.* vol. I p. 2
p. 203	*parvenus engender.'* Cohen, *Kimberley* p. 46
p. 204	*of the natives'* Hobson, *War in South Africa* p. 235
p. 205	*of his existence' ibid.*
p. 205	*and savagedom . . . drawers of water . . . the Outlanders.'* Manchester Guardian 28 Sept. 1899
p. 205n	*control of the Jews'* 21 Oct. 1892 quot. in C. Holme, 'J.A. Hobson and the Jews' in *Immigrants and Minorities in British Society* ed. C. Holme (1978)
p. 206	*Jew power ibid.*

p. 206   *full of Jews . . . by the Jews'* quot. *The Pro-Boers* ed. S. Koss (London, 1973) pp 55–7

p. 206   *the Jews . . . Jews operating . . . or Houndsditch'* 6 Feb. 1900 Parliamentary Debates, 4th Series LXXVIII, cols 795–7

15 : WAR FOR THE RAND

p. 208   *Rhodes is a coward'* 3 Apr. 1897, Lewsen vol. III p. 264

p. 209   *Moscow . . . 1812* see C.W. van der Hoogt, *The Story of the Boers* (New York, 1900)

p. 210   *irrevocable failure'* O. Schreiner,

p. 211   *Pitt say to this?'* Cartwright, *The Corner House* p. 176

p. 211   *with their months'* R. Kipling, 'The Absent-Minded Beggar'

p. 213   *industry and judgment'* Clarendon, *The History of the Great Rebellion*, pp 861–4

p. 213   *North Americans.* E. Waugh, *A Tourist in Africa*, pp 157–8

p. 214   *South African mining'* Rothschild to Rhodes 5 July 1901 CJR, C:27

p. 214   *thieves and swindlers'* Parliamentary debates 7 May 1901

p. 215   *Christ-like about Beit* quot. Cartwright, *Corner House* pp 259–60

p. 215   *on the Rand* J.B. Taylor

p. 215   *gold in the World* quot. O. Letcher, *The Gold Mines of Southern Africa* (1936) p. 139

p. 215   *Jew-imperialist conspiracy'* J.A. Hobson, 'How the Press was worked'

p. 216   *(and Champagne)* H. Belloc, 'On a General Election'

p. 217   *rub them in* H. Belloc, 'Verses to a Lord'. Reproduced by permission of Gerald Duckworth & Co. Copyright © 1970 Estate of H. Belloc

p. 217   *rule South Africa.'* quot. D.J.N. Denoon, 'Capitalist Influence and the Transvaal 1900–06' *Historical Journal* II (1968) p. 310

16 : USUALLY IN PARK LANE

p. 218   *and white labour* Hobson, *War in South Africa* p. 230

p. 219   *laziness of the race'* 14 Apr. 1903 NEL B/As.1

p. 220   *upon the Kaffir'* F. Cana, *South Africa from the Great Trek to the Unions* p. 217

p. 220   *by labour unions* 23 Aug. 1902, Duminy p. 243

p. 220   *in this country . . . white country'* quot. *The Times, History of the War in South Africa* v

p. 222   *in Park Lane'* quot. in R.S. Churchill, *Winston S. Churchill* II 1901–1907 Companion Volume p. 523n

p. 222   *South African Jews'* H. Belloc and C. Chesterton, *The Party System* (1911) p. 56

p. 223   *done by us?* Churchill to Selborne 16 June 1906 *Churchill* II *Companion Vol.* p. 548

p. 223   *in many ways identical'* Duminy p. 420

p. 224   *in the mines* Phillips to Wernher 3 June 1907, Fraser p. 179

p. 225   *with great disfavour* Knollys to Churchill 24 Apr. 1907, *Churchill* II *Companion Vol.* p. 656

p. 226   *at Wernher's name.'* Gutsche, *op. cit.* p. 224

p. 227   *fear of the wind* Cohen *Kimberley* p. 112

p. 228   *in a distressed age'* Jane Austen, *Persuasion* ch. 1

p. 228   *that is final'* Wernher to Phillips, 29 Apr. 1910, HE 144

17: SETTLING THE SCORE

p. 232   *you were second'* J. Welcome, *Neck or Nothing* (1970) p. 115
p. 233   *illicit diamonds' ibid.* p. 193
p. 233   *ruffian he employed'* Winning Post, 17 Dec. 1904
p. 233   *Eton College. Prodigious!' Winning Post* 24 Mar. 1906
p. 234   *on my part'* Welcome, *op. cit.* p. 222
p. 234   *disgraceful transaction' ibid.* p. 227
p. 234   *the brother Joel.'* Jackson, *op. cit.* p. 250
p. 236   *in their hands'* NEL PR 11/1106/2
p. 237   *in the main street.'* Court report in Cohen papers, Cullen Library, Univ. of
          Witwatersrand
p. 237   *was his reputation' ibid.*
p. 237   *said quite enough' ibid.*
p. 237   *to tell deliberate lies' ibid.*
p. 238   *decline the proposal* quot. Roberts, *Diamond Magnates* pp 301–2
p. 239   *of Sir Joseph Robinson.' Cape Times* 7 Nov. 1929

18: HOGGENHEIMER

pp
241–2   *Dalai Lama of Kimberley'* E. Jessup, *Ernest Oppenheimer: A Study in Power* (1979)
          p. 58
p. 247   *formerly occupied.' ibid.* p. 217
p. 247   *of De Beers' ibid.* p. 78

19: REVOLT AND REACTION

p. 251   *the entire world.'* E. Glanville, *Through the Red Revolt on the Rand* quot. N. Herd,
          1922: The Revolt on the Rand (Johannesburg, 1966) p. 19
p. 251   *from Eastern Europe'* Phillips to E.A. Wallers 6 Apr. 1922, Fraser p. 341
p. 255   *on my advice.'* Bailey to Churchill 26 Nov. 1920, Martin Gilbert, *Winston S.
          Churchill* IV 1917–22 (1977) *Companion Vol.* p. 1255
p. 256   *it is a misfit'* Cohen, *Kimberley* p. 211

20: LEGACIES

p. 257   *British–Jewish capitalist influence'* T.R.H. Davenport, *South Africa* (1977) p. 235
p. 257   *the methods of fascism'* quot. 'The Boer War' in A.J.P. Taylor, *Rumours of War*
          (1952) p. 158
p. 259   *today, my boy?'* DSAB
p. 259   *vast material concerns' ibid.*
p. 264   *the Zionist scheme . . . in his purse'* M. Gitlin, *Vision Amazing* (Johannesburg,
          1950) pp 20–21
p. 265   *conquer and rule . . . known to be loved'* E. Waugh, *A Tourist in Africa* pp 157–8
p. 265   *how not to do it'* K. Clark, *Another Part of the Wood* (1974) p. 221
p. 265   *back to Lenin's tomb.'* Personal information

EPILOGUE: THE GOLD-REEF CITY 1985

p. 266   *horizontal ridges.'* N. Gordimer, *A World of Strangers* (1958) p. 109
p. 268   *poor drudges underground.* G. Orwell, *The Road to Wigan Pier* (1937) pp 34–5

It ought to go without saying that these references are as full and accurate as they might be. Unfortunately, 'industrial action' – doubly inapt phrase in this case – at the British Library, while the book was in proof, made final verification of some of them impossible.

Index

GEOFFREY WHEATCROFT was born in London in 1945, and read Modern History at New College, Oxford. In 1975, he joined the *Spectator,* eventually becoming its literary editor. Between 1981 and 1985 Mr. Wheatcroft was a free-lance journalist, and his frequent reports from Southern Africa appeared in the *New York Times, Harper's,* the *New Republic* and elsewhere. He also writes about the other subjects that interest him: politics, travel, opera, sports (especially racing), cookery, and wine. A columnist now for London's *Evening Standard,* he is at work on his next book, a biography of the Wagner family. He lives in London.

PRAISE FOR
Chris Pegula's Books

"From Dud   h humor,
em   ng
fr   es
(h   ."
—   ng

"F   u-
m   n,
                                               m

"I   book   gh
all   m

"V   as
yo   m

"I   ys
su   et
so   of
ju   -
ut   T

"A   n.
C   y
ev   rt

DIAPER DUDE

The Ultimate Dad's Guide
to Surviving the First
Two Years

Chris Pegula

with **Frank Meyer**

A TARCHERPERIGEE BOOK

tarcherperigee

An imprint of Penguin Random House LLC
375 Hudson Street
New York, New York 10014

Most TarcherPerigee books are available at special quantity discounts for
bulk purchase for sales promotions, premiums, fund-raising, and
educational needs. Special books or book excerpts also can be created
to fit specific needs. For details, write: SpecialMarkets@
penguinrandomhouse.com.

Library of Congress Cataloging-in-Publication Data
Names: Pegula, Chris, author.
Title: Diaper dude : the ultimate dad's guide to surviving the first two
years / Chris Pegula with Frank Meyer.
Description: New York : TarcherPerigee, [2017] | Includes bibliographical
references and index.
Identifiers: LCCN 2016048500 (print) | LCCN 2017002838 (ebook) | ISBN
9780143110262 (alk. paper) | ISBN 9781101992616
Subjects: LCSH: Fatherhood. | Child rearing. | Parenting
Classification: LCC HQ756 .P44 2017 (print) | LCC HQ756 (ebook) | DDC
306.874/22—dc2
LC record available at https://lccn.loc.gov/2016048500

Printed in the United States of America
1   3   5   7   9   10   8   6   4   2

Book design by Katy Riegel

Contents

{DEDICATION}

To Meredith,
You are my love and light.
Thank you for inspiring me to be
the best dad possible.

Kai, Juliette, and Cole,
Parenting you has been the best job I've ever had.
I love you.

Introduction

"BE THE CHANGE that you wish to see in the world."

The words of the great Mahatma Gandhi ring through my head as I drift off to sleep. I wish I could say I was leafing through some of the master's proverbs from some highbrow collection bound in dusty ol' leather given to me by some college mentor at a key transition time in my life. But the reality is that I had just been watching *Iron Man 3*, which stars Ben Kingsley, who *played* Gandhi in the 1982 movie, which reminded me of the aforementioned quote. That's how my brain works. It's like connect-the-dots up in here. Aaaaand I might have had a few beers before bed . . .

Now it's four a.m.

Barely coherent, I roll over in bed to find Meredith, my

wife, not there. Bleary-eyed and confused, I step out of bed and place my foot directly into a puddle on the floor.

"It's happening!" Meredith exclaims. "My water just broke."

I'm in shock.

"What? Are you sure?" I respond.

A brainless response, I know, but my anxiety got the best of me. Like she's *not* going to be sure if water has poured out of her.

"But how is that possible? You still have four more weeks to go."

These are the actual things I said, folks. I know they sound stupid now, because my wife Meredith has told me so many times, but at the time I was just trying to be logical in the face of chaos.

Okay, take a deep breath. What should I do next?

Squish.

I gaze down to see my feet are covered in my wife's liquid.

All righty, then.

Meredith grabs her robe, crams a bunch of stuff in her bag, and heads for the car.

"You coming?" she asks.

I stand there paralyzed, not knowing what to do next. This is our second baby too, so you'd think I would've picked up a few things along the way the first time around, right? But no. My mind? Blank. My preparation? Zero. Sure, I could have packed my bag for the hospital a few days ago like Mer asked—no, begged—me to. But I am a man, dammit! I look logic in the face and laugh! Haha!

Somehow I manage to pull myself together and throw on a clean pair of shorts, T-shirt, and sneakers. I squeeze a gob of toothpaste in my mouth and grab my toothbrush while dashing out the front door.

A few minutes later when we screech up to the hospital entrance, I leap out, grab a wheelchair, and gently plunk Mer into it. As I begin to push her inside, an obnoxious voice bellows from behind me.

"You can't park here," the surly one commands. "This is patient loading and unloading only."

"Look, buddy, this *is* the patient and I *am* unloading!" I bark back like a badass. But then quickly followed up with the more realistic, "I am soooooo sorry for my outburst, sir. Please give me a moment to get my wife inside and I will come back and move my vehicle." I can clearly only stay mean for about ten seconds.

Deep in sweat, I manage to wheel Mer to admissions, where I quickly hand her off to the staff as I run to figure out where I am going to park my car. *What the hell? This is Los Angeles. Why don't they have valet? They have valet everywhere else, why not here in Tinseltown?* And just as that thought cements itself onto my brain, I see the sign *directly* in front of me that reads "VALET: 8 A.M.–8 P.M."

Ugh. It's still only five a.m.

I take another deep breath and start to laugh.

By the time I get back and find my wife heading to the delivery room, she is knee-deep in pain. What follows is a bit of a blur, but to say it is a physically exhausting emotional whirlwind would be a vast understatement.

The contrast of emotions one experiences during childbirth is almost impossible to explain. My best description would be to imagine yourself on the tallest, fastest, scariest, and most thrilling roller-coaster ride in the world. The anticipation, nervousness, excitement, fear, and exhilaration invading your body is overpowering, even dizzying at times, and can send anyone into a state of utter exhaustion and nausea. And we men aren't even the ones actually having the baby!

Yet all that pain and angst disappeared the moment I saw my precious little one for the first time. When the doctor held up Juliette and I laid eyes on her, with her furrowed brow and melodic scream, I was hooked for life. That moment is emblazoned in my brain forever, as are the births of our other two children, Kai and Cole. These are among the life-changing moments a father experiences that he never forgets.

Almost as quickly as all this joy hits, the doctors whisk the baby away for tests as your family and friends gather to celebrate. But you two just want to get some rest and hold the baby, so soon enough everyone gets the hints and takes off, leaving you and your weary partner alone to care for this new baby you have been longing to meet for the past nine months.

Huh? How one can be left alone to care for a newborn while in a facility filled with highly qualified staff trained for this very purpose is beyond me. Yet once they brought the baby back from the litany of tests (like she's some sort of lab rat), there we were, left unaccompanied with our newborn while in a complete daze, a state of shock, really. This was our second child but we still just sat there slack-jawed when the nurses patted me on the back and said, "Well, good luck with the baby and we'll see you in the morning!"

Good luck with the baby?

What the f—?

That moment was when Meredith and I were officially inaugurated into our new roles as mother and father of our second baby, little Juliette.

I didn't sleep a wink that first night. I can still hear the gurgles coming from my little burrito-wrapped daughter, who slept in a clear plastic bin angled to prevent the fluid, which she ingested while baking in Meredith's belly, from choking her to death. Seriously, that first night alone was the most frightening experience ever. I occasionally catch myself reliving it today, more than fifteen years later. Thankfully, we had expert help just outside our hospital room door. *But what happens when we are discharged?* I wondered. Now we had two children, one a newborn and the other two years old, and I was baffled. The sheer mystery of it was crushing. Thoughts and fears were crashing around my brain. *How am I gonna deal with all this?*

Dude, relax. It's not all traumatic and stressful. Most of parenthood is an absolute blast, especially in these first two years. But you need a little help, and that's okay.

Certainly one of the key ingredients in becoming a good dad is *support*. Support from your wife or mate, or from your family, friends, or fellow parenting community, is key. My hope is you will also find it in the pages that follow, through my experience and guidance. I have three kids of my own—Kai, Juliette, and Cole—so I know of what I write.

Be aware: there is no exact rulebook on how to be the best dad in the world. Consider this more of a playbook on how to be the best dad *you* can be. The effort and energy you put into this will be returned for many years to come. Or, more frighteningly, the lack of effort you put into being a good father will surely haunt you, your partner, and your pocketbook, for the rest of your life.

But parenting is not only about the relationship between you and your child. Your relationship with your partner plays a big role in good parenting as well. Keeping a strong foundation in your relationship with your partner will set the tone for the relationship with your kids. Keep in mind that you are a walking and living example for your children. How you interact with your partner sets the tone for how they will eventually treat their own relationships in the future. And how you act within your world will set the tone for how they interact with people, treat property, respect authority, and so on. Marriage is not easy. It takes work. Once you add kids to the mix, then it's a whole new ball game.

So the task at hand is to navigate through this new phase of your life, fatherhood, while keeping your marriage and sanity on track. Just like in your relationship, when you had to ask yourself at some point, *What kind of husband do I want to be?* You need to ask yourself, *What kind of father do I want to be?*

What kind of dad *do* you want to be? Do you want to be just like your pop, or the complete opposite? Put some thought into the type of person you want to be and the type of child you hope to raise. For me, I wanted to raise each of my sons and my daughter with the awareness and skills to be the best person they can be while being conscious of how their actions affect the community around them. That's important to my wife and me.

So ask yourself . . .

Who are you?

What kind of kids do you want to raise?

I OFTEN hear dads saying how they promise to parent their child better than their father parented them. It's an interesting statement to make and one we usually don't really address until we become parents ourselves. History is an important teacher. Take a moment to reflect on the impact your father had on you as a child. How has it affected you today? What did you wish your dad had done differently? Can you do better? If your dad was amazing, how can you too be amazing? Maybe you had a mom or aunt who was more of an influence than your dad. What did they do that made you a better person? These are questions to reflect upon as you are about to become a dad yourself. I can guarantee you there are no perfect dads. We will all fail at some time. The trick is to learn from it and grow.

Yes, there has been a major shift in the dad role.

I recall an incident when I attended my first New York City trade show with my company, Diaper Dude. It was the East Coast debut of the brand and I was nervous and excited to get feedback from potential customers. The idea of a diaper bag for dads was new to the retail industry that I was targeting, so I had some work cut out for myself. I remember one boutique owner who laughed when she saw the company name, Diaper Dude. She remarked that her son would never be caught dead carrying a diaper bag, since he wasn't "that type of guy." It struck me as odd that she was okay with the fact that her son was not expected to partake in the responsibility of taking care of his child. I was shocked.

What exactly is "that type of guy"? Not me, I can tell you that much. But it was a common depiction of men in our society in the '40s, '50s, and into the '60s, and still is to some degree today. The traditional American man went off to work to put food on the table, while the woman stayed home to raise the kids. But today women can have successful careers and still be mothers, and this dynamic has created a shift in the parenting roles of moms and dads.

For instance, here in Los Angeles when the economy tanked around 2008, a lot of dads lost their jobs, and moms had to step back into the workforce. I know a lot of couples where dad decided to raise the baby while mom went back to work. Many men are becoming stay-at-home dads, since their spouses are the main source of income.

"It's a new era for dads," my friend Bill's husband Scout

points out. "It's the reverse of how we all grew up. When we take our daughter to school, there's just as many dads dropping off as moms. The dads volunteer at lunchtime, help out at events, drive carpools."

It's true, dude. Gone are the days of the man being the sole breadwinner of the family, leaving his partner home to raise the kids. Today it's becoming more common for parenting to be a joint effort, with men assuming a more active role in the day-to-day responsibilities. In fact, as more moms thrive in the workplace, it's forcing us men to reexamine the parenting dynamics and, thus, more of us are becoming stay-at-home dads.

A recent study by the Pew Research Center found that the number of stay-at-home dads has nearly doubled from 1989 to 2012. In 1989, there were approximately 1.1 million stay-at-home dads.[1] In 2012, that number increased to 2 million. The 2008 recession definitely played a part, since many fathers had lost their jobs and found it tough to find new ones. While this study specifically focused on stay-at-home dads, imagine the number if it included working dads who share the parenting role equally with their spouses.

In 1949, anthropologist George Murdock defined the "nuclear family" as characterized by "common residence, economic cooperation and reproduction [containing] adults of both sexes, at least two of whom maintain a socially approved sexual relationship, and one or more children."

Adults of both sexes? Socially approved sexual relationship? Economic cooperation? Umm, that disqualifies half my friends and my children's parents. The definitions of a typical American "family" and "parent" are evolving. Rest assured, this is a good thing. The more we can open our minds, accept people for who they are, and break stereotypes, the better off society will be. It's fascinating to witness this role reversal in modern parenting.

"My wife was always the breadwinner, and was always gonna make more money than me because I was a teacher," *DadNCharge* blogger Chris Bernholdt told me recently. "So when we got married we sat down and decided that if one of us was going to stay home to raise the kids, it was going to be me. With me staying home, she could really pursue her career and support our family."

No matter who is the breadwinner in your family, the choice to be involved, and to what degree, is entirely up to you. While some jobs might take you away from your family, there is always the option to change careers. It might seem challenging and scary, but a career shift is an option and another way to allow yourself to prioritize your participation as a parent. It's something to consider, if you haven't already. Being a parent does involve sacrifice at some point or other.

"It's no longer a purely gender-based discussion. It's about what are you good at," my friend Charlie points out. "So if you want to be an at-home dad and take care of the domestic responsibilities, it's something that you can do. There's more technology now, you can telecommute so you can be more flexible with your career, and be a work-at-home dad or a stay-at-home dad if your spouse can work. It's really about being able to select for yourself what your role is, because before there were really confining, specific ideas about what dad did."

"My son went into Career Day recently," Chris says, beaming. "When they asked him what he wanted to be when he grew up, he said, 'a stay-at-home dad.' I thought that was so cool."

That is cool. The times they are indeed a-changin'. And you don't have to sit idly by and watch the world change around you. Hop in the driver's seat, dude.

Remember, "Be the change that you wish to see in the world."

Month-to-Month Guide

Month 1

For you and baby, this first month is all about getting acquainted with each other. Keep in mind that your newborn will basically just be eating, pooping, and sleeping most of the time. The sleeping will be erratic, but soon everyone will adjust to the new routine. Keep an eye out for baby's first smile, which happens around six weeks. The main form of communication at this stage is through crying, which often means he's hungry, needs a diaper change, or wants to cuddle with you. Close contact is key at this early stage. Don't freak out when you notice that dark green, even black, tar-like substance in your baby's diaper. That's called meconium. The first few poops are the mucus and bile ingested while baby was in mom's belly. Once that exits your baby's system, if he is breast-feeding the poop turns grainy and yellow. How fun it is to talk poop.

Month 2

Your baby's sucking reflexes are strengthening and you might notice her sucking on her fist or fingers. This is a great skill they develop to comfort themselves. She is starting to gain more control over her body and you will notice more head control when she is on her tummy. She still sleeps a lot during this age, but not necessarily throughout the night. Hang in there, it's not much longer before you get some shut-eye.

Month 3

Your little dude is gaining strength at this point. His hand coordination is getting better and he might be able to reach and grab on to toys. His head muscles are also getting stronger and he might be able to hold his head up while on his tummy. Your baby's getting more active and might begin to learn how to roll at this age (he will rock later). He is eating more, which translates into longer sleep time (up to six or seven hours at a time). You'll notice his personality emerging and communication skills improving. He'll be babbling away, and when he likes something he will let you know with a smile or a turn of his head toward you. On the other hand, if baby is experiencing something he doesn't like, you'll know by his crying, fussiness, and/or flailing limbs in displeasure.

Month 4

If you were to compare your baby's weight now to her birth weight, chances are she pretty much doubled it. She is getting even more control over her hand motion. Beware: her skills now allow her to explore items she can grab and place in her mouth. Make sure there are no small items within arm's reach that she could choke on. You need to be vigilant about this, dude. And on a happier note, if you haven't experienced a full night's sleep in four months or so, this could be the time it starts!

Month 5

Your baby's muscles are getting stronger! He might be getting close to sitting upright on his own without toppling over, if he isn't sitting up already. Rolling over is another skill he's getting good at by now. Your baby's grasp is getting stronger and he's now grabbing items and transferring them from one hand to the other. He might even be able to hold his own bottle or sippy cup at this point. Look out, world!

Month 6

Your doctor will most likely suggest introducing solid foods around this time, if he hasn't already. Cereals with breast milk or formula will most likely be the meal of choice. Your little one is getting closer to crawling and trying to prop himself up on all fours, rocking back and forth. He will notice familiar faces like yours and your partner's, as well as family, friends, and frequent visitors. Imitating sounds is also something you'll hear, so make a game out of it if you can. If you assist him in standing he will enjoy bouncing up and down, since his legs can now support his weight (with a little help).

Month 7

Baby's personality is starting to show. It will become more evident what she likes and doesn't like. Sitting up while playing with toys is becoming common. She also will probably enjoy a game of peekaboo. You might notice some teeth about to pop out. Excessive drooling and fussiness is a sign teeth are a-comin'. But don't stress if they don't arrive just yet—all in good time.

Month 8

Separation anxiety starts to rear its head at this point, if it hasn't already. Don't be surprised if you notice baby feeling shy in the presence of others. Most babies are crawling by now, but it's fine if yours is not there yet. If there is crawling going on, you might see your little one begin to pull himself up using furniture for support. Best to make sure loose objects like cabinets or bookshelves are secured so they won't topple down. Your baby's vision is almost what it will be as an adult. How cool is that?

Month 9

Your little one is quite the talker at this stage, babbling away and making funny sounds. Her coordination is improving and you might notice her ability to bop between crawling and standing with awkward ease. She is as curious as can be, so be sure to do a clean sweep of the room to make sure there are no small items within arm's reach. She most likely understands "No" by now, but won't always obey the command. As a father of teens, I can assure you this does not change!

Month 10

Baby most likely understands simple commands like waving goodbye or clapping hands by this stage. Separation anxiety and stranger anxiety are still present. You will notice improved hand-eye coordination and find that your little one gets a kick out of dropping items and watching you pick them up. His motor skills are improving and he's doing fun stuff like pulling himself up, babbling nonsensically, and imitating your behavior like brushing hair and talking on the phone. Oh, perfect, a comedian.

Month 11

Your child might already be close to cruising along with the help of support from furniture or you, dad. But don't worry if he is still in crawling mode. That's totally expected too. A timeline developmentally for walking is usually eleven through sixteen months. But if your little one's an adventurer now, curiosity can get him into trouble once you turn your head, so keep watch. Temper tantrums might begin to rear their heads as your baby discovers what he likes and dislikes, wants and does not want. Remember to keep your cool and be mindful of how you use your words. After all, *you* are the example!

Month 12

No longer a baby, your toddler is growing and learning quickly. Still a curious adventurer, he will most likely skip a nap at this point and only take one a day. You might notice a desire to do things on his own, like eating with his fingers or attempting to use a spoon. He might also help you choose what socks or shirts to wear in the morning. Now is a good time to introduce a lesson in manners by saying "please" and "thank you."

Months 12–15

Kids love to gives hugs and kisses at this stage. They also love climbing stairs and furniture, so keep an eye out, since they are still very curious. Hand-eye coordination is sharpening and your toddler will get a kick out of singing songs like "Itsy Bitsy Spider" and "Patty-Cake." She can usually point to parts of the body when asked, as well as follow simple directions like handing you a toy when asked.

Months 16–19

Doors are a fascination at this stage, so watch out as your toddler is having fun opening and closing them. Make sure he doesn't pinch his fingers! Your toddler might be able to connect pictures to names when asked, so be sure to pick up some age-appropriate picture books and read them together. This stage continues to be a blast and you will love witnessing his vocabulary skills grow anywhere from twenty to fifty words.

Months 20–24

Your toddler is learning day-by-day. She is interested in helping you around the house, so invite her assistance when you are folding clothes or putting dishes in the dishwasher. You might notice some unfavorable behavior during this stage, like hitting, biting, and kicking, but this is part of development. It's important not to overreact, but use a kind tone to explain that the behavior is not okay. Toddlers this age are simply testing limits. She is getting better at kicking a ball and is pretty good at throwing it underhand as well. By two she might be able to throw a ball overhand. Watch out, dude, you might have a baseball player on your hands. Curiosity and awareness of body parts is natural, so don't be freaked out by her self-exploration of her body. Your child will be getting closer to speaking two- to three-word sentences. Be sure to continue reading together as much as possible to encourage learning new words as well as identifying objects and animals.

CHAPTER 1

The Arrival

OUR FIRSTBORN, KAI, was born at exactly 7:11 a.m. I remember that night as if it were yesterday. After twelve hours of waiting and wondering, Kai finally arrived without a hitch. His lungs were robust, his scream powerful, and his reaction to entering this world unforgettable. "Waaaaaahhhhhh!" It was like a fire alarm. There was no mistaking the intensity of this little dude when he arrived into our lives. If his volume alone was indicative of his feelings, Kai was as mad as hell that we took him out of his warm digs and was not averse to voicing his displeasure. Loudly. If there were a crystal glass in our room at the time of his arrival, I am positive he would have shattered it. But once I cut the cord and had the chance to hold him in my arms, he was at peace. He was where he was meant to be. Life was good.

After my wife, Meredith, had the chance to lock eyes with Kai and caress his skin for a brief moment, the nurses stepped in to perform their inaugural duties and he was whisked off in preparation to be handed over to his parents for good. It was amazing to watch their ease and confidence as they bathed and weighed him, suctioned his airways, pricked his skin to determine his blood type, took his temperature, scrubbed his scalp, clothed him, and finally swaddled him up. (Swaddling calms the nerves and makes babies feel secure, but more on that in a few.) These moments occurred quickly and effortlessly. I tried desperately to keep up and take notes on each task. (If only the iPhone had been invented sixteen years ago!)

Having alone time with your partner and newborn is a must. Try to create a calm and peaceful environment as you introduce yourselves officially as mom and dad. Take this time to absorb this new dynamic.

"I was exhausted, so I was happy you dealt with the hospital and were more the eyes and ears than I was," Meredith confessed to me recently. "You later told me that you were faking it, but you made me feel like you knew what you were doing."

Blood Type

Among the tests your newborn undergoes after birth is determining the blood type. To this day I keep forgetting what my blood type is, much less my kids'. It is wise to store this information in a place you remember and can easily access.

The universe is powerful and can work in your favor if you let it. Take the time to breathe in this new gift delivered into your lives. You'll have plenty of time later on to celebrate with your friends and family as they congratulate you on your new arrival. Be warned, though: having a little space is something you should address ahead of time with relatives and friends who might arrive without notice. Your partner has just been through the ringer. Now you are her buffer from the outside world and the anxiety, frustration, and germs that can work their way into your sacred space.

The First Few Nights

The next few nights are a whirlwind and you won't be getting much sleep. Your partner, on the other hand, can use all the sleep she can get, so encourage it. Help her make it so. Depending on whether the delivery was vaginal or by

C-section, she is going to be in pain, and allowing her to rest will assist in the breast-feeding experience if she goes that route.

My wife's exhaustion just wiped her out and she was hardly in any shape to start breast-feeding. I, of course, was clueless as to what to do. That is, until Corky, the hospital lactation specialist, grabbed me by the arm and instructed me on the ins and outs of "latching onto the boob." This was not your typical interaction with a stranger. After all, I had just met this woman, and we were talking about my wife's breasts. To give her credit, Corky knew how to speak the language and relate breast-feeding to guys. Using the analogy of holding a football, she executed the play on how to encourage Kai to latch on and feed from the breast. Eventually the awkwardness slips away and the football hold becomes your child.

Pee-Blocking Devices

When it's time to change your boy, unless you're incredibly fast, you are bound to need assistance to prevent getting showered on. Check out the Pee-pee Teepee. Yes, that's a real thing and is exactly what it sounds like. They are dome covers made of combed cotton designed specifically to keep urine from spraying in your eyes. When our youngest, Cole, came home from the hospital, we had our nurse help us out, since we had two other children at the time. The first time she changed Cole's diaper he let loose and got her right in the kisser. Well, at least they say that urine is sterile!

Changing a diaper was not so foreign to me. Having five siblings, I had encountered a dirty diaper or two among my numerous nieces and nephews. But changing the diaper of my son was a different story. Let me advise you to keep a blockade of some type to protect the wild sprayings of your new son's tool. More than once I learned the hard way to block, duck, and cover. But with practice comes perfection, and now I can dodge pee with the best of 'em. (Hmmm . . . that sounded awkward.) I'm sure you will be able to diaper your dude with ease soon, but until then, be patient. If you have a girl, then be sure to get a rundown from your nurse or doula on how to clean a dirty diaper. Among other things, you'll learn to go from front to back when wiping clean, to avoid infection.

☛ **PARENT HACK**
Having trouble with those first few cleanings?

Cleaning those first few tar-like poops (meconium) can be tricky. Try using a little coconut or almond oil on baby's bottom to make cleaning easier. It gets baby's tushy just moisturized enough so that the doody should wipe off with ease.

Swaddling is an essential skill you must master. Being snugly swaddled actually mimics the security that your newborn felt while in the womb and will save you from those moments when baby is struggling to get some sleep. Swaddling is an art unto itself, and practice makes perfect. Use these guidelines to help you turn your little one into the perfect burrito. Practice with a doll first. Don't wrap too loosely or else your lil' one will wiggle right out of his blanket. Don't wrap too tightly or she'll be uncomfortable. Kids can get squirmy, and unless you've perfected the art, you might have a hard time at first. Check to make sure your baby is not overheating. If you notice your baby's getting warm or sweating, unwrap her or turn on the air to cool her down. It takes time to get it right, but you'll be surprised how quickly you will be able to swaddle with your eyes closed. (Pro tip: don't actually close your eyes.)

SIDS

Sudden infant death syndrome is the death of an infant before one year old that can't quite be explained. Often it's the result of accidental suffocation or strangling that occurs if items are left unattended in the baby's crib.*

By the end of the day of Kai's birth, I longed for some shut-eye. Meredith and I chose to keep him in our room to sleep while in the hospital. We were paranoid about the baby being switched or stolen like we'd seen on TV. (I know, but we were watching a lot of Lifetime movies and *Law & Order: Special Victims Unit* at the time.) In hindsight, having Kai sleep in the nursery might have been a great choice, and would've allowed Mer and me to get some much-needed rest. Instead I spent the night wide awake, constantly staring at his belly to make sure he was breathing. The fact that there is a *slight* chance that SUDDEN DEATH could happen made me absolutely neurotic. Not knowing much about it made matters worse.

---

* The Centers for Disease Control recommends always putting your baby to sleep on his back on a firm sleep surface.

Kangaroo Care

Kangaroo care is a skin-to-skin touch method that promotes bonding, helps stabilize baby's weight, assists with sleeping, reduces stress and pain, and boosts mental development.

My wife and I were firm believers and used it with our second son, Cole, as we felt it helped him feel secure and added an overall positive energy to the sterile hospital environment. It's especially interesting to note that Cole has a strong thirst for knowledge, started to read early at the age of two, skipped fourth grade, and is a wizard with electronics. While I can't prove kangaroo care was solely responsible, I think it helped.

I watched from afar as I called my parents to share the news with them. Today you are fortunate to be able to text or post your news via social media. Just be sure to get an approval from your partner ahead of time before you go a-tweetin'. Perhaps this is a conversation to be had before you even head to the hospital. The last thing you need is for her to be upset with you for posting a photo she did not approve of . . . especially on THE BIG DAY.

Bringing Baby Home

After spending forty-eight hours in the hospital, Meredith and I were excited to be heading home. I, for one, was scared as hell. The thought of being left alone to care for this new being was terrifying. Throughout the day, I couldn't help wondering why they require a license to drive a car, yet no requirement is necessary to have a kid. A wheelchair is standard protocol for mom when she is departing from the hospital. Meredith welcomed the assistance, as she was wiped out from the past forty-eight hours. It was finally time to head home with our pride and joy.

H-h-hiccups?

Hiccups are quite common in newborns. Overfeeding is a common reason hiccups occur in babies. When your baby eats too much or too quickly the diaphragm muscles spasm, creating hiccups. When you are feeding baby with a bottle, try taking a break to burp her before giving the remainder of the bottle. If you feed twice as often and half as much you might notice a difference. Remember, you don't want to underfeed your baby at the end of the day, you just want to help eliminate a potential nuisance. Pay attention to the amount of air intake your newborn makes while eating from the bottle. According to pediatrician William Sears's book *The Attachment*

Parenting Book: A Commonsense Guide to Understanding and Nurturing Your Baby, holding the bottle at a forty-five-degree angle will move the air to the bottom of the bottle, preventing too much air intake when your infant is eating.

By this time, I hope you have already installed your car seat, but you'd be surprised how many of us install it incorrectly. Don't let your manhood get in the way of asking for assistance. Safety is your first concern. The most important thing is to make sure that neither the seat nor the harness is loose. Another big mistake is to turn the car seat face-forward too early. According to the American Academy of Pediatrics, you should not face your child forward before the age of one or before they reach twenty pounds. This is not something you want to mess with. Take your car to a specialist to have the car seat installed correctly, or check safercar.gov.

The first few nights were like daddy boot camp . . . but since it was me and my family, it was a lot closer to *Stripes* than *Full Metal Jacket*. I was so impressed with Meredith when we got back home. As exhausted as she was, she snapped right into action and was so immediately confident as a mother. It was as if she was born to do it. For me, it was not that easy. I *wanted* to feel confident as a dad but

I found myself relying a great deal on my wife's instincts and following her lead. Yet I knew I had to find my way through uncharted waters myself.

For one thing, I was still freaked out over the thought of sudden infant death syndrome. I did the research and learned that having your baby share your room is fine, but not in your bed. You might have seen numerous images on social media of dads sleeping next to their baby, which melts the heart of every mom who comes across it. Sure, it looks sweet, but be careful when it comes to sleeping with your newborn. Believe it or not, simply rolling over while sleeping can accidentally cause suffocation. Better safe than sorry. There are numerous brands that carry co-sleepers (a hybrid between a bassinet and a crib) that attach or sit right next to your bed. We bought one from Arm's Reach, and it made nighttime feedings so much more bearable. Make sure to keep loose objects and pillows out of the crib. While these plush and cozy items might look good to the eye, for safety's sake, get them out of the crib.

☛ **PARENT HACK**
Put your baby to sleep with white noise.

Getting our first little one used to sound helped him sleep
easier when his siblings came along. A white noise machine
should do the trick. Now our kids can sleep soundly through
a thunderstorm.

I was also incredibly nervous about making a mistake.
The first time I held Kai in my arms I was shaking like a
leaf. I was paranoid that I would break him, smother him,
or drop him. I was hyperfocusing on making sure his
neck and head were secure and safe. The neck muscles
need to strengthen, so at first his head will wobble out of
control. Pay attention to your newborn's lack of strength in
neck muscles and make sure to support and protect baby's
head. Be aware of the soft spots on the head as well.

"Hey, what's the deal with those soft spots?" you ask.
There are two delicate places on your newborn's head, called
the anterior and posterior fontanels. They are gaps between
the bones of your baby's skull. The fontanels allow the baby's head the ability to squeeze through the birth canal
during birth. The gaps remain open under the skin to allow
your baby's brain to grow. Those gaps are protected by

sturdy membranes and eventually fuse together as miner-als build up inside your little one's head. While it's import-ant to be cautious when handling those areas, it will also be practically impossible to avoid touching them, especially when you are washing his head. The trick is to be gentle.

Keep It Clean

The umbilical cord falls off anywhere from seven to twenty-one days after birth. Keeping it clean and dry is important. In the past, it was recommended to clean the stump with alco-hol but the American Academy of Pediatrics recently changed its position on this, as treatment without alcohol actually healed faster with no more risk of infection. Giving your new-born a sponge bath and allowing the stump to dry is really all you need to do. Folding down the diaper to keep it from irri-tating the stump is important, as well.

Eventually my comfort level increased, and before I knew it I was mastering the one-hand hold and multitask-ing with the other. Once I felt comfortable, I held him like a football, tucked close to my side, nestled in my arm. (Just don't spike that ball!) Of course, it helped that he was only five pounds fifteen ounces. I actually took some photos of him in his crib next to a jar of peanut butter and a spoon, just to record how small he was.

And also, slowly but surely, I mastered the art of changing diapers and bathing and feeding our son, while at the same time paying attention to the needs of my wife, who had undergone an episiotomy. Her physical limitations gave me no other choice but to step up to the plate and take over. But she was a great coach along the way.

Becoming comfortable with your newborn can be challenging at first. Each of our kids were born well before their due date. So you can imagine the concern when someone like myself, six feet tall and 190 pounds, attempts to be gentle with an infant. But trusting yourself is the first step in building confidence. The second step, of course, is not dropping the baby.

Under the Blade

I'm not here to judge your decision on whether to circumcise your baby. For me, I didn't want my boys to ask why theirs looks different than mine, so I had them circumcised. But if Mer and I were to have more children today, after reading more research about the pros and cons of this procedure, I might have chosen a different route. Especially after witnessing the actual process in person. Watching my newborn son being strapped into a plastic restraint system and listening to his piercing cry as the doctor performed this ritual is not an experience I care to relive. Many argue over health risks and hygiene issues. I say, do your homework and make your own decision.

A large portion of your time will be dedicated to charting the eating, peeing, and pooping of your baby, so keeping a log of these activities will help, especially once you get to your first doctor visit (around two weeks after you get home). Newborns will lose some weight in the first five to seven days of life. A 5 percent weight loss is considered normal for a formula-fed newborn. A 7 to 10 percent loss is considered normal for breast-fed babies. So tracking this info will be a wealth of knowledge for your newborn's first doctor visit. A good thing for mom to do is to keep a notepad strapped to her side to record the baby's feeding and pooping schedules. Charting this is smart, as it enables parents to learn when it's time to feed and when to expect a poop. It also alerts you two to how much downtime you have so you can catch up on sleep or run some errands. Another way you can assist in this task is by setting the alarm on your phone to remind your partner of the feeding schedule. Check for apps (iBaby Feed Timer is popular) to help track your baby's feeding schedule or simply track it on your own in the notes section of your phone.

One thing you don't want to do is jump on the Internet every time there's a problem with your kid. Even though we have such easy access to information at our fingertips, do not immediately jump on the Net to diagnose your child's condition. You will only set yourself up for a freak-out session. While you might find some quick-fix remedies, you'll also be exposed to diagnoses and wacky stories that will set off every red flag and alarm in your system. The Internet can be one big panic button when it comes to self-diagnosing. First, try to get in touch with your child's physician. Usually *he or she* can assess the situation over the phone and spare you the drama you might discover online. If an *actual doctor* believes there is a life-threatening condition, they will surely suggest you head straight to the emergency room.

Homeland Security

Now that you're home, you'll need to assess your battlefield of potential weapons via a security sweep. That's a fancy way of saying you need to get down on all fours and crawl around looking for potential disasters and baby threats. While your little one is not mobile until at least six months old, it's best to get this task achieved before it's too late. By assuming this charming doglike position, you are able to scan the area from your child's perspective. It's amazing what you will discover. From sharp-edged furniture to enticing electrical outlets, your home is a world of mystery and wonder to a mobile baby. Securing furniture that is wobbly, removing items from countertops, like candles or frames, that are loose and easily accessible to your curious adventurer, are a must.

Make sure to inspect your window treatments, and if you have cords from curtains or blinds that are lying low, be sure to secure them high out of reach from your little one's hands, as they can serve as a strangulation device for your child. Installing locks on toilet seats so your child doesn't accidentally fall into the bowl and drown, as well as securing your dishwasher and refrigerator doors from accidentally crushing your child's fingers, are a definite must. (It's been nearly two decades since we had babies in the house and I still forget that some of the drawers in our home remain babyproofed.)

The interior of your home is not the only place you need to be concerned about when it comes to your child's safety. Be sure your yard is safely enclosed, as well as your pool if you have one. Most states require your pool to have a gate with a lock that cannot be accessed by a child. You can always double up on security by installing a pool cover, preferably one that remains secure even if your child happens to step on top of it. Covers that are used solely to retain heat can actually contribute to drowning, since they are lying loose on the water.

Setting the water temperature in your home lower than 120 degrees is recommended, to prevent scalding your child when he is washing his hands or taking a bath. A big concern is keeping them out of reach of medicines and toxic cleaning products. Speaking from experience, I can say that you do not want to be holding your child's hair over the toilet bowl after administering ipecac syrup to induce vomiting. It's just not fun.

☛ **PARENT HACK**
Can't find the bib during feeding?

During meals our little dude always turned into a mess because we never remembered to have a bib handy when food was a-flying. Take a self-adhesive hook and attach it to the back of baby's high chair to hang the bib on. This way, a bib's always within reach.

If you are a DIY type, you can hit your local hardware store or Amazon to purchase the necessary babyproofing items you will need to make your home safe. If you're no handyman, there are many services that will come to your home and assess and install the necessary babyproofing your environment needs. Do not let your ego get in the way if you are not capable of installing the gadgets yourself. The important thing is to do the best that you can to make your home a safe haven for your baby to explore, grow, and learn. It goes without saying that you should make sure anything you buy meets current safety standards.

Here are ten childproofing tips to get you started:

1. Install layers of protection over pools and spas. Ideally you want a barrier completely surrounding the pool, including a four-foot-tall fence with self-closing, self-latching gates. A pool alarm ain't a bad idea either.

2. Put safety latches and locks on all cabinets and drawers in kitchens, bathrooms, and other areas to help prevent children from gaining access to medicines, household cleaners, or sharp objects.

3. Use safety gates to help prevent kids (or drunk adults) from falling down stairs or entering certain rooms.

4. Doorknob covers and door locks help keep children away from places with hazards, but make sure the

devices allow the doors to be opened quickly by an adult in case of emergency.

5. Apply anchors so furniture and appliances can't fall over, especially if you live in high-risk earthquake areas.

6. Install anti-scald devices for faucets and shower-heads.

7. Make sure you have smoke alarms on every level of your home, inside each bedroom, and outside sleeping areas. Change the batteries at least once a year and check 'em once a month. Use carbon monoxide alarms too.

8. Window screens are not strong enough to stop a falling kid. Use window guards and safety netting to help prevent falls from windows, balconies, decks, and landings. But make sure at least one window in each room can be easily used for fire escape.

9. Corner and edge bumpers help prevent injuries from falls against sharp edges of furniture and fireplaces, so stick 'em on every table and desk you got, dude.

10. Outlet covers can help prevent electrical shock and possible electrocution. Make sure they are large enough so that children cannot choke on 'em.

For more overall safety information on babyproofing, check out cpsc.gov.

Going Back to Work

Early on in my career, as I was building the Diaper Dude brand, I made it a priority to be as involved as possible in parenting. Starting a business out of my home afforded me that luxury. While it was convenient to be able to leave "the office" whenever Meredith or Kai needed me, it was also a source of stress knowing that I could be called upon at any given moment (like in the middle of an important conference call). During that era, my kids didn't realize I even had a job. I was around almost all the time. Eventually, as they got older and more verbal, they started to catch on. In time, it became a challenge for me to work from home, as the kids wanted my constant attention. But I couldn't afford a separate office space at the time either. So my solution was to run my company out of our garage . . . but I didn't tell the kids. You'll figure out what works best for you and your family. Don't be afraid to get creative.

Burping

Burping my kids was one of my favorite bonding times with them. Call me juvenile, but bodily functions crack me up. Having my tiny tot let out a roaring burp was hilarious to me. There are a few ways to achieve a burp from your little one. Over the shoulder is the most common technique. Be sure you prep your clothes with a burp cloth to avoid puke staining them. Rest your baby on your shoulder and gently pat her back and alternate with a rubbing motion up and down. If you find that after four minutes or so that she does not belch, then chances are it ain't gonna happen. Try laying your baby on her stomach on your lap and attempt to get a burp from that position, or sit your baby on your lap while supporting the head and pat gently on her back. Let the hilarity—and gastrointestinal relief—ensue.

Every morning I would say good-bye to my family after breakfast, get into my car, and drive around the block to the alley, where I pulled back into the garage. Quietly I would sneak into my office space so the kids would not hear me, and get to work. If they spotted me and figured out I was actually home all day, all bets were off and I was screwed! They just loved having me at their disposal, so getting work done around them was impossible. Of course, the day finally came when my kids outsmarted me. I thought it was nap time and got lazy. Instead of driving back to the front of the house pretending I just got home from work, I decided to sneak into the house through the back door. Well, as soon as I entered, the door alarm chimed, and within seconds I heard footsteps running to discover me. The jig was up.

Finding the time to bond during the first few weeks of your child's life is so necessary for you, your partner, and your child. There are many advantages and benefits your child will reap when dad is part of the picture and inter-acting early on. A recent article in *The Economist* cites that the early interaction of dads in the child-rearing process benefits their children.[2] Fathers who take paternity leave are more likely to take an active role in child-care tasks. This early interaction has longer-term benefits for a child's learning abilities. A 2014 study by the University of Oslo found that paternity leave improved children's perfor-mance at secondary school; daughters, especially, seem to flourish if their dads had taken time off.[3] So if studies sup-port the benefits of dad's involvement, there is no reason for you to miss the first few weeks of your newborn's life if your job will allow.

Having had the luxury of spending the formative years at home with my children, I encourage you to take advan-tage of the chance for this as well. Make it a point to find out early on if you are eligible for paternity leave and plan accordingly with your partner. Many employers are required by federal law to allow their employees up to twelve full weeks of unpaid leave. But not all of them. Each state is different. Recently, though, some cities have been requiring employers to give new parents fully paid bonding time with their babies, and a few progressive companies are even offering up to two weeks' paid paternity leave. Hopefully these concepts will become the norm.[4]

Strategize when you are going to head back to work, dude. While the first few weeks after the baby is born are a whirlwind, you will get a good sense of how your life will function with your newborn addition. Working out your schedule with your partner before heading back to work is highly suggested. Meredith felt overwhelmed once I got back into my work schedule. She even remarked how jealous she was that I was able to interact with the outside world while she was experiencing cabin fever caring for Kai. This was actually a good example of when being open to my partner and being willing to take over was a huge benefit. Meredith and I came up with a daily schedule where I could relieve her and let her get some much-needed downtime to regroup and slowly integrate herself back into her life again. She was stoked that I listened to her and was willing to work with her, rather than getting defensive and seeing it as an attack. Sixteen years later, I can genuinely say I am happy I made the decision to be available as much as possible during this early stage.

Vaccinations

In 1989, Michael Jordan made history by hitting an unbelievable jumper with six seconds left in game five of the Eastern Conference first round against the Cleveland Cavaliers, giving the Bulls the win by one point. Jordan's shot was recorded as one of his greatest clutch moments and was even re-created in a popular ad for Nike's Air Jordan sneakers in 2006. If you were one of the fortunate spectators to witness that moment, it was spectacular. Unlike Jordan's famous shot, which was celebrated as one of the greatest moments in NBA history, it is commonly recommended that your newborn be subject to a *series* of shots for the next few years that will leave little to be celebrated (and hopefully leave no recollection whatsoever for either you or your little one).

Vaccinations have become quite the controversial topic. Actress Jenny McCarthy got into a boatload of trouble when she voiced her opinion on the link between her son's autism diagnosis and his vaccinations. While I am not medically trained, and the opinions on this topic are merely my own and those gathered from friends and colleagues, it is *you* who has the final say in whether you choose to proceed with vaccinating your child. Don't go with the flow, don't go with a quote from a magazine article, don't go by what some friends tell you. I highly suggest gathering as much information as you can to make an educated decision. Here's a chart of recommended vaccinations, dates, and doses, but check out cdc.gov/vaccines for more info.

	HepB	Dtap	Hib	Polio
Birth	•			
2 months	•	•	•	•
4 months	•	•	•	•
6 months		•	•	
12 months	• (6-18 months)	• (12-18 months)	• (12-18 months)	• (6-18 months)
15 months				
18 months				

	PCV	RV	MMR	Varicella	HepA
Birth					
2 months	•	•			
4 months	•	•			
6 months	•				
12 months	•		•	•	
15 months	(12-15 months)		(12-15 months)	(12-15 months)	•• (2 given from 12-23 months)
18 months					
19-23 months					

Meredith and I struggled quite a bit when it came to vaccinating our children. Some families in our immediate circle felt that since many of the diseases had been irradicated, there was no need to vaccinate. Take polio, for instance. Unless one was traveling to a part of the world where there was a polio outbreak (or traveling back in time, which is even less likely), vaccination seemed useless. Polio used to be very common in the United States until the vaccine was introduced in 1955. It was one of the most dreaded childhood diseases of the twentieth century. However, thanks to that vaccine, the disease is practically nonexistent in this country now. In many countries, the vaccine is required before children can even attend school. Just think, if, back in 1955, no one had vaccinated their child, polio would still be a huge problem today.

Measles, on the other hand, is a different story. In 2015, there was an outbreak of measles reported at Disneyland in California from a traveler who most likely was exposed to it while overseas and was infectious while visiting the park. Those who broke out with the measles virus were all unvaccinated.[5] It was a scary time for all parents, including those who hadn't vaccinated their children. Ironically, one day I was riding in an Uber to pick up my car from being serviced at the dealership. Somehow my driver and I got into a discussion about the measles outbreak in Cal-

ifornia, since I lived near the high school where an infected teacher was employed. It turned out that the teacher
was a relative of my Uber driver. For a moment, I pan-
icked. Thankfully I'd had my measles vaccination already,
but what if I hadn't and had caught it because she was
exposed prior to picking me up?

Do your homework on vaccines, talk to your doctor, and
make an informed decision. Failure to protect our children from known diseases can contribute to their resur-
gence. Ultimately, Meredith and I decided the risk was too
great to not vaccinate. When we decided to move forward
with the vaccinations, we did find that splitting up the timing of certain vaccines—such as measles, mumps, and
rubella—made us feel more comfortable, since our children wouldn't be overloaded with more than one vaccine
at a time, giving their immune systems time to recover
and hopefully experience fewer side effects. We discovered
this possibility by talking with our physician. Your doctors
are available to help in these moments, so make full use of
their knowledge and expertise as you decide what's best
for your family.

Crying vs. Colic

There was a moment when Kai was just a baby and Meredith and I were taking a walk in the neighborhood with him in the stroller. He was very sensitive and this time we tried everything we could to soothe him, but no go. We held him, we sang to him, we rocked him; it was exhausting. Sometimes a change of environment can do the trick, hence the walk in the stroller around the hood. Our walk started out with Kai crying relentlessly. A woman happened to be walking by and commented that we should not let our baby cry. I had to hold Meredith back from decking the lady square in the jaw. If only she knew the great lengths we had gone to, to get this kid calm. This was meltdown time, so to have someone comment on our parenting was really hard to digest at that particular moment.

I don't care what anybody says about their perfect baby. All babies cry. A newborn is adapting to a completely new environment while you and your partner are adapting to a whole new lifestyle dynamic. It's especially challenging since your baby has no way to communicate to you except through crying and "oohs" and "ahhs." But those are important responses that cannot be dismissed. Crying most likely means your baby is uncomfortable and wants her needs met, like a diaper change or a feeding. Be mindful of this and get into the habit of recognizing how your baby's demeanor changes when you take action. This is your first stage of communication with your baby. Look for cues, clues, and subtle (and not-so-subtle) signals that your baby is uncomfortable or distressed. Seems elementary, but not everyone gets this from day one.

Finding ways to soothe your newborn or toddler is like finding the Holy Grail. You don't realize how trying and nerve-racking moments like these can get until you are in the thick of it. So arming yourself with some tricks of the trade to help soothe your crying child will make parenting more enjoyable for you, your partner, and your child. For instance, make sure to check to see when your baby was last fed. Since your baby cannot yet communicate with words, crying is the only way to let you know she/he is hungry. Depending on your stance with pacifiers and thumb-sucking, you can choose to offer one or the other as a soothing technique. While I can't guarantee your child will avoid orthodontia in their teenage years, it's worth the money spent later for both your kid's and your own sanity.

Motion is a great quick fix to get your baby calm. We used to have a swing in our home that all three kids loved to hang out in. Sometimes a car ride did the trick and would knock out Kai within minutes. Nothing beats the warmth and touch of mom or dad cuddling, or wearing your baby in a carrier. Whatever you choose, make it a ritual that includes calmness and relaxation. Remember, the idea is to calm your baby down, not overstimulate him. If you find that motion just makes him more restless, then quit while you're ahead.

But some children will just not be soothed by these methods. Some babies suffer from colic, and colic is not just crying. Colic is a different story entirely. To put it simply: colic is a pain in the ass.

☞ **PARENT HACK**
When babies get gassy, they get upset.

Lay baby on her back and rotate her legs in a bicycle-pedaling motion, from elbow to knee. This helps relieve the gas . . . but expect a few baby farts to the face.

About one-fifth of all babies develop colic, usually between the second and fourth weeks. They cry inconsolably, often screaming, extending or pulling up their legs, and passing gas. Their stomachs might be enlarged or distended with gas. What is the cause of colic, exactly? There is no definitive explanation, but most often it means that the child is unusually sensitive to stimulation and cannot "self-console" or regulate his nervous system. Sometimes in breast-feeding, baby's colic is a sign of sensitivity to a food in the mother's diet. It can also be a sign of a medical problem, such as a hernia. Generally, the colicky crying stops by four months, but it can last until six months.

Are there any remedies for colic? Consult your pediatrician first, but here are some tips worth trying:

- If your partner is nursing, she can try to eliminate milk products, caffeine, onions, cabbage, and any other potentially irritating foods from her own diet.
- If you're feeding formula to your baby, talk with your pediatrician about a protein hydrolysate formula.
- Don't overfeed your baby. Try to wait at least two to two and a half hours from the start of a feeding to the start of the next one.
- Walk your baby in a baby carrier to soothe her. The motion and body contact will reassure her.
- Rock her, run the vacuum in the next room, or place her where she can hear the clothes dryer, a fan, or a white-noise machine. Steady, rhythmic motion and a calming sound might help her fall asleep. However, be sure never to place your child on top of the washer/dryer. (It's insane that I feel the need to even say that, but I read the papers . . .)

- Try a pacifier. While some breast-fed babies will actively refuse it, it will provide instant relief for others.
- Lay your baby tummy-down across your knees and gently rub her back. The pressure against her belly can comfort her.
- Swaddle the baby in a large, thin blanket so that she feels secure and warm.

The Postpartum Experience

Not long after the birth of our youngest son, Cole, I was slated to undergo sinus surgery. I had been subject to numerous sinus infections and this was the only way to cure my chronic condition. My doctor suggested recovering away from my home (which meant at my mother-in-law's) in order to avoid the children accidentally injuring me by jumping on top of me and hitting my nose. He knew how active I was as a father and how "spirited" my kids were, and was concerned my kids could do some harm, since they were just toddlers. It was a brutal recovery and my emotions were erratic. I was irritable, suffering from extreme discomfort, and did not want to be seen by anyone, since my face was a mess, all swollen from the surgery. When I finally got through the difficult stage of this experience, I was still feeling the side effects of the trauma I had endured.

With that in mind, I can only imagine the amount of discomfort and irritability Meredith and other moms feel post-labor. The fact that they also need to be available to care for their newborn the moment after birth is a true sign of the strength and courage women possess. I, for one, do not take this trait for granted. It is important to keep this in mind for the next few months when it comes to your partner's post-delivery state. It is tough to recuperate from having a baby, especially if your baby is colicky and shrieking all the time. Rest, peace, and quiet go a long way, but sometimes it goes deeper than that.

"I would get these big emotional waves," my friend Amara said of her post-childbirth emotional state, "these moments of eeriness and being overwhelmed at the very idea that I've brought this human into the world and now have to do my very best and nurture her into her life, and the responsibility of it all."

Some of the main contributors to a woman's disposition during and even after pregnancy are hormones. During your partner's pregnancy, her levels of estrogen and progesterone increase substantially. These are largely responsible for the shift in moods you both might have experienced during the previous nine months. Once she delivers the baby, though, those levels drop rapidly back down to their normal pre-pregnancy levels. This dramatic change plays a large part in the postpartum depression (PPD) your partner might experience over the next few months. She will appreciate your sensitivity to this matter and, moreover, your ability to assist with daily child-care responsibilities as you both transition into this new role. Be patient with your partner as she adjusts to being a mom. As men, we are conditioned to find solutions to problems and might try to offer quick fixes to our partner's problems. This might *not* be one of those times. You might find that simply listening to your partner's struggle and validating her feelings will help her feel understood and comforted. If her condition is one that needs medical attention, you'll need to step up and get her some assistance. Encouraging her to speak to her ob-gyn for advice is the first step. She or he can usually recommend a great therapist for your partner to see. It's important not to take this matter lightly.

Many women experience postpartum depression after delivery. In fact, over 950,000 women a year are diagnosed as having PPD.[6] Postpartum depression is suffered by a mother after childbirth, typically stemming from the combination of hormonal changes, psychological adjustment to motherhood, and just plain exhaustion. PPD can occur anywhere from three to six months after birth.

"You don't really hear about the negative aspects of motherhood and how that initial transition can be very hard to deal with," says my friend Daphna, who is a marriage family therapist. "Many women, such as myself, experience postpartum *anxiety* more than depression. There is a component of tremendous fear involved that runs the show because of the newness of the experience, and that can lead to depression. Everything is turned on its head, and fear is so big you feel like you can't function or cope.

"I assumed I would be madly in love instantly, and instinctively know how to feed and take care of this baby," Daphna explains. "How hard can it be? But it was so much more complicated than I could have anticipated. Everything felt so painful, so scary, difficult and out of control. It was as if I had done something with my life and my relationship with my husband that would never shift back. It scared the crap out of me. And that fear was so paralyzing that it impacted the way I was taking care of my child and dealing with my husband, who was no longer my husband, but a dad now. It was a postpartum *reaction*. Everything was so new that it stopped feeling familiar, and it was really hard to navigate."

Most of the time, PPD can be treated with therapy by a licensed therapist, and, if necessary, medication prescribed by a doctor. In some cases, PPD can be a more serious problem, especially for women who have a family history of depression or anxiety, experience severe PMS, felt depressed during pregnancy, had a difficult pregnancy or delivery, have a troubled marriage, lack supportive people in their lives, or are caring for a chronically ill baby.

Keeping an eye out for signs of PPD is important. Some symptoms include mood shifts as well as anger, anxiety, guilt, and/or hopelessness.

Other signs of PPD include:

- Sadness
- Irritability
- Feeling overwhelmed
- Crying
- Reduced concentration
- Appetite problems
- Trouble sleeping

Keep in mind that it is completely normal for your partner to experience bouts of anxiety, weepiness, and irritability immediately after delivery, due to the sudden drop in hormone levels once the baby is born. This state is often referred to as the "baby blues" and is common in up to 70 percent of women who give birth.[7] These symptoms usually fade away within a few weeks after birth.

Rest assured, not every hormone surge is so harrowing. My wife had the complete opposite experience. "I was just in love with my babies," Mer reminds me. "I just wanted to cocoon myself, and didn't want to be with anybody else. I was so happy I had the opportunity to be a mom, and that we were alive and well. It was a really magical time."

Interestingly enough, dude, *you* might also go through a similar hormonal experience. Hormone levels fluctuate in men as we undergo stress. Financial pressure coupled with the new responsibility of taking care of your newborn can contribute to a tremendous feeling of hopelessness and anxiety. Don't let your ego get in the way if you are feeling the pressure of this new role. It's important first to acknowledge if you are feeling overwhelmed and to seek help immediately. Sometimes the simple act of talking about it with close friends who are already parents can put things into perspective for you. Hopefully you and your partner have the skills to communicate openly so you can bring up how you are feeling without stressing or overburdening her with your feelings. Speak up, and encourage your partner to, as well.

CHAPTER 2

The First Few Months

THESE LAST FEW weeks have been pretty crazy with lots of new information and experiences coming at you full force. Keeping the team-player mentality is crucial as a parent, especially when it is all so new. Work out a schedule together with your partner. If your circumstances are such that you are the sole caregiver, call upon whatever family and friends you can to help out so that you get a break every so often.

If your partner is breast-feeding, chances are she will at times feel like a slave to your child's feeding schedule. My wife was exhausted, so as soon as she finished feeding Kai, I took over to burp, change his diaper, and put him to sleep. Having my assistance seemed like a godsend to her.

She found it difficult to even find the time to jump into the shower if I wasn't around. It's the little things . . .

☞ **PARENT HACK**
Sleep when baby sleeps.

Sleep deprivation makes bonding with baby challenging. When baby goes to sleep, mom and dad should try to rest too. Whether it's ten minutes or four hours, take it when you can!

If you've seen the Bill Murray flick *Groundhog Day*, you will understand how similar the first few days after your baby's birth are to that movie. The cycle of repetition is downright stupefying. "You just have to embrace it and try to find the joy in the repetition," Meredith once told me. You wake up and perform the same tasks you did the day before over and over again, so having a partner to share in that experience definitely helps keep your sanity. Tag-team whenever possible.

But in all this madness, have you been able to set aside time to bond with your newborn since you got home? Pretty amazing, ain't it? If not, don't stress about it, but start carving out some time to really connect with your child. Even if it's just before you put her down to bed, or in the morning before you are off to work.

Actually, you might have already begun bonding with your newborn and might not even be aware of it. Cuddling with your baby is one of the most rewarding times spent bonding with your children. Cuddling promotes security, safety, and love. It's important to realize that interaction with your baby is an opportunity for them to learn and grow. The freer you can be in your interaction with your child, the more enjoyable and beneficial the results. Of course, newborns are too young to understand specific words, but easy conversation can lay the groundwork for language development. The tone of your voice can communicate ideas and emotions. Speaking to your baby by describing what you see, hear, and smell around the house and outdoors is a fun way to start this process. Just be simple in your selection of words.

☞ **PARENT HACK**
When the baby feeds, feed mom a snack.

Breast-feeding moms are working extra hard and are often tired. Simple snacks like cheese and crackers, a protein bar, or a smoothie will make her feel better.

Reading to your baby is a good way to bond and set the tone for teaching. Choose black-and-white hardcover books for newborns (since that's what they can see best), and ones with bright, large, colored images like Eric Carle's *The Very Hungry Caterpillar* or the ever-popular *Goodnight Moon* by Margaret Wise Brown for babies that are older, like three months and up. Or if you really want to warp your kid's mind early, go straight to Dr. Seuss and Shel Silverstein. Bottle-feeding (formula or breast milk) is another opportune time to take advantage of bonding with your newborn. Whether you choose to hum a tune, tell a story, or simply take in the moment with your baby, using this time to bond is mutually beneficial.

Welcome to the Daddy Game

The first month after bringing home your baby is consumed with mixed emotions. The excitement and reality of actually being a dad, showing off your newborn to family and friends, and joining the ranks of other parents in your circle of friends are just a few of the positives. But let's not overlook the feelings that are not so easy to endure, like fear and angst. For generations, men have adhered to the tradition of bottling up their emotions, determined to be pillars of strength at all times, and essentially fearing emasculation if we display an ounce of vulnerability or sensitivity.

There are many reasons why we allow fear and anxiety to get the best of us. The uncharted waters you are about to encounter as a dad is amazing, but frightening as well. It takes time to adjust to your newborn's rhythms. I was a wreck the first few nights after we came home with each child. The first is always the scariest, because it's all so new. Juliette wasn't so bad, though I was completely and ridiculously unprepared, if you recall. But Cole was a preemie and was hooked up to monitors his entire three-and-a-half-week stay in the neonatal intensive care unit. There was a monitor that alerted us if his breathing stopped. The alarm would sound so loud when it went off it scared me shitless. And it went off quite often, so shitless I remained. Yet when it was finally time for Cole to come home I had grown so used to the monitor's annoying yet reassuring shriek that I wanted to bring the damn thing home with us!

☛ **PARENT HACK**
When baby gets clean, so does mom.

While mom is feeding the baby, set up a shower or bath for her. That way, when you change the baby, mom can clean up as well. This is where you get to be a rock star!

The one thing that kept me sane during all the worries and woes of parenthood was my wife. Sharing my fears with Meredith made me realize how off the deep end I was allowing my fantasies to take me. She would bring me back to reality and help me find my confidence as a father. The times when even she couldn't make me feel secure, I reached out to our circle of doctors, nurses, and other parents for advice. The one thing I learned is to take initiative, seek answers and help. There will be times you want to pick up the phone to call the doctor because you are worried about a cold your baby just picked up. Hey, if it calms you, then give him a call. That's why doctors get paid the big bucks. But you don't need to call the doctor *every* time your kid sniffles. Kids get sick. They just do.

Attend as many pediatrician visits as you can so that you can ask questions about what's in store the next few months. The more proactive you are, the better you will feel. Get to know your baby's pediatrician so you will feel comfortable calling as many times as you see fit to ease your anxiety. As a parent, you will always be vulnerable to fear and angst, among many other emotions. Make a plan of action with your partner ahead of time so that if and when the time comes that your little one is in need of medical care, you will be prepared and spare yourself the unnecessary stress and fears that arise from making that dreaded trip to the emergency room. Preparation is a main ingredient for successful parenting.

Introducing Baby to the World

From the moment your baby is born, you will be inundated with visits from family and friends. As exciting as this is, it can get a bit overwhelming. Not to mention the fact that you will want to protect your newborn from germs and viruses that might be circulating among the crowd. So you'll need to strategize how to handle expected and unexpected visits from family and friends. This is another conversation to be had ahead of time with your partner so you can both be on the same page. It might even help to discuss this with family and friends prior to the birth of your newborn. The last thing you want is to encounter those awkward moments of turning family away when the timing is not right.

Depending on if your parents, in-laws, and immediate family live in town or not, you might have to accommodate incoming family or at least have to host some early visits during a time when you'll be exhausted and hardly in entertaining mode. So addressing visitation times with incoming family and guests is best done *before* you are crazed and in the thick of newborn hysteria. Today you have the advantage of contacting everyone at the push of a button via email, Facebook, or any social media platform you like and giving the world a heads-up that you are going "offline" for a few. Pop-ins can be painful, so try to head them off at the pass.

☛ **PARENT HACK**
Baby got burned?

Soak a cloth or paper towel in vinegar. Apply it to the burn until the area feels cool. It will ease the pain and prevent a blister, baby!

If your relatives can afford to stay in other accommodations, that would be ideal. Much less chance you will murder them in their sleep. Exhaustion, irritability, and hunger often play into the dynamic when it comes to those first few weeks of parenting. Having your space invaded with no ability to escape is challenging and uncomfortable for any new parent. You can always suggest to your relatives that they'd be more comfortable in their own space with privacy at a local hotel or Airbnb. You owe it to yourselves to be that specific, especially if your family dynamic is a bit much to handle.

Try to find visit times that work for *you* so that when your parents or in-laws come to dote on your little one you two can take a little break. You can use these visits to catch up on some much-needed sleep, or try to organize your newly chaotic schedule of baby-care needs. If your family and friends are willing and able, put them to work, dude!

Big Brothers and Sisters

Throughout all of this, make sure any other siblings are getting attention too. Jealousy is very common among new brothers and sisters. Prepare your other children for the big day their new little brother or sister arrives. Make them feel special when they come for that first visit. We had Kai bring Juliette a stuffed animal as a welcome gift from him when he came to the hospital to meet her for the first time. The smile on his face when he presented his sister with her toy was magical. He was eager to touch, hold, and play with his new sister. While it might seem daunting at first to let a toddler handle your newborn, you will need to find a way to creatively have him feel empowered and involved.

Meredith was great at introducing Kai to Juliette. First, we explained the rules of washing hands so he could help to keep his baby sister clean. Having Kai touch Juliette's toes and hands while Meredith was feeding her was a great way to get him involved without handing her over completely. Once Juliette was fed, changed, and swaddled, we had Kai get into Mer's bed. I sat myself right next to him as we ceremoniously handed over Juliette to her big brother Kai.

Be mindful that there is the possibility your other child will feel hesitant at first upon meeting his sibling. That's normal. Allow your child to take his own time to adjust. The last thing you want to do is force him into an uncomfortable situation. Talk to your older child about becoming a big brother or sister. Honor his feelings and let him know it's all right to take his time getting used to this addition to the family and the new dynamic. Try asking him to make artwork to display in their sibling's nursery. Be sure to include alone time with the older sibling when you bring your newborn home so that he can feel just as important as his brother or sister. Inviting him to watch you feed, change the diaper, or even rock your newborn to sleep will make him feel a part of the process. Just keep in mind that it's normal for him to have uncomfortable feelings as a newly minted sibling. After all, that's why the term *sibling rivalry* came about. The best you can do is to communicate clearly and simply with your older child about how important he is as a big sibling, and how important he is to you. As a helper, he will feel a part of the experience and most likely feel more welcoming to having a brother or sister.

Entering the Friend Zone

Introducing family to your baby and your new lifestyle is one thing, but friends can be a little more complicated. Some people are just not comfortable with babies—not yours, not anyone's. And other friends will simply not be comfortable with you becoming a parent. They won't say anything, but you'll start to get a feeling for which friends are truly down with all this daddy business and which ain't. The ones who are into it can be helpful during these early days. The ones who seem weird about it all, you can reconnect with later.

There are also always friends and family who are delightful to have around, but just don't know when it's time to leave. They don't read the signals or take the hints, as well-meaning as they may be. There is no reason to have visitors linger when you are exhausted and yearning to catch a few moments of rest. Having a plan that addresses this issue will help to avoid unnecessary stress. A little trick Meredith and I used when visits were getting too long was saying a code word to inform the other it was time to encourage our guests to leave. All she needed to do was say "red dog," and I knew it was time to move 'em on out.

Baby's Best Friend

Speaking of man's best friend, what about the furry creatures in your life? If you have pets in your household, you will need to take the proper precautions to prepare them to meet your newborn. Establishing boundaries from as early as the pregnancy months is recommended. Training your pets not to enter the nursery, establishing a quiet place for them to retreat if there is overstimulation once you bring your newborn home and, most important, keeping their food away from arm's reach once your newborn enters the toddler phase are just a few of the ways to prepare your home for your newborn and pets. Bringing home a blanket or piece of clothing from the hospital, so that your pet can get familiarized with your newborn's scent, is a good tip. If you feel your pet is a potential threat, then you will need to involve a trainer or consider other options like having it stay with family or friends until your newborn is older. No matter what, it's not acceptable to leave your pet alone with your child.

Late-Night Calls

Establishing a plan for those much-dreaded, middle-of-the-night doctor calls is worth doing. Bring this up at your initial visit for your newborn's checkup. Ask what the procedure is for emergency night calls so you can feel prepared once that time arrives. Prepare an emergency contact list on your fridge or in your smartphone with numbers to your pediatrician, poison control, and relatives who can be contacted to help out. Try apps like Heal, which connects you to a doctor right away.

Get in the habit of washing your hands before you attempt to handle your newborn, especially after playing with pets. Meredith strategically placed bottles of Purell around the home so I never had an excuse not to be prepared to offer assistance when needed. Specifically, having a bottle placed at the entrance of your home and using it to greet your guests is a good idea. It might seem a bit neurotic, but they'll get over it. Peace of mind and your baby's health are the goal here. You also want to be sure that every guest is free from colds, sore throats, and the like. Most people are clueless when it comes to being sick around a newborn. In this case, *you* are the guard dog.

Sleep and the Lack Thereof

Sorry to break this to you, dude, but sleep? . . . it just ain't gonna happen for a while. Oh, how we unknowingly took for granted the one act that we long for the most once a baby comes into the picture. When I think about the numerous all-nighters I managed to pull off in college and during my bachelor days, I don't know how I ever survived them. Today you couldn't pay me enough to miss my eight to ten hours of sleep. So if you're reading this before your kid comes into the picture, get all the sleep you possibly can now, because once your little one shows up, you're in for a real treat, my zombie friend.

Let's talk about how we can make the best out of a tough situation. First off, planning *how* you are going to handle the night shift with your newborn is important. If your partner is breast-feeding, hopefully you can get some extra zzz's in while your newborn eats. But once feeding is finished, stepping in and taking over to burp, change the diaper, and put baby back to sleep is very noble and highly advised. Again, tag-teaming in this manner will benefit you both. The more time you give your partner to rest, the better her body will respond to producing milk. Make sure to allow yourself to get some shut-eye too while she feeds, as it'll put you in a better mood once your shift sets in.

Meredith and I were ardent about making sure our firstborn Kai was with us at all times, including bedtime. We chose to have a bassinet placed right next to our bed so we could stand watch and be alerted to any issues that might arise during the night. Eventually, as he grew out of the bassinet, we opted for a co-sleeper. It attaches to your bed so your baby can have their own space yet still be at arm's reach. For our second child, we chose to have her sleep in her crib from day one. It just made life much easier, since we now had two under the age of two to deal with. Interestingly, Juliette slept longer and better from the start than her brother Kai. (Today, however, they both are talented at sleeping till midafternoon if left uninterrupted by trivial things like going to school. How times will change!)

Establishing a bedtime ritual is smart and recommended. Turning down the lights, lowering your voices, playing relaxing lullabies, all contribute to setting the tone for sleeping. Reading a book and rocking your little one to sleep was one of the ways I enjoyed putting Juliette to bed. Kai, on the other hand, was not as easy. I remember numerous times lugging him in his car seat for a ride up and down San Vicente Boulevard until he managed to fall asleep. Back then, that was the thing to do among my parent friends. I laugh now thinking about how I'd see fellow parents driving the opposite way in the hopes their little one would be fast asleep by the time they reached the driveway. It was pretty comical. Once he finally managed to fall asleep, the challenge was getting him out of his car seat and into his crib without waking him up. You'll figure out what works best—car ride, a walk in the stroller, a bouncy chair, or another trick that puts your little one to sleep (and will eventually become a tip you pass on to other newbie dads).

Massage

If you mastered the art of massage while your partner was pregnant, you can now test your skills on your newborn. Babies respond positively to touch when applied in a calming, relaxing way. Lightly massaging your newborn can help stimulate the production of serotonin, which ultimately calms and soothes your baby. It's best to pick a time to massage when your baby is calm. If you attempt to massage her when she is fussy, you might be overstimulating her instead of relaxing her. Using baby oil, or even olive oil, lightly rub your infant's shoulders, back, legs, and arms (being sure to avoid the spine). Turn down the lights, pick some soothing music, and let your fingers do the walking. To perfect your skills, log onto www.infantmassageusa.org to find a class taught by a certified educator or infant massage (CEIM) practitioner near you.

Ideally, setting your baby up in his own room from as early as you feel comfortable is the way to go. There is no right or wrong way when it comes to sleep. Some friends I know insist on their baby sticking to a sleep schedule no matter what. That's fine, if that's your decision. Hell, I knew one couple who hired the *actual* Baby Whisperer herself to whisper their baby to sleep. Now, that seemed a bit much to me, but to each his own. Others prefer their newborn to adapt and sleep while out and about, so they don't need the comforts of their home surroundings. You'll need to decide what works best for you. And if you end up having more than one child, you will have the knowledge from your first and apply it to your second and/ or third, provided you haven't lost your mind and memory (and keeping in mind that babies have their own individual preferences too). Hang in there. Before you know it, they will be sleeping well into the afternoon as teenagers, and you'll have the opposite problem on your hands.

When the Poop Hits the Fan

Diaper explosions are a horrific and sadly unavoidable part of parent life.

A diaper explosion can best be described as just that, an explosion of poop that essentially ends up all over your child either from an incorrectly applied diaper or a bad case of the runs, or both. Whenever one of our children encountered a DE, I usually ended up throwing away their stained clothes while hosing them down in the backyard.

☞ **PARENT HACK**
Vomit/Poop Cleaning Tip #1

Clean up everything you can first. Then make a paste of baking soda and water, and spread liberally on the mess. Let it dry overnight. Vacuum it up, dude!

Making sure to properly clean their car seat, baby carrier, or stroller is very important. Those items are way too expensive to disregard if you can clean them off. Make sure to do a thorough job, though, as you don't want any residue!

But dude, don't be a cheapskate either. Sometimes, once something has been infected, shall we say, you just gotta throw it away. Sure, it was painful to toss some of my children's nice clothes at times, but it's not worth them catching some infection, or just living in smelly, soiled clothes. Buying those inexpensive one-pieces sold in packs makes throwing them away sting a little less.

Going on the Go

Another art form to master is dealing with the deed while you are away from home. Be sure to stock up on extra clothes and wipes in case of accidents. One of my biggest phobias is public bathrooms and kids. Public bathrooms are like walking into a germ museum. Not to mention that some men's public restroom stalls can be downright *disgusting* and completely unfit for a child in every possible way. It's one of the few times I hate how curious kids can be, since they tend to touch everything in sight. I usually try to get in and out as quickly as possible.

Having a portable potty seat that folds is a godsend, coupled with a plastic cover so that you can keep the potty seat clean and discard the plastic covering once your child is finished. If not, then clean the living hell out of the stall toilet seat before you place your child on top of it. On a side note, I have gotten really good at holding my breath. I would like to say those days are long gone, but I go to enough sporting events and concerts to know it's a talent that lasts a lifetime.

☛ **PARENT HACK**
Vomit/Poop Cleaning Tip #2

If you don't have a good stain remover on hand, just MacGyver that shit! Combine one part Dawn dishwashing detergent with two parts hydrogen peroxide. Works best for poop stains on clothes, puke stains on carpets, and random juice and pee stains.

Finding a restroom with a changing table can be a hassle and a royal pain in the . . . rear . . . when you are left with no assistance to change your child. Meredith and I once got yelled at for changing Kai in the booth at a restaurant we were having lunch at in New York City. No one was around us at the time, so we felt if we quickly changed his diaper it wouldn't be an issue. Well, the manager didn't feel the same way and lost *his* shit on us. I asked him what he expected us to do, since they did not provide any changing tables in either of their bathrooms and there was no way in hell I was going to put our child on the bathroom floor. We made sure to cover the area we used to change

Kai with one of his blankets, as well as the changing pad
from our diaper bag. Let's just say, we never went back to
that restaurant again. Thankfully, today times are changing and more restaurants are becoming more family-
friendly, but it would serve you to find out ahead of time if
they do have a changing station in case the need arises.
Better safe than sorry.

Cabin Fever

I've always had a hard time sitting still so, for me, being
cooped up indoors is torture. The first few weeks at home
with each of our newborns was definitely a test of patience
and will. To go from your daily rituals—work, exercise, or
errands—to putting everything on hold to focus on this
new dynamic in your life takes a great deal of tolerance
and fortitude, coupled with the unbridled love and excitement for your new baby. Preparing for the possibility of
cabin fever will help you to avoid the stress and strain on
your relationship as well as your inner emotional self, not
to mention the overall emotional state of your household,
which can negatively affect your little one.

"I was shocked that my husband was able to read the
newspaper at breakfast so soon around our first baby," re-

calls a recent new mom. "I was constantly gazing into our baby's eyes, astounded by this new creature, and I'd look over and he's drinking coffee and reading the newspaper. I was like, *Oh, my God, how could you be doing that? How could you not be staring at this baby 24/7?* It took me a little while to realize that's what *he* needed, to carve out a slice of his own time amidst the chaos of the new family life. For fifteen minutes, he enjoyed his coffee and paper and was in his own world. I realized he was actually a great example for me, and that I needed to find the time to do the same thing."

In an ideal world, you and your partner would have a chat about cabin fever during her pregnancy. But if that conversation got left by the wayside, as many do, you will need to take the driver's seat and navigate this potentially rocky road. Create a schedule (by the hour, day, or week) where each of you take some time to regroup and reset. It will be difficult at first to feel entitled to your alone time, especially for mom, since your newborn is largely dependent on her if she is breast-feeding. You'd be surprised how a simple walk around the block or a drive through the hood can help you reboot and ultimately be a better partner and parent.

Once I began to get back into my work schedule, Meredith started to get a taste of cabin fever. Soon she was commenting on how I got the joy of interacting with the outside world while she was cooped up inside tending to the daily rituals of feeding, diapering, and putting Kai to sleep. I was baffled as to how she came to this assessment. I saw my work schedule as stressful and all-consuming, especially since I felt guilty spending time away from our newborn son. In hindsight, I understand how she came to this determination. I took for granted the simple, everyday interactions with the mailman, the neighbor, or the local barista. I soon realized how challenging it must be for my wife to have sole responsibility while I was at work. Especially for a first-time parent. Addressing this as early as possible will help eliminate feelings of resentment and frustration with each other. Establishing a relief schedule will help even more and perhaps even avoid cabin fever altogether.

Breast-feeding in Public

Breast-feeding in public has become quite the controversy over the years. Today there are solutions to those wandering eyes that make moms feel uncomfortable with the natural act. Bebe au Lait is one company that is leading the pack. They have a great selection of nursing covers for your partner so she can tend to your little one wherever she may be. This is another topic that's subject to personal preference, but my two cents is that nursing is a natural act, and not an offensive one. A little tolerance goes a long way.

Feeling cooped up, or even trapped, can happen in a relationship even without bringing kids into the picture. Giving your partner some downtime to reboot, be it a trip to the nail salon or a walk around the neighborhood, is a great way to be that team player.

You don't want to find yourself getting into a competition with mom over who does more work with your newborn. You *both* will be doing a lot! But you *both* need to have some downtime as well. So have a conversation about this beforehand. Remember, supporting your partner in taking some time to take care of herself is part of parenting as well. Parenting is a round-the-clock occupation. The ability to disconnect once in a while is valuable so you can be complete and available when the time comes for you to reconnect. Give her a night off to reboot, dude. A change of environment is important and useful. Just as connecting to your child is important, the room to recharge can be just as valuable.

Complications: Jaundice, Reflux

If I had a nickel for every time I complained to my parents about my headaches, stomachaches, or shortness of breath when I was kid, I could probably retire now at the ripe old age of forty-four. I am amazed at the patience and calm my parents displayed as they persevered to get to the root of my constant array of minor maladies. I'm happy to report that all of my ailments were easily treated and eventually disappeared as I matured through adolescence. Despite the numerous tests and doctor visits I endured, not once did I sense the fear or stress my parents might have been undergoing. Their ability to maintain their composure and strength was a standard I strive to achieve in my own parenting journey. But I am the first to admit, it ain't easy!

Today, as a parent of three, I've experienced my fair share of medical scares with my children, beginning pretty much as soon they were born. Kai showed up three weeks early, weighing five pounds fifteen ounces. Not to be outdone, Juliette beat him by a week, making her debut four weeks early, weighing in at four pounds eight ounces. And then Cole trumped them both. At thirty-two and a half weeks, he entered this world a lightweight at four pounds three ounces. *Bam!* While all three children were born healthy and strong, the circumstances that contributed to each child's early arrival left me feeling scared, hopeless, and out of control.

Cole's premature birth in particular was the most impressionable. Having gone through the birthing process twice before, we knew that the moment you come home with your child is memorable, to say the least. This time, the fact that we came home empty-handed was unsettling. We actually felt like we experienced a loss. Though Cole was under the best possible care he could receive in the hospital, our hearts were heavy having to spend evenings away from our precious newborn. Not to mention the stress of having to explain to our two other children that their new brother needed an extended hospital stay. (According to the Mayo Clinic, only 12 percent of pregnancies result in premature birth.) Support from family and friends, especially when the unexpected occurs, is key during times like these.

Jaundice

One condition that can occur after birth is infant jaundice. Jaundice is a yellow discoloration in a newborn baby's skin and eyes. Infant jaundice occurs because the baby's blood contains an excess of bilirubin, a yellowy pigment of red blood cells. Infant jaundice is a pretty common condition, particularly in babies born before term (thirty-eight weeks' gestation) and in some breast-fed babies. In infants it usually occurs because a baby's liver isn't mature enough to get rid of the bilirubin in the bloodstream. In some cases, an underlying disease might cause jaundice. When Juliette had it, the doctor recommended exposing her to sunlight. It sounded weird, but she ended up undergoing a few sessions of light therapy, which was totally noninvasive and painless, and it worked!

Reflux

Reflux is another issue that can create quite an alarm in both new and veteran parents. Happily, it's quite treatable, and once diagnosed by a doctor it can usually be alleviated with a few simple measures.

For infants:

- Elevating the head of the baby's crib or bassinet
- Holding the baby upright for thirty minutes after a feeding
- Thickening bottle-feedings with cereal (do not do this without your doctor's approval)
- Changing feeding schedules
- Trying solid food (with your doctor's approval)

For older children:

- Elevating the head of the child's bed
- Keeping the child upright for at least two hours after eating
- Serving several small meals throughout the day, rather than three large meals
- Limiting foods and beverages that seem to worsen your child's reflux
- Encouraging your child to get regular exercise

If the reflux is severe or doesn't get better, your doctor might recommend medication to treat it.

CHAPTER 3

Your New Life

YOU MIGHT BE excited and amped about becoming a dad for the first time. Or maybe you're scared out of your wits about this new journey. Both are completely justified reactions. "The process of parenting is finding your confidence and alternately being scared shitless," my *HowToBeaDad* blogger friend Charlie stresses. "You want to do a good job, you know a lot is riding on it, so you rise to the occasion. Finding out I was going to become a father, and then being there for the birth and being a participant in that process, set me up to say, 'I'm in and I'm going to do whatever it takes.' The fearlessness came from a willingness to experience anything."

Of course, your schedule gets more complex when kids

come into your life, but that's just par for the course. Sometimes I reflect on my life pre-kids and think how much simpler things were back then. But compared to life now, you couldn't pay me to go back. My world is definitely more fulfilling with a family than it was without, but it's not without its chaotic moments. When life inevitably gets hectic, you need to figure out how to manage yourself amid the chaos.

"If only there were more hours in the day" is a phrase I say all too often, especially these days with three active teenagers. Finding the time and energy to keep up with your work, your relationship with your partner, and your relationship with your child can seem daunting at first, but it's necessary. And it gets even more challenging when you add a second or third child to the mix. So having an open and positive attitude keeps you on track and allows you to enjoy this journey even more.

I'm not sure you ever find that perfect balance between family and career, but the choice is yours on how you choose to handle things when they get to be too much. You choose how you react to stress. If you react poorly, then it's time to start working on it. For instance, when I'm on a deadline for work I tend to get really anxious or "snappy" (as my wife would say). Until I finish my projects and my deadline is met, I am told I am not the most pleasant person to be around. But it wasn't until she brought this to my attention that I actually made an adjustment in my attitude.

☞ **PARENT HACK**
Teething baby?

Put a dash of apple sauce on the middle of a cloth (not a paper towel), and then roll it up and freeze it. Remove it from the freezer and let baby suck on that tasty treat till the pain goes away.

An important lesson I've learned is to actively take steps to reduce my stress through efforts like exercise, meditation, or by simply being aware that I am under pressure at the moment and can be easily set off. I make it a point to run at least three miles up to five times a week just to keep my sanity when I'm not training for one of my marathons. If I don't take the time to get my run or workout in I end up cranky and my family calls me on it. I owe it to my wife and family to take care of myself so I can be in a better place for them, and they know me well. You'd be surprised how the simple acknowledgment of how you are feeling in the moment can really make a difference. How you choose to behave based on this information is another story. For me, I value my family relationships. So if my work gets put on hold because my son wants me to throw some baseballs, then I'm going to choose to spend time with him.

Ferris Bueller said it best: "Life moves pretty fast. If you don't stop and look around once in a while, you could miss it." Don't be the guy who missed it.

Bonding over Daily Tasks

Have you ever passed a pet store in your local mall and observed the adorable golden retriever or Labrador puppies? You can't help but notice yourself desperately trying to connect with them. Without even realizing it, your voice rises three octaves and you start acting goofy and making funny faces in the attempt to get attention or a reaction from those captivating little buggers. The feeling those innocent little puppies evoke from you is a lot like how your child makes you feel the moment he enters your life.

Allow yourself to connect to that feeling those puppies incite in you as you begin to explore your relationship with your newborn. We are so often distracted by the seriousness of life, work, and relationships, that we bury that goofy, childlike sensibility deep down inside. Fight the urge to judge your behavior when it comes to interacting with your little one. Allow yourself to make silly faces, speak in silly voices, and say nonsensical things to your newborn. Getting to the point of feeling free and connected is what's most important. There is no right or wrong way to achieve this.

☞ **PARENT HACK**
Baby likes to play with his dirty diaper,
leaving a mess?

Get a pair of footy PJs and cut off the feet. Put the PJs on baby backward, with the zipper in the back. Now your little one will not be able to get to his diaper and make a mess.

When you're changing diapers, feel free to sing a song, make it a game, or tell your newborn a story. I used to playfully explain each and every movement in the diaper-changing process as it was occurring, so my kids would know what was happening.

"Let's take off that silly, dirty diaper. Can you say, 'Bye-bye'?"

(Wipe as baby is waving. Toss diaper in hamper.)

"And now we are making Mr. Poopy go away! Can you say, 'Bye-bye, Mr. Poopy'?"

(Baby waves. Bang—a smile.)

Well, something like that.

Simple interactions like that begin to establish a bond between you and your newborn. One of my favorite moments of bonding with each of my kids was bath time. Making sure the water temperature was not too hot, I would chill out in the tub with Kai on a bath cradle that allowed him to float on top of me while enjoying the relaxing warm water. This is another great moment to share a story, sing a tune, or even take in some classical music. And let us not forget the power of a rubber ducky (thank you, Ernie from *Sesame Street!*), submarine, army diver, or even Nemo.

Taking walks, wearing your baby on your chest, bottle-feeding, reading books, and bedtime rituals are all great times to establish as your personal bonding time with your baby. It might take a few weeks to establish your rhythm, but once you get the hang of it, it becomes second nature.

Parenting Styles

If only we could all agree on a parenting style that would create the same outcome: a world full of compassion, love, generosity, and support. I have faith that the up-and-coming generation has this same desire. Just think if decades from now a generation can look back on us as the ones responsible for starting a movement of parenting that makes life better for our children. It's all in the way you decide to help mold your little one along the way.

There are many parenting styles, but psychologists believe they fall under three main categories: authoritative, authoritarian, and permissive.

Authoritative parenting is a strict style. Parents enforce rules and expect their children to follow without any hesitation. Failure to follow usually results in punishment. This type of parenting leaves no room for question or challenge by the child. I would categorize my dad as an authoritative parent. Whenever I would question or challenge his rules with a, "Why not?" his response would be, "Because it's the law" or "Because I'm the father." I know this sounds a bit harsh, but he actually said it in a way that would make me laugh.

Authoritarian parenting, on the other hand, leaves room for explanation and understanding when rules are challenged or questioned. This type of parenting is more supportive and nurturing toward the child, as opposed to the punitive, authoritative style. I strive to be more like this type of parent, especially since I was raised more from an authoritative background.

Permissive parents are the type who deem themselves more as friends to their children rather than authority figures. They usually have very few demands to make of their children. Rarely is there discipline in this type of parenting. We used to have neighbors whose son would call his parents by their first names at two years old. While I found that hilarious, the boundaries were definitely blurred between parent and child. "Mom" and "dad" were not even part of their child's vocabulary. Instead he called them by name, Jeff and Carrie. When I observed their interactions, it was clear to see how there were no rules to play by, which is confusing to children. While it appeared that Jeff and Carrie were really laid-back, at the end of the day, they lacked control and direction. As parents, it's our job to guide and direct even when our children don't like us for it. Today, every once and a while, my daughter Juliette attempts to call me by my first name instead of dad. Sometimes I call her out on it and other times I simply ignore her and she gets the picture. Secretly it cracks me up, but as a parent I need to stick to my guns in order to be taken seriously.

A more recent type of parenting has come into the picture, called uninvolved parenting. This parent is neglectful of the needs of their child and usually unaware of key players in their child's life, like teachers and friends. This type of parent might be heavily focused on career and have hired help such as a nanny to replace their role and raise their child. This parent is usually distant emotionally from their child as well. We all know a few parents like this. It really sucks when it's one of your friends too, whom you otherwise respect, but you can tell they don't care much about parenting or just prioritize themselves over their child. Not a fun place to be if you're the child.

☞ **PARENT HACK**
Worried baby will fall out of bed?

Many kids get squirmy when they sleep. Try placing a pool noodle under a fitted sheet to make a barrier. This should keep your kid from falling out of bed, which will let you sleep easier.

A more specific and immediate style to add to your parenting choices is attachment parenting. Dr. William Sears's *The Attachment Parenting Book: A Commonsense Guide to Understanding and Nurturing Your Baby* breaks down the principles of attachment parenting quite well. He explains how this method encourages a strong early attachment to your child and advocates a parental responsiveness to your baby's needs. Attachment parenting seeks to encourage parents to establish a secure, nurturing connection to their baby so, in turn, their child will grow into a strong, secure, and independent child. Some examples of this type of parenting include baby-wearing and skin-to-skin touching. Positive discipline and constant, loving care are also characteristic of this parenting style. Nighttime parenting is also encouraged. Co-sleeping is part of this style and something Meredith and I took advantage of when our first was born. The great thing about this type of parenting is that dad can be involved in most every way possible.

"With Kai we were really into attachment parenting," Mer recalls. "I wanted us to be present, and make sure his needs were being met so that he could feel safe in the world and move on to the next step of development. Being attentive and in tune with the baby is very important at that early age, because they are not communicating beyond crying and cues, so you don't really know what they want. You have to be open and learn."

Three . . . Two . . . One . . . Go!

You need to be on top of your baby's every move once he becomes mobile. It won't be long before your baby will be crawling, scooting, or running around your house while picking up anything and everything he can get his hands on, with the end goal of putting it into his mouth. Meredith, who is a germophobe, was flying to New York with our youngest, Cole, when he was two years old. She spotted Cole playing with the tray in the seat back in front of him. She leaned over and very calmly explained that there are germs on the tray and we want to make sure you don't put your hands in your mouth after touching said tray. Cole looked at her and in a matter of minutes dismissed her request by sticking out his tongue and licking the tray. Kids. Gotta love 'em.

A recent offshoot of attachment parenting is helicopter parenting, a style of parenting where the parents hover over the child and monitor every move. They fear the child is not capable of making decisions on their own. Some researchers claim that technology makes helicopter parenting more common, since parents can have nonstop access to their child, thanks to texting and cell phones. Kids become more dependent on their parents in a way that stifles independence and emotional growth. While helicopter parenting is more commonly applied to parents of high-school- and college-age students, it's good to get it on your radar now so you can avoid being guilty of piloting that chopper.

RIE (Resources for Infant Educarers) was founded by Magda Gerber and is yet another style of parenting. Respect is the basic principle of RIE. RIE encourages parents to treat their baby as a unique human being, not as an object. By observing your baby, you learn what they need. Parents who practice the RIE method are encouraged to look and listen closely before responding to their child's needs. An example of this would be not to jump in when your baby is fussy or crying, but to observe and discover what she's trying to tell you. During caregiving times like diapering and feeding, you encourage your baby's involvement. RIE encourages your child to participate instead of being a passive recipient. Time for uninterrupted play and freedom to explore is also a basic principle of RIE. Trusting your baby's competence is encouraged. By trusting the little one to be an initiator, you learn what she is ready for.

"Helicopter, free-range, there's so many titles. How would I describe my style? Dive-bomb parenting, maybe?" my blogger pal Charlie says with a laugh. "Accepting your own work as a parent is hard. It's something we should all be working on, but it's ongoing. Parenthood has been around since it all started and we're still trying to wrap our heads around it."

While it's interesting to read up on all these different styles, I actually question the idea of labels and subscribing to any one particular way of parenting. Try going into your new role with an open mind. Sure, mistakes will be made. There is no one right way to be the "best" dad or mom.

"I find the things I need to work on the most are the things I get the most frustrated with about my kids," Charlie adds. "It's an ongoing process to learn to work on the things in my own life, which are the same things I need to work on as a parent."

Trust your instincts. When the end result leads to creating a family filled with love, interaction, and discovery, then chances are whatever path you chose was the right one. "We had to learn along the way from things that worked and didn't work," Mer reminds me. "The first few years you're not going to see the consequences of your choices, but as they grow up (or you have more children) you start to evaluate your methods and adjust your approach."

Dudes 'n' Dads

If you're married, I'm sure you can relate to the shift you have experienced in your "dude" relationships. The spontaneous meet-ups at your favorite dive bars are less frequent, since you now have your spouse to consider before heading out to hang with the guys after work. While you can still maintain a strong bond with your buddies, your priority has naturally shifted, since there's two of you to consider now. Once your little one enters the scene, this dynamic becomes even more complex. Your free time becomes more and more scarce as you assimilate to your new role as a father.

☞ **PARENT HACK**
Sick of cutting everything into tiny pieces with a knife?

Try using a pizza cutter! Roll, roll, roll—done!

The challenge of balancing personal time, couple time, work time, and family time can seem impossible at first. The first month was a huge blur for me. The last thing on my mind during it all was a social life. But as I got the hang of my new routine, it dawned on me that I did indeed have a life outside of the house that it might be nice to connect with again. Yet would my friends even want to hang with me anymore? Would they wince if I brought the baby along? Could I even get out of my house without the baby?

Dad's Night In

If your buddies are close and meant to be in your life, they will stick it out. Most will. If not, you'll find new friends along the way. If the opportunity presents itself, you might even want to give your friends a heads-up that you'll be MIA the next few months as you figure out your new routine. Just because you're not available to hang for a little while doesn't mean your social life is totally kaput.

Life does not end when you become a dad, dude. You adjust. You get family and sitters to cover you, so you can get out and hit the town once in a while.

"At first, you go out less because the baby is so young, and don't even really wanna leave," my coauthor Frank Meyer says, "but after a while you just pick your nights out more carefully. At some point, I just had to come to the realization that I wasn't going to be at every party and cool concert anymore. But that's okay. I went to plenty in my heyday, and there will be plenty more down the line. You just ease up for a few years."

As life moves forward, you will be amazed at how quickly you form new relationships with fellow dads you meet at the park, on a walk, or while running errands with your little one in tow. Don't be weirded out if you are an introvert and have a tough time meeting people. Let your baby do all the work. Next time you're out with your little one, if you pass by another dad, give him a friendly nod. Before you know it, you'll be chatting away about sleep patterns and junk. (Athomedad.org is a good resource for dad meet-up groups.)

"I've met many of my closest friends over the last few years at kid functions, playgrounds, and birthday parties," Frank adds. "It's not like you bond with every single parent you meet, but you definitely spot the like-minded ones from across the field. *Hmmm, looks about my age, wearing a Black Flag T-shirt, kid doesn't seem like a nightmare...* Soon enough you are chatting with some stranger and might have met a new friend. Or at least someone cool to pass the time with during daddy duty."

Who's Watching the Baby?

They say it takes a village to raise a child, and while it is not entirely clear if that old proverb originated in African or Native American culture, both cultures value the importance of community when it comes to raising children. Western culture is quite different from Native American and African tribes, but this phrase is still very much applicable and valid. Bottom line, no matter where you are, this parenting stuff is hard and you might need some help!

Finding the right day care and figuring out the right age to start is a lot to digest. While some experts suggest waiting until your child passes the age of eighteen months, starting off earlier can work too, since children are less aware of the transition. If you're lucky, you can afford for one parent to stay home with the baby while the other goes back to work. But many families do not have that luxury, and need to sort through the options to find the best child-care solution for their needs.

Financial hardship, not being geographically located near family to call upon, or having strained relationships that prevent you from reaching out to them for help might leave you with no other choice but to seek day care at an early age. But, man, oh, man, it's expensive! And sadly our government is not up to speed with the rest of the world when it comes to child-care assistance. While most U.S. states allocate at least three months of government-assisted paid leave for mom, countries like Sweden offer both parents up to 480 days of paid parental leave, with 60 of those days reserved for dads. C'mon, America! Get with it!

Do your homework and come prepared with a list of questions when you are interviewing the day-care staff. A few key questions include:

- How long has the center been in business?
- What are the center's accreditations?
- Are you licensed?
- How flexible are you with pickup and drop-off times?
- Do you supply diapers or is that up to the parents?
- Can parents visit?
- How are the children grouped together? By age?

For ideas of more questions, check out the "Daycare Interview" on Babycenter.com.

Once you make your decision, be sure to allocate time for your child to adjust to the new surroundings. Plan on hanging around for a bit at first to ensure your child feels secure and confident with his new environment. It might take a few days of you staying the entire time, but use that to your advantage to observe how the facility is run. Your best bet is always to lean on recommendations of friends or coworkers who might already use a day-care facility, to feel more confident about your choice

My friend Julie found it was easier to put her son in day care early on, by twelve months old. "Certainly older children entering day care reap the benefit of socialization, but at that age they really haven't developed separation anxiety yet," she explained. "And it helped us get used to leaving our baby with someone else, which was important. It was a home day-care situation and we became great friends, like family."

"My wife was able to take the first year off work, but eventually she too had to go back to work, so we sent our daughter to a local day care," Frank told me. "We really liked the lady that ran it, but she had four to six other kids to tend to as well. Her teenage daughter helped out, but not all the time. Needless to say, at some point, Bella was on a swing unattended, fell, and almost bit her tongue off. She had to get stitches all the way through. It certainly wasn't the caregiver's fault and could have happened to anyone, but I couldn't help thinking that if she had less kids to take care of, she might have been watching and caught the fall."

Keeping the best interests of your child at the forefront of your decision will make both you and your child feel secure about the choice you make. Listen to your gut when it comes down to the final decision. It's tough, I know, but this is just a sliver of the difficult decisions you will be faced with as a parent. Whether it's a relative or a paid employee, finding that village is important and necessary.

I am thankful for the many wonderful teachers our children had the advantage of experiencing in their formative years. Thanks to social media, I am able to still maintain a connection to those who have been a part of our early child-rearing days.

iParent

The baby-tech industry is exploding. I recently exhibited my company Diaper Dude at one of the biggest annual shows in the United States. The show was geared toward the retail industry so buyers of stores in the United States and abroad could come check out the newest products in the baby and child category. There was one item that was showcased that blew everyone's mind. It even went viral, with over twelve million views when a video was posted of the demo in action. The product is called the Pockit Compact Stroller, from GB Child USA. Essentially the stroller packs up so small it is recognized by the *Guinness Book of World Records* as the world's smallest folding stroller! The convenience and ease of this product's functionality is amazing, yet is it really necessary?

It's easy to get sucked into the latest and greatest gadgets on the market. Whether it's the Bugaboo stroller, the latest ultra-modern crib purchased by Kim Kardashian, or the funky mamaRoo infant seat that bounces up and down or sways from side to side, the product selection on the market today is endless. But our grandparents' parents survived without the latest and greatest tech, so my motto is, "If you can afford it and it makes your life easier, then go for it. But if it is going to stress out both you and your bank account, forget about it!" A parent is not considered qualified based on accoutrements.

Baby Gear

Wow, how the times have changed since my kids were born. As exciting as all this baby tech can be, it can get overwhelming. Keep in mind, it's not a competition with other parents to score the latest and greatest, so there's no need to break the bank. But if you are looking for the basics and some style at the same time, check out a few of my favorite brands that are sure to assist, impress, and help pave your way into stress-free parenting.

Of course, I would be remiss not to mention the diaper bag designed for dads, my own brand Diaper Dude. With selections from messenger-style, backpacks, and totes, you will be sure to find a style that suits you. Even better, Diaper Dude is part of the True Dude family. In fact, 5 percent of Diaper Dude sales supports the critical work of our charity partner Futures Without Violence and their program Coaching Boys into Men. True Dude is on a mission to redefine what it means to be a dude in the twenty-first century. It's a great feeling to know your purchase goes toward helping make our world a better place for our children. Check out www.truedude.com for bags and more about our social efforts.

Baby-wearing is the act of wearing your child in a carrier. It is a practice that has been around for centuries and has become more common in the United States in the past few decades. ErgoBaby is a brand I love. Their award-winning ergonomic design keeps your baby safe and comfortable, while staying close to you. Their sophisticated design allows you to wear your little one in multiple positions. With a large selection of styles and colors, ErgoBaby will not disappoint. Check out the store at ergobaby.com for more information.

☛ **PARENT HACK**
Baby is still teething!

Fill an ice-cube tray with formula, breastmilk, or water. Grab the nipple of a baby bottle, put it in the tray, and freeze it. An hour or so later baby's got a popsicle straight from mom.

Shopping for a stroller is very much like shopping for a new car. The brands to choose from are fantastic and the styles are more sophisticated than ever. From the ever-trendy Bugaboo stroller to the athletic-focused Bob strollers, you can search the Internet for one to suit your style. The aforementioned GB Pockit Stroller is one of my favorite new baby devices.

4moms is a brand designed to bring ease to parenting with innovative designs. From infant seats and tubs, high chairs, play yards, and strollers, the 4moms brand is both technologically advanced and stylistically impressive.

Remember, the most important thing to keep in mind is whether or not the product will assist your needs while reflecting your style and keeping you stress-free. For the best selection of parenting gear on the market today, check out Buy Buy Baby. The main thing is to go with your gut. You don't need to buy a bunch of crap you don't need, so keep it simple, dude.

Old-School Fun

A friend once confided in me that he didn't feel skilled when it came to playtime with his newborn. He felt inse-cure and was worried he'd be seen as silly. I was just the opposite. I *loved* being goofy and carefree. But that is just my personality. Don't let your head or ego get in the way when it comes to fun with the kid, daddy-o. Playtime with your child is crucial. It's important early on and when your kids are older. To this day, my youngest still challenges me to a game of *Call of Duty*.

Be creative and have fun with your newborn and/or toddler. Act out a bedtime storybook with fun, imaginative voices, or sing a tune while your infant is having some tummy time. Besides being fun, playtime helps to stimulate the baby's motor skills and social development, doctors say. It's also a good excuse to act like a kid yourself. So the next time you have the urge to pull out your iPhone and check in on Facebook or Snapchat, instead focus on playing a game that will benefit your little one's growth and development as well as your relationship.

Speaking of your smartphone, one of the biggest drags about all this tech is that it takes away from simply experiencing real life, without the need to post, capture, and upload it along the way. For instance, for me, a big benefit to being part of a large family was always having enough players for a game of kick the can. My siblings and I spent countless hours occupied with good ol'-fashioned fun, surrounded by nature. Unfortunately, nature has a huge competitor these days: the Internet. I have mixed feelings about this. While electronic innovations surely are the future and something we all need to embrace and excel at, are we leading our children into a nonsocial world with limited interaction with nature and even other people, for that matter?

There is a legitimate component of stress that comes along with parenting. Finding ways to eliminate stress, or at the very least deal with it in a healthy manner, is a top priority. So it's no surprise when you see a parent reaching for an iPad to distract a toddler so the parent can attempt to finish a task at hand, be it a conference call for work or an everyday chore like throwing in a load of laundry. I even see it happen at the dinner table when I'm out dining with my family.

At first it can feel like a godsend to have your little one behaving peacefully, but are we setting ourselves up for an issue later on when our children grow up with short attention spans and few communication skills? Before the iPhone era, we functioned perfectly fine without handling a crisis the very second it occurred. Do you need to be on call twenty-four hours a day? Everything has to be dealt with immediately? Really? Because it's only been in the last decade or so that we even had that option, and our parents, and their parents, got along just fine without a miniature computer in their hand to text their every thought.

Put Down the Phone

Try to avoid being on the cell phone too much around your newborn. Infants need attention. That little device that is glued to your hand has to go bye-bye for a while. It is a distraction, plain and simple. Sure, you want to stay in touch with the rest of the world, but if you can find it within yourself to put down your phone for a portion of the day, you will be a hero. Your baby does not want to feel like she is in competition for your attention, and neither does your partner.

On the other hand, exposure to computers and electronics at an early age can be likened to the exposure to multiple languages early on. Kids pick up second and third languages much easier than adults. I can't tell you how many friends and colleagues are amazed at how advanced their toddlers are with working their iPhone or iPad. From this perspective, I find it fascinating and beneficial. In fact, I rely on my eleven-year-old son Cole to assist me with anything electronic, since he is a whiz when it comes to that department. He built his own computer at age nine just from watching a YouTube video. In fact, I am hoping to retire off of my son Cole's tech talents in a few more years. (Is that bad?)

Get outdoors with your family every chance you get, dude. Grab a mason jar, fill it with some grass, and catch some fireflies. Just punch a hole in the lid so they can survive. (Rotting dead insects and animals are shockingly ineffective ways to win your child over, for the record.) While we can't deny that our future is going digital, we can choose how much and when to expose it to our little ones.

CHAPTER 4

Strength in Numbers

WHEN YOU START having kids, your relationship gets put to the test. Sure, you will experience all the joy and excitement of starting a family, but that newness wears off once exhaustion sets in, and then suddenly you're questioning this new lifestyle choice. Raising children together truly does test the strength and foundation of a relationship. If you are on the same page about the type of family you want to raise, and what type of parents you want to be, then you're off to a good start. Communicating along the way and presenting a united front to your children is key, so a good move is to have a heart-to-heart talk with your partner about what lies ahead. Perhaps you can

come up with a plan, some ground rules, on how to deal
with those stressful moments that might set you both off.

"It's easy to build resentment, but you have to step back
and see the bigger picture," my friend Adam advises.
"There are days when I come from a hard day at work and
I just want to chill and my wife hands me those kids like
she's handing me a baton in a race. So I get resentful for a
minute, like, 'Hey, why are you passing the kids off to me?
I had a tough day too.' But then I look around and see one
child is trying to jump off the couch, one is trying to beat
himself with a battery from some unknown location, and
the third kid is crying for a bottle. Madness. It's like herd-
ing cats. And I realize, oh, yeah, her day *is* tough too.
Maybe tougher than I could ever imagine. So I better roll
up my sleeves, jump in, and say, 'Let's go.'"

Listening is just as important as being heard. In an
ideal world, both of you get the opportunity to unload your
stress and vent to each other about your day. But there
needs to be a give-and-take. There are days when Meredith
can handle me unloading to her and it rolls off her back.
Other days, I know from a simple look on her face to not
even bother. In the past, that might have made me resentful, but today it's a different story. Our relationship has
gotten to the point where we simply understand not to take
things personally. We can read each other's cues. That's

why discussing this ahead of time will save you the unnecessary stress and strain on your relationship, not to mention your bank account, if communicating from the get-go
helps you avoid therapy. But if you have the extra cash and
are willing to put in the time, couples therapy can be a
game-changer.

Having a friend to unload to has been a saving grace for
me, as well. Even if your best buddy is not a parent, having
another outlet to unburden your troubles to will do won-
ders. We all want to be validated and heard. If you can find
that from friends, then do it. Grab lunch, hit a happy hour,
or even just make a phone call. Being able to get things off
your chest will lighten that load. Then go home and be
that soundboard for your partner.

Once you get into the parenting groove, you'll master
how to navigate the hard times that inevitably arise. You
can develop this skill together, you just need to be on the
same page. So start that dialogue early. And the next
time you get home from work and walk through the front
door, don't freak out if the first few words out of your
partner's mouth are, "They're all yours." You got this,
dude.

Discipline Starts with Yourself

It's important for your partner and you to discuss your approach to discipline early on so you can be on the same page. You want to nip bad behavior in the bud, so being a team is crucial at all times. If you do disagree with your partner on a specific issue, never express it in front of your child. Have a private talk so that you and your partner can honestly discuss your differences and remain a united front.

That being said, the good-cop, bad-cop dynamic is practically unavoidable in parenting. My parents did it with me, and sometimes we do it with our kids. As long as you two have decided up front how to play off each other in front of your child and aren't actually fighting and disagreeing, go for it.

There will be times when you will experience frustration, irritability, and anger. One of the most challenging parts of being a parent is monitoring your reactions when in the company of your children. Kids are like sponges. They absorb everything, even when you think they aren't listening. *Especially* when you think they aren't listening. Meredith is constantly eyeballing me to keep my voice down when I am sharing a story that might be inappropriate for our kids to hear. "The kids hear everything," she says. Sometimes they understand what you are saying and other times they are clueless, but they hear you. So get into the habit of keeping private conversations private and working out any tension you might encounter with your partner *away* from the kids.

Emotions are tricky, and when we're stressed out we sometimes end up reacting before we think. But keep in mind that you can choose the environment you wish to raise your child in and how involved you want to be. Make it a point to show your child how you and your partner can work out differences. Show them how to resolve issues, not blow up over them.

Try to avoid unleashing your anxiety via sudden outbursts of irritability and anger. Kids need stability. Children need to know their parents are in control, and that they're safe with you. Having an outlet is important to help channel and release your anxiety in a healthy and productive way. If you are the physical type, visit the gym or go for a run. Work off that excess energy. Yoga is brilliant when it comes to focusing your energy and letting go.

Using Your Words

Spring break was the perfect opportunity for our family—along with my father-in-law, his wife, my sister-in-law, and her family—to go on a family trip. We usually headed somewhere warm and beachy. One particular time we headed to Cabo San Lucas. If you've ever visited Cabo, or Mexico in general, you know you have to wait in line to get through customs. Ahhh, the joys of waiting in a long, slow line with three children under the age of seven in a foreign country. Cole, our youngest, was about two at the time. Somewhere he managed to pick up the phrase "stupid fuck." Yes, you read that right (chip off the ol' block, that boy). However, since he couldn't quite pronounce an *f* properly, he made it sound like "puck." Cole managed to entertain himself by approaching strangers and calling them a "stupid puck." Of course, Cole's siblings were extremely amused by this, which only encouraged him more. Meredith was horrified and pretended to be his nanny. It was hard for me to keep a straight face.

I'm the first to admit I have a problem when it comes to language and dropping *f*-bombs. But when I became a dad, I knew I had to clean up my act.

Watching your language around the family is always in your best interests. Choosing the appropriate TV shows, movies, music, and even friends and family you hang out with should all be with the consideration of what's appropriate for you and your child. Of course, he'll eventually be exposed to bad language (and eventually bad behavior, and other bad things the world has in store for him). But why rush the process?

And when the inevitable day arrives when your child uses the *f*-word in your presence for the first time, my advice is to check yourself before you react in anger or disapproval. If there's one thing you can assume with kids, it's that whatever you tell them not to do will be the only thing they want to do. If you tell them not to go somewhere, that's the only place they will want to go. Whatever is taboo is what they will become obsessed with. This doesn't mean you should ignore bad behavior. But evaluate your level of response. *You* are the only one who knows if something is a big deal or not. You can diffuse the situation entirely by how you choose to respond. Calm, cool, collected, yet informed is always a good way to approach things with kids. And with life in general.

Fighting and Tension

As parents, we have an obligation of not only taking care of our own emotional health, but our family's as well. A simple yet powerful piece of advice I received early on was this: Think before you react. This phrase can be one of the most challenging lessons to achieve, but it is powerful nonetheless. Do your family a favor and lead by example. They say the brain is a mirror to the child's developmental experience. Children imitate what they see and hear, and learn from what they experience around them. If your child is exposed to kind behavior, chances are they will be kind. The same holds true for exposure to aggressive behavior and violence. Keep negative emotional outbursts, threats of any kind, and overall hostility of any kind away from your child's world entirely, and you've got a much better shot at raising a well-behaved kid.

Because your child is keyed in to your emotions, it's important to establish a calm, communicative way to express yourself in your relationship to both your partner and your child. There's nothing worse than when my wife is upset with me because I overreacted or couldn't understand her point of view. If my stubbornness gets in the way, it just alienates us and makes me feel worse for forcing her to indulge in my emotional hurricane. I've learned it's best to take a breath before verbally striking back. We both feel like crap if we don't treat each other with respect. Getting into the habit early on of managing your stress will not only make you a better partner, but a better parent as well.

I saw my parents arguing many times as a child. I wasn't even sure what they were fighting about. I just recall how unsettled I felt when I saw them emotionally charged and speaking in a tone that felt unsafe. Look, if I am being honest, my folks were big yellers. On one occasion, my mom got so upset that I thought her head would explode *Scanners*-style. I am not kidding! I was probably five or six at the time. To this day, that moment still remains burned in my memory. While it was scary and painful to experience, I use it as a reminder for myself today to remain calm when engaging in emotionally charged moments, be it with Mer or the kids.

Tantrum Tales

Picture me this: The entire Pegula clan circa 1980-something on a trip to Kmart in my parents' brown and white Buick station wagon. We had the model with the tailgate that opened like a door so you could get in from the back. Three kids sat in the way back and the others in the middle row. As usual, my brothers were successfully getting under my skin. After numerous attempts to block them out, I lost my shit and attacked them in the car. *Wham! Bam!* The car screeched to a roaring halt and my dad, with numerous threats and a raised voice, finally pulled into a parking spot and did his best to smooth over the backseat chaos. Meaning, he yelled at us. Again. *This* was a Pegula family outing.

The car doors flew open and everyone did their best to exit as quickly as possible while avoiding getting hit by the flailing appendages that took over my tantrum. To a passerby we must have looked like a friggin' clown car at the circus. To make matters worse, I decided I was not going to get out of the car no matter what the consequence. While everyone split up for their destination inside Kmart, I held my dad hostage, unwilling to budge. Unfortunately, his size and strength overpowered my stubbornness, and before I knew it I was being dragged through the patio furniture aisle. Being wholeheartedly committed to my current emotional state, I managed to kick, scream, cry, and blurt out every expletive I could conjure up in retaliation for being dragged around the store.

Over twenty years later, I managed to be the recipient of such behavior, in what my father has since described as "sweet, sweet payback." Kai would lose it every time we went to his favorite candy shop if he didn't get all the sweets he wanted. It became *a thing*, so we decided to limit the number of trips to this store per week since it was getting out of hand. One day he asked if we could go, and we said, "Not this time, buddy." Well, the boy went completely berserk and had a fit. There was no calming him down. Karma, anyone? How my dad managed to keep his cool during all my fits even half the time is beyond me. Experiencing the same behavior myself made me feel guilty for my own past transgressions.

Tantrums are one of the most challenging aspects of raising kids. But they are an inevitable part of parenting, so it's smart to arm yourself with tools that will allow both yourself and your child to survive the next breakdown.

Preparing for the inevitable breakdown in your parenting future is a solid idea. Depending on your demeanor, you might be able to let it roll off your back, but don't stress if, like most parents, you get the sweats, hot flashes, or even some dizziness every time you imagine the judgmental voices of those surrounding you. A key to keeping tantrums from escalating is to keep your calm. Chances are your anxiety is about to hit the roof. It's normal to feel out of control when your child's behavior is full-throttle. Dealing with your own feelings first will help when dealing with your child's current and future tantrums. Take a deep breath and don't give in to the feeling that all eyes are on you. After all, it's your child having the tantrum, not you. Removing your child calmly from the situation if you're in public is your best bet. Your action will demonstrate to your child that their behavior is not appropriate. It will set the stage that you mean business when they behave this way. Be mindful; it takes practice and probably will happen numerous times before your child catches on and understands the consequence.

Being mindful of triggers that set your child off into tantrum mode is needed as well. For instance, if you are about to enter the grocery store and you know your child will want to buy something and this time you are choosing not to let them, you need to prepare them ahead of time. If after preparing them they still end up in tantrum mode, calmly and quickly gather your things and head back home. While this is not a convenient move if you need to actually buy some groceries, leaving immediately is the best action you can take. I've been there numerous times in the past, and as annoying and frustrating as tantrums can be, it's best to take the steps to avoid them and teach your child the proper way to behave. I recommend checking out www.empoweringparents.com for more tips and knowledge on how to handle this behavior.

Following through is a central step in curbing your child's tantrum. If you give your child a consequence, you damn well better follow through with it. I've been guilty of making empty threats in my parenting years, and it came back to bite me every time. Not only does this add stress to the parenting dynamic, but it also seeps into your relationship with your partner if both of you are not on the same page. She will be resentful of you if you are a pushover (which I admit I was and have worked tremendously on changing). Bottom line: your child will not take you seriously if you don't back up your promises. Children need to understand that there are consequences for their actions, that you will discipline them if they break the rules, and that you will follow through.

Communication is your best tool. According to Jane Nelsen's *Positive Discipline*, as parents we often feel the need to complete our duties, like grocery shopping or other daily errands. If you are on kid-duty during these tasks, you'll most likely encounter resistance from your two-year-old. Yet you forge ahead in order to accomplish your task at hand, because, let's face it, you need to get it done. However, Nelsen suggests preparing your toddler by being specific with what you expect from their behavior. If they are not able to agree and abide by your request, you need to be willing to leave at once and abandon your original plan. Obviously this can be a huge inconvenience, but following through with consequences is a tool that will help in shaping the expected behavior during this period.

Remember, you are not here to be your child's buddy. You are his father. You can be a loving, caring person to him, as friendly and cool as can be, but when it comes to disciplining your kid, you need to be firm.

There is no one-and-done, easy-fix solution in dealing with tantrums. It demands practice and patience. And if it seems like none of these methods to ease tantrums are helping, and your child is still unable to control his emotions, consider seeing a doctor and possibly a therapist. It's not impossible that your child could be suffering from attention deficit disorder, or another issue, and that's an entirely different path on your journey.

CHAPTER 5

First Steps

As PARENTS WE get to witness our children taking their first steps. It's a very significant moment because after that they just keep on moving forward. Eventually they run right out the door! Whether it happens as early as nine months or as late as eighteen months, the process of walking is gradual. Once the walking starts, the curiosity increases. Your toddler is now officially on the road to nonstop discovery, which means you have to be on your toes to make sure their curiosity doesn't get them into trouble!

You are probably used to keeping a close eye on your toddler since he started moving around. It's shocking how quickly kids master the art of crawling. It seemed like I would turn my head and in the blink of an eye they would be clear across the room. They were friggin' speed demons. I thought it was exhausting then, but once they took to their feet, it really got challenging. There is a fine line between allowing your child to explore their curiosity and keeping them safe. Hopefully, by that time you will have babyproofed your home and addressed potential dangers like stairs and loose furniture. If you haven't, you better get to it stat!

Having some areas to allow your little one to run free and be relatively safe at the same time is a blessing. Heading to our local park was a great way to burn off some energy with my kids while getting in some fun bonding time. They loved playing in the sand, crawling up and down the jungle gym, and most of all slipping down the slides. It's always a blast seeing other dads and moms get into the fun with their own children too. You get to relive your childhood, to a certain degree. Although I definitely don't have the stamina I had as a kid.

Houston, We Have a Walker

Fast-forwarding to present day, my oldest son recently turned sixteen and is now legally allowed to drive. Kai's driving was a huge milestone for the entire family. It seemed like just yesterday that he was learning to take his first few steps. Now he is an independent young adult navigating his way through high school . . . and city streets! It's a moment that terrified and excited us at the same time.

The first few days of Kai's learning to drive were anxiety-provoking, to say the least. I admit I was a pretty annoying backseat driver, every second with my hand in front of me as if to protect me from flying through the windshield, and I would incessantly comment on his speed, his stops, and his choice of music. (I haven't let up on that last one, by the way.) It was very difficult for me to trust that he could be a safe, responsible driver. But finally, after a few weeks of practicing with Kai on the road, my anxiety and fear slipped away. I was actually quite comfortable with his independence. I never thought I would come to peace with the fact that I was not in control of our oldest son anymore. It was a bittersweet moment.

☛ **PARENT HACK**
Your toddler keeps sliding around the slick
surfaces?

Hardwood, tile, and even carpet can be hazardous for a toddler when they are on the move in onesies and PJs. Just grab a hot glue gun and write her name on the bottom of her socks, indoor shoes, or slippers to create a little grip.

Much like my anxiety got the best of me when Kai was learning to drive, you might find yourself becoming a backseat driver once your little one begins to master their first few steps. While we need to encourage their efforts, it's important to allow children to learn on their own. There will be many moments of imbalance and falling down, hopefully without injury, but eventually they will master the steps and before long will be zipping around your house or yard with you close behind. When you approach this phase of parenting, it's important to keep the following things in mind as your child experiences this next major milestone.

Encouraging your child with each step they take is important. Trusting them to learn on their own is tough. For the past year or so, your little one has been under your full control and constant supervision. You carried them, you wore them on your chest, you pushed them in the stroller. Now you are giving them the freedom to discover the environment around them on their own two feet. It's a bit challenging to give 100 percent control to your child. If it were up to me when my kids were learning to walk, I would've padded them with a cushioned body suit to prevent injury whenever they fell, like those human-dog-toy guys you see training pit bulls and Dobermans in the movies. But falling is important for them to experience so they can learn how to get stronger and become more confident at walking. You can't fly before you fall.

There are times when your kid will take a fall and your gut reaction will be to cry out and rush to their aid. But your outburst and over-the-top reaction can be scarier to a child than the actual injury itself, as your vocal and visual cues can actually increase the drama. Once you learn to hold back emotions and outbursts, you'll find many of these situations are not as bad as they seem and are easily diffused by a calm, metered approach.

Another mistake parents often make is to say that their kid is fine when he falls, whether they know it or not. Almost immediately after a spill dad says, "Awww, you're okay!" Well, how do you know your child is all right after a nasty fall? You didn't even bother to ask him first, you just told him. Look, dude, you ain't a doctor, so how do you know if junior is really uninjured? But in order to ease *your* anxiety, you immediately blurt out that phrase as a way to comfort yourself when you should be comforting your child. Acknowledging *the fact* that he fell and explaining what just happened to your child is much more useful. *Asking* him if he is all right instead of *telling* him he is fine is very different. Choosing the right dialogue is important. It wasn't until after I learned this fact from our preschool teachers that I started to see a difference in my children's

behavior when I made the choice to give them more freedom to fail.

We want to encourage our kids to be confident and secure. If we project our anxiety and fear onto them, we are
not doing our job of shaping them into the beings we hope
they will become. It's a catch-22. Our fear shows how
much we love and care for our children, but it can also be
paralyzing. A first step to supporting and encouraging
your child is learning to censor your reactions. It might be
difficult at first, but if you remind yourself to embrace
these moments with support and encouragement, you will
discover a very different reaction when your child begins
to master his steps.

Mastering these first steps is exciting to watch. At first,
your child will learn to pull herself up and lean on anything she can grasp to support her balance. Definitely
double-check your bookcases if you have any and make
sure they are secured to the wall. Since your child is just
discovering her mobility, now is a good time to revisit the
babyproofing in your home. Before long, your kid is going
to be a speed demon. If putting a gate in front of the stair-
way cramps your style and you choose not to have one, you
better damn well be sure you are all over your kid when
she takes off for the stairs. I don't know anyone who would
choose style over safety, but if that happens to be you, I

suggest changing your mind and opting for safety and security.

☛ **PARENT HACK**
Door slamming got ya flustered?

My kids love to slam doors, but sometimes they smash their little fingers! Cut a pool noodle in half and slide it on the door jamb. This will prevent the door from closing completely.

Stoves are another source of danger, especially when your child is learning to walk. Curiosity is part of the process and your stove is a wonderland of knobs, buttons, and tasty boiling delights that will entice every curious adventurer to explore. Check out the babyproof stove guards at your local baby store. There are covers that attach to the knobs to prevent your child from turning the stove on. There is also a guard that prevents your child from reaching over and touching a hot burner. Again, while these items might put a damper on your design aesthetic, my motto is better safe than sorry. These are temporary and worth the safety they provide. Don't worry, your pad will be cool again soon enough.

A good friend told me how his son just loved banging the doors in their house. It was something he mastered soon after he perfected his footsteps. How delightful. You might want to consider investing in soft doorjamb stoppers. These will prevent your little one from slamming your doors, prevent his fingers from getting pinched, and prevent you from going insane.

Remember, insanity is no bueno.

Get Some Class

Now that your kid is mastering the art of walking, this is a good time to consider enrolling her in daddy-and-me-type activity classes. There are numerous programs today where instructors provide balancing activities, tumbling, dancing, and singing, all age-appropriate activities for your developing toddler. These are great ways for both baby and parents to burn off excess energy while learning how to socialize and have fun at the same time. Often you can end up finding a new network of dad friends to hang with too. Creating a community of buddies who are going through the same thing as you is a plus. Take advantage of it if you can.

☞ **PARENT HACK**
Don't let baby lock himself out.

Kids love to play with doors, but they might lock themselves
in—or out—of a room. Try stretching a rubber band across
one door knob to the other. It will prevent the door from actually locking.

If you are lucky enough to be able to include your child in
work activities, give it a go. Kai used to love when I brought
him along with me to walk a few dogs, back in my dog-
walking days. Sometimes he'd enjoy a ride on my back in a
carrier while hanging on to a leash or two. Other days, he
would revel in the act of walking the neighborhood streets
showing off his canine buddies to everyone we encountered.
Bonding moments like this are ones I cherish forever, and
you will too. Taking the time to bond with your child and
demonstrating patience as they go through growth and
change are keys to being a solid father and building rewarding experiences.

Being patient with your child is important as they are
mastering this new discovery in their lives. Often we are
rushing to get from task to task in order to get as much
done as possible, but once your mobile toddler is part of

the picture, you need to accept that it just might not be possible to get as much done as you used to. Even visiting a department store can take more time, since they will now want to explore and flex their newfound independence on their feet.

Just a Little Patience . . .

It's so easy for parents to lose it with their kids, but the repercussions that behavior will have on your child are far worse than the relief you might feel at that moment when you let it fly. Not to mention that eventually your guilt will sneak up on you and you'll feel like crap for losing your shit because your kid just dumped his cereal all over the table simply because he felt like it. Hey, just because you happen to be having "one of those days," don't take it out on your child. Find an outlet for your frustration, to blow off that steam, so that you can be levelheaded when you communicate with or discipline your child. As parents, we have a duty to guide our children to become the best possible human beings. We must lead by example. If they see us losing it, how can we expect them to keep it together?

☛ **PARENT HACK**
Toys too friggin' loud and annoying?

Try putting tape over the speaker of the toy to help reduce the volume. It's much better than kidnapping or dismantling Peek-a-Boo Elmo.

So what are some ways you can work on not losing it when your kid is driving you nuts?

Breathe and think before you react. This is a big one, so I'll repeat it. Breathe and think before you react. It sounds so New Agey but, dude, it is solid advice. Just take a moment to check in with yourself. Getting familiar with your emotional temperature is the best way to do this.

Meditation is another great tool. I admire Meredith for her dedication to meditation. She is great at it. I, on the other hand, am so high-strung that unless I am about to go to sleep, meditating is virtually impossible. However, being a marathon competitor, I run for hours at a time. Running has become my meditation. I can't tell you how good it feels to be able to focus for as little as ten minutes or as long as four hours. It's not about the length of time as much as it is about the frequency and dedication.

Exercise is a given when it comes to burning off excess emotion and stress. It also sets an amazing example for your child. If you can include your partner in your routine, even better. You can also use that time to check in with each other. Nothing wrong with killing two birds with one stone.

Exercise!

Exercise is a great way to relieve your stress, in addition to its overall health and wellness benefits. Exercise pumps up your endorphins, improves your mood, and acts as meditation in motion. Whether you are a runner, lift weights, cycle, or do yoga, dedicating the time to take care of your health is part of your parenting responsibility. It is also a great example to your children as they grow older and witness your focus on staying in shape. If getting to the gym or a class seems like a hassle, no worries. You can even stream some great workout classes right to your iPhone or iPad, so you really have no excuse. (Sorry, dude.)

Taking every opportunity to *stay present and enjoy the parenting experience* is something we parents should all aspire to. While we can't be perfect, we can make doing our best at being patient with our children a priority. So next time you feel your blood pressure rising, or find yourself overreacting to your child's stubborn behavior, or even feel conflict arising with your partner, remember: breathe and think before you react. If you can achieve this much, then you are gonna rock as a dad.

Let's Talk About Sex

One of my favorite childhood rituals was getting the mail from our mailbox. Sounds like a silly favorite memory, but I grew up in the country, and our home was situated on a hill about an eighth of a mile off the main thoroughfare. In order to get the mail, we could either pick it up by car on the way to our home, or I could walk an eighth of a mile to the mailbox, or get it my favorite way . . . by riding my Honda XL80S motorcycle! Each of my siblings had their own cool motorbike, but since I was the youngest I was only allowed to have the slightly lamer XL80 model. It wasn't a piece of crap, but it wasn't exactly the most macho bike either. Fortunately, as I got bigger I inherited the hand-me-downs of my siblings, which in our family included badass motorcycles. So to ride my two-wheeler down that dusty highway at top speed, combined with the excitement of leafing through my favorite *Archie* and *Jug-head* comics for the latest hilarity and practical joke items, was analogous to a kid waking up to open presents under the Christmas tree.

As I was feverishly rummaging through the mail, I would often come across a postcard from a resort in the Pocono Mountains called Cove Haven. Cove Haven can best be described as a getaway for honeymooners and lovers where they can enjoy the lush landscape of the Poconos while relaxing in a bubble bath in a *huge* champagne glass. In fact, on the front of the postcard addressed to my parents was pictured a man and woman naked *in* said champagne glass filled with bubbles. You can only guess what that image does to a sheltered adolescent tween growing up in the countryside. I was always confused as a child as to why my parents wanted to spend time *away* from the family at a location like *this* to sit in a giant glass. Of course, now, as an adult and parent of three children myself, I get it. I mean, they weren't just sitting around in giant goblets looking at the view!

With that in mind, the attention you give to your relationship is something that is going to impact most all dynamics of your life. My motto is: if my wife is happy, I am happy. Which in turn makes my kids happy. Hence, we are all happy. Sounds like a no-brainer, right? But happiness takes effort, and this is where I'll remind you of how important it is to put in the effort to make yourself as happy as possible.

What makes *you* happy and what makes your partner happy might be two totally different things. Most guys are happy when they are getting some, but depending on where your partner is emotionally and physically at the time, sex might be the farthest thing from her mind. Not to mention the fact that finding time for sex can be virtually impossible with a newborn or toddler in the house. Even if you have a spare minute, you both might be wiped out and find you don't even have the energy to do the deed. Dude, this is normal. As with any transition in life, you need to find your rhythm. It might not be easy at first. She might be in the mood when you are not, and vice versa. But being sensitive and tuned in to when she is ready is most important.

There are a lot of factors to wrap your head around, considering that your partner is still going through emotional and physical changes, having given birth within the past year. I'm not here to give you tips on how to score with your partner. You already know how to do that. But what I would like to suggest is that you be communicative and open with your partner about sex. Depending on your relationship, you might feel awkward talking about scheduling a time when the two of you can get it on, but it is better than feeling like there is no interest or time because life is too hectic since your baby came along. *This* is where the real work happens in your relationship.

Dialoguing is the best way to express your needs, your desires, your emotions. Your partner should/will appreciate your desire to work on this part of your relationship. Understanding where she is coming from and speaking honestly will get you both into a habit that will stand the test of time. But if you are silent and take a backseat to this topic, you both can be interpreting signals and concocting stories in your mind that are the furthest from the truth. Actually listening to how your partner feels about sex, and what she is experiencing, is mandatory.

Some women lose the urge completely for sex for quite a bit of time. The hormonal transitions her body has been experiencing can contribute to her lack of interest. For instance, if she is breast-feeding, then the last thing she might want is you messing with her breasts.

"The whole time I was breast-feeding, I was like, *Stay away,*" my wife reminds me. "I thought, *I am a milk machine, and that's my job now.* So I just wasn't feeling very sexual or physical during that whole stage. They say it's about six weeks for your body to recuperate from childbirth, even more if you've had a C-section. But eventually you start to relax and the feelings come back."

Some women say the loss of their sexual urge can actually make them feel depressed and scared. "I was worried for a while, actually," Meredith confessed. *"Is that it? Do I no longer care about sex?* I was relieved that it eventually came back."

But not everyone feels that way. "I was ready to get back into it pretty soon after I wasn't sore from having the baby," another mom told me. "My husband was a little concerned about it hurting me, but he was ready when I was ready too. I actually got much *more* in touch with my natural rhythms based on what was happening with my hormones. I listen to my body now more than ever."

Whatever your partner's experience, it might take some effort to help her feel the urge again. Use this time to revisit the dating period you shared before the two of you got serious. In addition, help out as much as you can to ease the pressure of taking care of your newborn. Throw in a load of laundry, clean up the kitchen, and make dinner. These simple tasks will reduce a ton of pressure your partner might feel. Women are wired totally different than us guys, so while these acts seem menial or inconsequential to you, they will go a long way toward inspiring a desire for intimacy from your partner. The less stress your partner experiences, the more likely you can start to build back your sexual relationship.

Once you rekindle that fire, scheduling alone time for the

two of you is important for reducing stress and keeping your relationship intact. Start to consider getting a regular night away if you can. Reach out to family or friends who can watch your baby, and spend some quality time getting back into your groove as a couple. Whether you call on a family member or hire a sitter so you can do date night, make it a ritual to find that time for the two of you to connect.

When the baby is really tiny and sleeping all the time, you can just bring the kid along on date night. Our Kai would sleep right through our night out at our favorite restaurant. If your baby is a deep sleeper, don't feel like you two can't take a break and grab dinner together.

"But it's also okay to *not* have the focus on the couple right away," Mer chimes in. "There's a lot going on those first few months, and there can be pressure to immediately get the relationship and sex right back on track, and I don't think you have to jump right back into it all right away. If you need some time before you start up with date night, so be it. Wait until you guys are up and running a little bit. You've got to get settled with your new family."

Hey, you might not have the opportunity to unwind in a champagne glass bubble bath, but you can certainly enjoy a glass of bubbly with your partner in a good ol'-fashioned tub. Your relationship is important, so put in the work and have fun together!

Vacations

My parents were quite the travelers back in the day. My dad used to rent an RV for the sole purpose of schlepping everyone to our vacation destination in one vehicle. Today I'm sure it would be considered illegal to hang out in that bed above the driver's seat, but back in the seventies pretty much anything was fair game. How my parents managed to travel with all six of us is beyond me. I can barely handle my three.

Mommy-and-Me Time

When our first was born, the last thing I imagined was feeling jealous over the time he spent with Meredith. But oddly enough it turned out to be a legitimate concern. Parenting is demanding, especially the first year. Be prepared for your partner to be highly focused on caring for your newborn. It has nothing to do with you. But having a conversation about your time alone together is a great way to let her know how you feel. Staying connected to your partner is work, but once you put in the hard work, you will soon discover that mommy-and-me time involves you too, not just your newborn.

Preparation is the key when it comes to traveling with little ones: having the right gear, enough supplies, and, above all, patience. Depending on your child's age, you want to make sure you do not leave home without the essentials (diapers, wipes, hand sanitizer, bottles, utensils, etc.). The worst thing you can do is forget one of these as you are about to board a plane. If you're driving to your destination, then you can always make a pit stop. But once that cabin door closes on the plane, there's no getting off. This might sound like common sense, but you'd be surprised how many parents forget the basics. After all, you have a lot on your mind packing for the entire family. You might want to consider making a list as a reminder of what you cannot afford to forget. Don't let your ego get in the way on this one. Preparation is always your best friend.

Another trick for getting through these stressful travel adventures with children is distraction—theirs, not yours. Distraction is your friend, my friend. Getting creative with ways to buffer the situation and change the emotional tone of the meltdown will help bring the boil down to a simmer. We were on a fight once where I spent about an hour in the bathroom flushing the toilet as a distraction just to get a smile on my daughter's face. She was losing it over not being able to play with the tray table of the seat in front of her. She was driving the passenger in front of her nuts. So I picked her up and made our way to the back of the plane. I thought at least getting her into the bathroom would separate her from the rest of the crowd. In there, she became fascinated with the flush of the toilet. So I spent an hour flushing the toilet over and over, to her amusement. As torturous as it was to be confined in such a small space for such a long time, I wanted to keep the peace with my neighbors on the plane and keep my sanity.

Traveling with kids can be a very pleasant experience or a total nightmare, so hedge your bets and do some due diligence.

Avoiding On-the-Road Rage

Stress is a major factor that can put a damper on your vacation. Here are a few pointers to help ensure that your quality time is high-quality time.

Don't make threats you won't follow through on. Any mafioso or loan shark will tell you this. Decide ahead of time (with your partner's input) what your limits and their consequences are, and make good on your promises. Just as it is at home, credibility and follow-through are key when you're on vacation. Paradoxically, kids like to know someone's in charge—in a loving, clear, and fair way. So don't be afraid to establish your rules of the road.

Get on the same page. The moment your kids detect that you and your partner are not in agreement on an issue or punishment, you are pretty much screwed. It's amazing how savvy kids are. It's like they detect fear and weakness. Like bears.

Don't wait until the night before to get everything together. A lot goes into planning a trip, and the packing alone could drive anyone insane. Make a list, check it twice, and start early.

Pack a few surprises. A great trick I learned to distract a child on a long flight is to pack some surprises like books, Play-Doh, snacks (healthy-ish?), or little toys. Wrapping them so your toddler gets to experience a little mystery will make the experience more fun too. You need to strategize when you actually present these distractions. If you have a three-hour trip, then stock up on a minimum of three small distractions. At every hour, or earlier if you see they're no longer interested and starting to get restless, have another distraction to present. You can even go as far as having them pick out from a bag of wrapped distractions. Be creative and have fun.

Meet the neighbors. A great way to break the ice with the fellow passengers around you on a flight is to prepare a survival bag for *them* too. What's a survival bag for a fellow traveler? It's simply a pair of ear-plugs, a small bottle of water, or even aromatherapy oil (preferably lavender to help calm their mood), and a kind note apologizing ahead of time for any misbehaviors your child might be guilty of. Your fellow traveler will be shocked and delighted by your effort.

Don't be disappointed if your first trip with your little one doesn't turn out as smooth and enjoyable as you expected. The most important tip you can remember is that children can sense our energy, so if you're stressed they will pick up on that stress. Set the tone and ask your partner to help out by reminding you to chill out if she senses you are getting stressed.

And when all else fails, a shot of whiskey can also do the trick (for you, not your kids). Just remember to be responsible.

Daddy's Little Tax Credits

The last thing on most new parents' minds when starting a family is the amount of money having a kid will save you on your taxes. What a bonus! Who knew? For all the money they are going to end up costing you, it's nice to know the government is willing to give you a break per child. The legalese tends to change unexpectedly, so I recommend going online to irs.gov to see how much savings your family can qualify for. Here're a few pointers:

- A qualifying child must be a U.S. citizen you claim on your taxes, who can't support themselves, and is your son, daughter, stepchild, foster child, brother, sister, stepbrother, stepsister, or a descendant.
- The Child Tax Credit can be worth as much as $1,000 per child under the age of seventeen, depending upon your income.
- If the amount of your Child Tax Credit is greater than the amount of income tax you owe, you might be able to apply for the Additional Child Tax Credit.
- This credit is limited if your adjusted gross income is above a certain amount (depending on your filing status).
- The credit is also limited by the amount of income tax and any alternative minimum tax you owe.

CHAPTER 6

Uncharted Waters

PARENTING FAILS ARE bound to happen more than once in your life. Nobody is perfect. "You got that right," *DadNCharge* blogger Chris chimes in. "During my first week of staying home, my son was in a high chair at the breakfast bar in our new house. He leaned back in the chair and fell out of the chair backwards. I had never heard him scream like that, so I rushed him to the ER. We waited hours, with my kid shrieking the whole time. He ended up breaking his collarbone. On my first week of being home taking care of the kids my son breaks his friggin' collarbone. Unbelievable."

Okay, they might not all be quite *that* bad, but shit happens, dude. Learning to be easy on yourself as you get familiar with this new and ever-changing landscape of parenting is crucial. This is a full-time job. You don't want to get burned out too easily, so approach each day with curiosity and a positive attitude. If you do, you will find that this parenting gig is actually pretty fun.

Relationship Reminder

Nurturing your friendship with your partner throughout this journey you are on together is also key, so that you can express frustrations and share milestones with each other. Giving yourself and each other the freedom to fail and learn above all is something most parents forget to do. Go easy on yourself and your partner when things get tough. You are a team, and to be a good team you need to communicate. Family is like a chain reaction. If you and your partner are good, your kids will be good. If you and your partner are tense, your family will feel tense. Parenting takes effort, and as long as you put in the time and energy you will reap the rewards twofold.

Respect the Bubble

These days, there's increasing awareness of the importance of personal space and bodily safety. I've come to believe it's never too early to introduce the idea of healthy boundaries to a toddler or young child. At our local public school, young students were taught vocabulary to use if they felt their personal space was being threatened. The students were to imagine a bubble around themselves. If conflict arose, the students would use the learned vocabulary to keep them safe and protected. For example, if one student was not comfortable having another student invade their personal space, they would tell that person not to burst their bubble. I love how this image simplifies the idea of preserving the feeling of safety for the child.

So how early is too early to start having conversations about boundaries and personal space? Only your comfort level can dictate what age is appropriate, but I suggest starting earlier rather than later. Waiting to discuss boundaries with your child until they are old enough to discuss sex education might be too late. It's perfectly acceptable to speak to your child about personal space in a manner that is on their level of understanding so they can start to learn what is appropriate and what is not.

Big Mama

A great resource for talking to your child about social, emotional, and behavioral issues is the Mother Company (themotherco.com). Founded by two moms, the Mother

Company aims to educate and create a generation of kind, compassionate, considerate kids. The site has a wealth of knowledge, featuring tools like videos and books that address age-appropriate social and emotional issues all children undergo. You will not be disappointed with the information this company has to offer, so go check it out today.

The biggest lesson you want to impart to your child about this is that they have control over their body. They have the right to say yes or no to anyone who wants to touch them, be it a hug or kiss (even from family). I used to feel so uncomfortable when one of my kids would shy away from hugging or kissing family members they were not used to seeing regularly. I felt like it was a reflection on my parenting, not to mention displaying disrespect to the person they did not want to show affection to. But Meredith made me realize that they have the right to decide whom they want touching them, and that is fine. It's actually natural for a child to feel uncomfortable being affectionate with someone they are not close to. Allowing them to listen to their gut and honoring those feelings will empower your child. What a great gift to give as a parent to your child.

What about the discussion of private parts? You certainly need to be age-appropriate in your terminology and

only need to discuss as much as you feel is appropriate for
your child's age. But do yourself a favor and use the correct
names for anatomy when discussing the parts of your
child's body. The last thing you want is your child refer-
ring to his penis as his "pee-pee" or "dingus" or some
other made-up name for his private parts. My grandma
used to call it a "piece of skin." She was an old-school Catholic woman whom I loved dearly, but calling it a piece of
skin was a bit confusing, to say the least. Teaching the
appropriate terms will once again empower you and your
child. It also avoids the feeling of shame around your
child's own body, and potential embarrassment when they
say something ridiculous around others that they learned
directly from you because you didn't want to say the word
"penis."

Children are curious little buggers, and you're bound
to come across a moment when they flash their private
parts because they think it's funny. Be sure not to overreact if this happens in your care. Simply explain to your
child that it is not appropriate to show our private parts in
public. His body belongs to him, and it is to be kept covered because it is private and personal. You'll want to avoid
scolding and shaming. Having the openness to communicate with him early on will keep the dialogue going as your
child gets older. Talking is healthy. Get into the habit of

doing it early with your kids. The more open you can be, the better your child will respond.

Birthday Madness

Most of us have at least some fond memories of birthday parties from our childhood. They might not have been fancy, or perfect, but often they were fun (or at least we remember them that way, decades later).

But here's a news flash. Birthday parties are not always great fun when you're the parent. Often they mean swarms of family and friends and a ridiculous amount of preparation—neither of which you might want to deal with. You're probably gonna have to bite this bullet the first few years and do the party thing. It means a lot to the people around you, despite the fact that at this age your child will have absolutely no memory of it whatsoever. None. But your wife will. And her mother will. And so will yours. Play the long game here, pal.

Today in L.A., birthday parties are on a whole other level than when I grew up in rural Pennsylvania. My kids have attended parties ranging from private VIP service at Disneyland and Universal Studios to gaming trucks and DJs bringing down the house. One of the most unusual birthdays I ever attended actually included a kiddie train, a live chimpanzee to be in a photo with you, and catered food that was served on fine china. Talk about setting the bar high! I made sure to explain to my kids that this was not the norm and this would be the last time they got to have an experience like this. It was overkill to the nth degree.

Speaking of birthdays, the first year is definitely an important and exciting milestone to celebrate. But as I'm sure you can agree, you don't remember your first birthday, so perhaps you can keep that in mind when you are about to plan your little one's special day. Take the pressure off yourself to make it perfect. The last thing you want to do is feel stressed out while taking care of everyone and everything at your party. It's much better to be in the moment and cherish your child's day. So with that in mind, here are a few thoughts that you might want to consider when it comes to celebrating your child's first birthday.

Safety first. Decorations always make a party festive, and balloons set the mood. But consider using Mylar over latex, as latex can be a choking hazard if the balloon accidentally breaks. Party favors are a typical thought as a thank-you to your guests for celebrating this special day, but they are not necessarily needed.

If you do decide to go that route, be sure to not include candy or other chokeable items. Instead, opt for a soft chew toy that is age-appropriate or even a hardcover children's book. Remember, it's safety first and the thought that counts. Some other thoughts to keep in mind are making sure silverware is kept out of reach and if you have a pet, be sure they are accounted for during the party. We chose to have our golden stay at a friend's place during the party hours. While she was the friendliest dog ever, you never know how your dog will react to overstimulation from your little guests.

Be allergy-aware. Be sure to find out if there are any food allergies to be concerned about. Wheat, peanuts, eggs . . . whatever it is, it's best to know ahead of time to prevent any problems on the big day. Ask your guests, and tailor the menu (and food prep) accordingly.

Bounce with caution. My kids loved bounce houses, and while one-year-olds might be too small for such an activity, it's definitely a great activity for kids two and up. Hell, I loved getting in a jump or two myself. But it goes without saying that kids smash into each other all the time in these things, so if you or your child are especially sensitive to pain, you might want

to reconsider letting them into the bouncy bruisina-
tor. Your call.

Honor the almighty nap. Keep in mind that during this
first year, your little one still naps quite a bit, or at least
you hope she still does. Keeping the length of the
party to two hours is ideal, three tops. There's nothing
better than a kids' party that's short and sweet.

Consider the joys of outsourcing. If you attend Daddy
and Me– or My Gym–type classes, you might explore
hosting your party with them. Businesses like these
are great, since they have professionals who host the
event and allow you to relax and partake in the festiv-
ities. The best part about that is you get to skip the
cleanup afterward. Who doesn't like that idea?

As your kid gets older, there will be plenty of time to
dazzle them and your guests with creative party ideas. But
for now, do yourself a favor and keep it simple. It can only
get better (or worse) from there.

Sugar Fix

Many parents use sweets as a reward for good behavior. I
suggest you do not. You don't need a sugar junkie in your

house. Rewards and consequences are definitely helpful for disciplining purposes, but it gets complex when you use food as a carrot (so to speak) on a stick. Teaching your child to have a healthy relationship with food is important. If you restrict sweets, the child will just rebel at some point. Moderation is the key.

When I was a kid, my parents were not very strict when it came to diet and sweets. I indulged in sugar every moment possible. Don't get me wrong, we didn't eat chocolate and junk food 24/7, but I don't recall ever being prohibited from eating a sweet whenever I felt the urge. For instance, after-school pickup usually involved a quick run through Mickey D's for a large order of fries and a hot caramel sundae with nuts. Oh, how I miss that sundae! We lived about twenty minutes away from my school, so picking up a treat often distracted us for the drive home and tided us over until my mom made her typical three-course dinner. If we had to race home soon, there was no need to fret. We had a special cabinet for snacks—a cabinet Meredith soon obsessed over—stocked with everything from chips and pretzels to candy and my favorite, Double Stuf Oreos. That cabinet remains stuffed with treats to this day and my wife is captivated by it every time we visit my parents.

Meredith had major restrictions on sweets growing up. Cereal was specifically a sore spot for her, as she could eat only healthy choices like Raisin Bran and Cheerios. I, on the other hand, had free rein to dive into Cap'n Crunch, Frosted Flakes, Count Chocula, Fruity Pebbles, or what have you. Once she discovered the special sugar treasure trove in my folks's house, it was tough to tear her away.

☞ **PARENT HACK**
Popsicles getting too messy?

Nothing more fun than sticky hands from drippy popsicles! Use a disposable coffee cup lid or muffin cup liner to create a li'l tray. Pop that 'sicle into that bad boy and now they are goo free!

I never imagined limiting my children from something I was exposed to as a child that I considered normal. But once I became an adult and changed my style of eating, I soon became an advocate for making sure my kids were eating healthy. If you start this practice while your child is young, you will be amazed at how she will choose to continue the practice on a daily basis. Your actions become examples, so *if you eat healthy, chances are your kids will want to eat healthy too.* It just takes some effort and education to get them started.

First off, never, ever use food as a bribe when it comes

to disciplining your child. It's a slippery slope. Associating behavior and foods can create habits that ultimately lead you down the wrong path. Kids are smart and will catch on to your tricks. If they realize good behavior results in sweets, then you might find yourself in a pinch the next time they start to act up. Children who become accustomed to sweets at an early age will most likely carry the urge for sweets into their adulthood, so doing your best to avoid the introduction until they are at least two years old is highly recommended.[8]

While Mickey D's was a saving grace for me as a child, today if I even mention that name to my kids they cringe and make the most unpleasant expressions. I'd love to take all the credit for their disgust toward fast-food restaurants, but most of the credit goes to Meredith and their school health classes. Obesity is a huge epidemic in our country, and fortunately giving children the awareness and insight to make smart choices really makes an impression. You don't have to wait until your kids are in school to begin the process. Once you start to read to your child, you can pick books themed on healthy choices and eating like *Good Enough to Eat: A Kid's Guide to Food and Nutrition* by Lizzy Rockwell. Sure, this might sound extreme, but why not get in the game early on?

Juice is another sweet that modern parents tend to

avoid, as whatever vitamins juices offer are often swimming in buckets of sugar. I was surprised when I learned that apple juice contained such high amounts of sugar. So if you're going to serve juice, at least water it down. It's no wonder my kids craved so much of it when they were young. Fortunately, today my kids are just not that into sugar. They are allowed to indulge at movies or parties, but their access to it is nothing like I had when I was a kid. Today I still have a sweet tooth for desserts, but have little desire for straight-up candy . . . well, unless we are talking about Swedish Fish, Milk Duds, or Junior Mints. Okay, maybe I never actually got over my sweet tooth!

Moderation is the way to go. Best to lead by example. At some point in their lives, your kids are going to find their way into the sugar sensation. Hopefully they'll see you as an example of healthy living, or even feel the need to correct your old-school, fast-food ways.

Kids 'n' Sports

I was always involved in sports as a kid. Sports are a major part of our family dynamic. But that wasn't always the case. Back when Kai was around three years old, Meredith and I signed him and his best friend Jared up for soccer lessons.

It was a weekly session that lasted all of an hour. But, man, that hour can last a looooong time, especially when your kid does not want to cooperate. Kai and Jared were very articulate for their age. They managed to have pretty intense, thoughtful discussions. They loved *Harry Potter* and moviemaking. When it came time for soccer, they decided that, rather than break a sweat and participate in such a rigorous activity, they would sit down in the middle of the playing field and simply talk about how great *Harry Potter* was and what next movie they were going to make. The only people who got exercise from this weekly outing were Jared's mom, Mer, and me. No matter how much effort we put into getting them interested in soccer, it was never going to happen. Eventually we got the picture.

Today, however, is a different story. One of my proudest moments is watching Kai, at sixteen years old, pitch from the mound with speed and accuracy while playing high school baseball. Watching him race for a pop-up hit to center field when he is playing the outfield or hitting a triple is up there as well. I would never have guessed this of a kid who at the age of eight tried out for his first baseball team and broke down into tears out of fear of not being good enough. The day that happened, a coach spotted his behavior and took him to the side for a pep talk that inspired Kai to drum up the nerve to give it a shot. That was over eight years ago. Today Kai is as passionate as ever about baseball. Thanks, Coach!

Juliette first got a taste of volleyball in fifth grade. She was a natural. Three years later she is playing both club and school volleyball. Her serve is ferocious. Watching her passion and skill as she plays is something both Meredith and I treasure.

Cole had his moments with basketball and baseball. Baseball was his sport early on, though. He was a great pitcher and catcher. He also has the speed to run like a bullet. But cross-country is where he found his passion. I loved cross-country myself and participated in it during my high school years. Its what's inspired me to run marathons to this day. Watching him run is always an inspiring experience.

The discipline and effort that goes into playing a sport is so great for kids. Besides the physical workout, they get to learn about the qualities of teamwork, dedication, and commitment, to name a few. They also learn how to work with other people to accomplish a central goal, not simply a personal goal. These are important traits that help define our kids. Learning how to win and how to lose is also important. But most important, it's how you play the game that matters. Sports are a great way to build up your child's confidence.

"I was never really a big sports guy growing up," my coauthor Frank admits. "But we put our daughter into soccer really early on. We didn't care if she had a future in it, but we liked the values it offered: being part of a team and working with others, setting goals and training to accomplish them, and listening to adults in authority. Turns out she also truly loves the sport, did indeed pick up many of the life lessons we hoped she would, and it's given her incredible focus and drive."

It is very important that kids learn how to listen to adult authority figures who are not their parents. In life, they will have coaches, teachers, bosses, sergeants, and/or team leaders they will need to respect and listen to. It's good to get them trained for this relationship so they don't see other adults telling them what to do as a threat. Be it team, individual, or combat sports, children need to learn early on that adults can impart wisdom, lessons, and values to them, and to accept these without feeling threatened.

At this stage of parenting, your child is just getting familiar with learning the basics when it comes to sports. Being a guide is the best way to allow them to discover which sport they might excel at and which sport they might despise. During our early parenting years, both Meredith and I feared Kai's disdain for soccer meant he would never be into sports when he got older. Clearly that was not the case. From baseball to football, Kai is obsessed with sports 365 days a year. Cole, on the other hand, couldn't care less, but that's okay. He still gives his best effort when cross-country season starts up. He knows the value Meredith and I place on sports and that following through with his commitment as best he can is all we ask.

Encouraging your child to play a sport will likely come when your child hits grade school, but until then it's not a bad idea to grab a ball and try for a game of catch. Children are young and uncoordinated enough at this point that it won't exactly be the most thrilling game, but exposing them early on will gain their interest. If you discover they don't care much at first, take their lead and don't pressure them to love sports just because you do. They will likely gain interest in sports as they get older anyway, so avoid making them resent sports because of the pressure you are putting on them. Allow them to find the enjoyment on their own. For now, it's just an introduction on your behalf to get them acquainted with something they might enjoy. If you don't have a sports star in the making, that's all right too. Your child will find their way to their interests, don't worry. Kids are full of surprises.

CHAPTER 7

Growing Up

As I ENTER this new phase of parenting (the "teenage terror years," as I've been calling them with a shudder), I reflect on how my parenting influences my children daily. Like you, this territory is constantly new to me. While I am excited to see how the next few years will evolve, I am also overwhelmed by the thought that my oldest will soon be off to college and beginning a whole new chapter of independence and discovery. It trips me out that he is two years away from the age I was when I left home for college. It seems like only yesterday that Kai was crawling around the floor in his own vomit, and now he's off to college to . . . well, he's probably still crawling around the floor in his own vomit, but still, I think you see where I'm going with this.

Sometimes I wonder, *How did we get here?*

☛ **PARENT HACK**
Your two-year old won't stop fidgeting?

Give your kid a big ol' cardboard box and some crayons, and watch their creativity flourish. Forts, artwork, submarines, lemonade stands; the possibilities are endless . . .

I couldn't have done it alone, that's for damn sure. My wife and kids have been such an inspiration in my life. They are the real reasons why I have found success in my business and in my family life. Their support and belief in me continues to motivate me to expand and grow, as a person and in my business. Of course, hard work, dedication, and persistence play roles, as well. But nevertheless, my family has been a crucial part of who I am today and what I hope to accomplish in the future. My role as a dad is just as important to me as my business. After all, being a dad is what inspired everything I created in my company. The fact that it adds support and encouragement to fellow dads all over the world is a total bonus.

But kids grow up, and the family dynamic changes. We

can't stay young and innocent forever, right? You, dude, are merely at the beginning of this journey. You still have a number of speed bumps ahead of you on this ride. Such as . . .

The Terrible Twos (?)

Man, it is so great to have a partner to help balance the scales when it comes to parenting, especially when experiencing the particular age of challenges, discovery, stubbornness, and independence known as the "terrible twos." Or "the terrific twos," depending on your kid. But for some reason the adjective *terrible* earned its way as the most popular title for this age. Is it really necessary to label this specific time as being so bad? For some families this age is a blast. For others it's challenging. But it's always exciting and fun, no matter what the temperament of your child may be.

"I'm Bored"

We love our kids madly, but sometimes we just do too much to entertain them. You are in the beginning stages of your parenting journey, so now is the time to set the path right. Encourage your toddler to play on their own. Being bored is okay. That's how kids find ways to entertain themselves. Set them up with building blocks, pots and pans, or any creative toys that will keep them safe yet allow them to be occupied without your help. Give them crayons, paper, and some inspiration. One of my favorite responses to "I'm bored" is, "Only boring people are bored." My son hates it when I say that.

So why is *terrible* used as a description for this particular age? Perhaps because it is in this stage of development that kids begin to test limits. Two-year-olds are mobile and curious. They love to touch and feel, to climb and put everything into their mouths. They also love, love, love the word *No.* They have limited language, so when they get frustrated if you don't understand what they are trying to communicate, they throw a fit. It's a natural reaction at this stage of development. The difficulty becomes when we as parents try to control and curb this behavior.

"Different kids have different personalities, so they all act differently in their twos," Meredith opines. "Without naming names, two out of three of mine were *much* more challenging. It's a personality thing."

☛ **PARENT HACK**
Got marker and crayon stains all over?

Try a dab of toothpaste. Just apply, scrub with a washcloth, and watch the stains fade away.

Maybe the *terrible* part should be less about labeling the child and more about how it feels to be parents. The baby you were in charge of is now defying you. Who likes to be ignored and defied? Nobody. You feel out of control. Meredith and I understood that in order for our children to become successful people, they needed their voice. They needed to say, "I need to feel separate from you, but I'm also scared because I'm so dependent on you." We wanted to support that.

Potty Training

Kids are usually ready to start potty training as early as eighteen months and as old as three years. There are signs to look for when your child is ready to try using the toilet on their own. First, if they are conscious of the times they need to pee, that is a sure sign they can begin the process.

If you find your toddler saying they need to pee or poop, then chances are they might be set to give training a shot. Check to see if their diaper is dryer for extended amounts of time, which usually indicates they are gaining control of their bladder. Being adept at taking off their clothes is another clue. Don't pressure your toddler into it. Take it slow and be patient.

Hygiene

Once your child is starting this phase, it's important to give them some direction on proper hygiene. Proper hand cleaning is a ritual you want to properly address. Depending on the age of the child, some parents find that a useful way to teach their child to wash their hands is by lathering up with soap and rubbing their hands together while singing the entire alphabet. Not such a bad idea, although it could take a while if your child does not know the whole song. But at least you know they will be giving their hands a good scrubbing. If not the alphabet, then try something fun like the Muppets' "Rainbow Connection," a Beatles tune, or something simple and fun you can sing together.

Taking the lead from your child when it comes to potty training is highly suggested. If you try to force it on them too early, chances are they are going to rebel and you will not get very far. Test the waters by having some simple discussions about going to the bathroom, and invest in some children's books that explain from a kid's perspective the whole potty-training experience. A few we loved to use to teach potty training are *Everyone Poops* by Taro Gomi and *Once Upon a Potty* by Alona Frankel. We even went as far as buying a kiddie potty and leaving it in the family room, so the kids could familiarize themselves with it and attempt to use it on their own. A way better approach than forcing it on them . . . though it led to some interesting conversation with guests about interior design.

Be warned, your child will challenge your patience. Try to take a deep breath, stay calm, and don't react quickly out of emotion. It's normal for your child to attempt to establish independence. We want our children to embrace independence once they get older. Just work the task at hand with patience and grace. Your reactions inform your child of the power they have when you lose control. Keeping a calm demeanor will inform them that no matter how they react you have the authority.

Actually, these taxing moments can be useful in the long run, as you are actually teaching your child how to react to things as they grow up. If you are the type of guy who flies off the handle and has a hair-trigger temper, then chances are your child will be that way too when they are older. Keeping your frustration in check and learning not to overreact are big steps toward being a great dad. Take the high road. You owe it to your family.

Autism, Developmental Issues

Every October, my family gets into Halloween mode. My kids can't wait to hit Six Flags, Universal Studios, or Knott's Berry Farm for the various Fright Fests, Horror Nights, and what have you. It's an evening filled with zombies scaring the living crap outta clueless guests as they wait in line to ride the roller coasters and explore the many haunted mazes. The creative mind that designed these mazes should be commended . . . or better yet, committed. The fusion of visuals and auditory performances by the zombie actors impacts even the most apathetic individual. It didn't take much to scare the crap out of me, but there was one specific maze that impacted me like no other. It was called Total Darkness.

Total Darkness was a completely pitch-black maze nightmare led by one intrepid leader of the group (me . . . *gulp*) while the other members (my kids and their friends) hold on to a rope that leads them through an agonizing five-minute pathway in complete darkness where fear of the unknown assaults your senses. We were all single-file, following the leader—me. I had no trouble displaying and vocalizing my fear every second of the journey. As I led my group through this terrifying experience, we were bombarded with flashes of demons, sounds of torture, and voices of evil piercing our ears. I'm not sure who was more scared, myself or my children. Imagine yourself walking blindfolded without a hint of sight or light, not knowing what was going to happen next. While this experience is vastly different from the parenting journey, the similarity of the emotional life is spot-on. In retrospect, the totally dark maze was less scary than the fear of not knowing what your future holds as a father.

As parents, from the moment of the child's birth we do everything in our power to protect and nurture our child. While we cannot predict the future, we envision a pathway rich with experience, knowledge, happiness, and safety. But sometimes we are thrown a curveball and that path is no longer so direct, and we must adapt to reach the goal we were originally seeking. My sister, Nicole, is a perfect example of someone whose journey became one filled with hills, valleys, twists, and turns. I am inspired and motivated by her unrelenting determination to make life better for her children, especially my nephew Cullen, who was diagnosed with autism.

Cullen is the second of four children, three boys and

one girl. Nicole's oldest child, Patrick, was born happy and healthy over eighteen years ago. Fifteen months later, Patrick became a sibling and our family welcomed Cullen into the world. Cullen too was born happy and healthy. Life for Nicole was going as planned.

Newborn years are tracked through milestone markers set by the medical community to gauge your child's development. While these markers are not set in stone by an exact date, you can usually assume progression is normal if your child falls within a few months of the marker at hand. For instance, crawling can occur anywhere from eight to twelve months of age. Some schools of thought suggest girls develop quicker than boys, so some markers might be affected by your child's sex. When it comes to walking, kids can start anywhere from twelve to eighteen months, so Nicole became concerned when Cullen was still not getting to his feet after a year.

"At thirteen months, he still wasn't crawling, so I kept talking to the doctor," Nicole recollects. "They said, 'He's a boy, they develop later than girls do.' My first child was walking by twelve months, so I had a point of reference."

Deep in her gut, Nicole knew something was off. She found it impossible to sit back and wait, so she sought the assistance of a physical therapist to engage Cullen into taking some steps.

After over two years of being told to relax and that nothing was wrong, despite Cullen still not talking yet, a doctor finally suggested they take him to a developmental pediatrician, just to have everything checked, and *that* was when they found out. Cullen was nearly three when he was finally diagnosed on the autism spectrum.

"We had no idea about what we were dealing with," Nicole vents. "The only thing I had ever heard about autism was the movie *Rain Man.*"

Now Cullen attends a local special ed school and, thanks to modern advancements in technology, is able to communicate using an iPad and apps. For instance, one app called Proloquo uses pictures that Cullen chooses to communicate for emotions or actions. For instance, if he is hungry he will go into the food category and choose what food he wants. Even more amazing is the iPad speaks the sentence for him. It's amazing that he is able to have a conversation with his mom, or anyone else, for that matter. Before this technology, it was a struggle to figure out what he wanted.

With Cullen, allergies also entered the picture and early on affected his ability to trust that eating was safe, since his reactions to milk, barley, and wheat were anaphylactic. At least two times, he was rushed to the hospital in an ambulance because of reactions. This is something many parents have to deal with, even if they do not have a special needs child.

Developmental issues, severe allergies, these are all fears we encounter and pray will never happen to our children. But if they do, taking action and trusting your gut are important and crucial. As a mother, Nicole's focus was to find a solution to keep her children safe. While at times it was heart-wrenching to undergo these experiences, her parenting skills continued to develop in her fight to make her children's lives as healthy and positive as possible.

Today, Cullen is adapting and learning how to prepare for adulthood thanks to the many wonderful organizations that support those individuals affected by medical disabilities such as autism, Down syndrome, and other medical conditions. Learning a trade like warehouse packing/shipping or janitorial duties are among the opportunities offered to create an income for those affected by these disabilities. But the most import thing for any child is having good parents or a solid support system that sticks by you. While I can only imagine how emotionally challenging parenting has been for my sister, I have been lucky enough over the years to witness her undying love and commitment to her family and watch her face every moment reflect courage and strength.

I think back to that night leading my kids through that

maze of darkness and can't help thinking how much life reflects that exact experience. I could've allowed my fear to paralyze and overcome my path during that maze or continue to move forward and face the obstacles that crossed my path no matter how terrorizing they appeared. As parents, we must always choose the path to move on.

Keeping Your Relationship Strong

Raising kids is a big responsibility. You and your partner spend so much time as a parenting team now that it can be hard to view each other as a couple anymore. Hell, it can even be tough to see each other as sexy, sensual creatures after all the dust settles. You just see a mom. And all she sees is an increasingly lumpy dad. The lack of sleep and the sheer energy that goes into keeping these tiny creatures entertained all day long is exhausting, least of all trying to work in some romance. Resentments build up, and before you know it at least one partner lashes out at the other, usually for a ridiculous reason. Don't let it get that bad, dude.

Big lesson: don't let your parenting distract you from being a couple.

Parenting is hard enough as it is, so don't add fuel to

the fire by letting it dismantle your marriage or sex life. During the first few years it's tough to find alone time with your partner. If you don't make it a priority, you are going to find that it's harder to stay connected. The relationship will surely suffer.

Meredith was great at carving out alone time for us to catch up, usually while the kids were sleeping. Every night can be an opportunity to spend fifteen minutes to check in with each other and learn about the other's day. When you are connected emotionally, physical intimacy is much more intense and effortless. The likelihood of being able to reconnect in bed will increase if you stay emotionally connected. The two go hand in hand. The trick is not to let becoming part of this parenting team dwarf your relationship as lovers.

However, Meredith and I have gone through many challenges in finding time alone, and not just when the kids were young. Even today, having teenagers in the home, we find it difficult to be alone. Fortunately, we get to spend our mornings together after the kids head to school and before I head to work. I long for those moments when it's just the two of us catching up and inspiring each other for the day. If time allows, we definitely take advantage of hitting the sack for some alone time, if you catch my drift. It's all about the effort and making it a priority.

You will discover down the line as your little one hits elementary school, middle school, and high school just how much the focus has been on your children. You don't want to suddenly discover that you don't know your partner the way you used to when you first met because you have been distracted with raising your children, work, and your social life.

While I'm sad to no longer have the infant and toddler days in my life, I am excited about the future Meredith and I have, both alone *and* with our teenagers. A whole new chapter in our lives has begun and we get to put the focus back on us as a couple again. But don't wait until it's too late. Make the effort now to stay connected. Besides benefitting your relationship, it will benefit your children from the great role modeling the two of you demonstrate as a couple.

The *D*-Word

Divorce can be an impactful experience for kids. It hits adults hard too, but we are used to life smacking us around. Children usually have no point of reference for this type of loss, beyond maybe a death in the family. While I never experienced divorce firsthand, I do know that relationships take time and effort. A strong, loving relationship keeps the family dynamic positive and healthy. But what happens if you and your partner just can't seem to make it work out? While I am a fan of exhausting all efforts to fix your relationship—attending counseling, reaching out to friends, working out the issues that have created an impasse in your relationship—there are times when you need to take a cold, hard look at your situation and make a decision that is in the best interests of you and your family. If repair is just not possible, then start working *with* your spouse on a plan to make sure your kid feels secure with your decision to divide the family.

"I come from a family chock-full of divorce," my co-author Frank reminds me, "which was all the more reason I was hell-bent on finding the right girl and staying married. No matter what. I was determined not to make the same mistakes my parents did. But after seven years we just weren't happy together. Deep down we both knew it, but I refused to admit it. So when out of the blue my wife mentioned that she wanted to try a separation, I was completely blindsided. It was a dark, dark time, to be sure. Eventually I realized she was right and it was the right thing to do, for us and for our daughter. Bella was around two years old at the time, and though it all seemed cataclysmic back then, she doesn't remember any of that now."

In her book *Parenting Apart,* Christina McGhee discusses in detail the effects of separation on the child, as well as suggestions on how to make the transition as seamless as possible when you decide to make the separation official. Depending on the age of your child, the efforts will need to be adjusted to meet the needs of the specific age, but be consistent in your behavior and involvement. Do your best to keep life in your home predictable by maintaining regular routines and normal activities for as long as possible as you work through whatever path you take. This involves routines like bedtime, meals, and consistency in day-to-day life activities.

Role Models

On a day-to-day level, if you can't actually minimize the conflict between you and your partner, then at least take it to the next room, or wait till the baby is asleep, or just keep your voices down. Just keep the stress away from your baby. This might be a challenge if things are highly charged between you two, but make your child the top priority. It's been said that children are like wet cement and whatever falls on them makes an impression. If your child is a toddler, you need to be even more sensitive, as they will internalize the reason you decide to split and feel somewhat responsible for things not working out between the two of you. They are still developing their cognitive skills and take things literally at this age, so the more consistent and fluid you and your partner can commit to raising your child together, the better off your child will be.

Just because children are young doesn't mean they don't know what's going on. Babies, even newborns, go through a developmental stage in which attachment to mom and dad is critical for their feelings of safety and security. If one or both of you are not present and receptive to this stage of development, you will notice issues arising like fussiness, irritability, discomfort, disruptive sleep patterns, and so on. Consistency from both parents is highly advised. This involves regular bonding time, frequent contact with mom and dad, and quality time with each parent. If you are able to make this a reality, you should find your child feeling emotionally, psychologically, and physically connected to you both, hence feeling secure and safe.

If one of you decides to move out, maintain a calm, collected relationship with your partner, and keep any tension out of the reach of your children. No matter what the reason may be for your split, children internalize things, so you don't want them thinking it has anything to do with them. Talk to your child and offer reassurance that they are loved by both of you with actions like hugs, kisses, or spending quality time with them. It's up to the two of you to make the transition as seamless and natural as possible, especially if they are not of an age where they understand what is happening or have limited verbal skills.

Kids need both parents if they can get 'em, as each brings different qualities to the table. Whatever issues you had as a couple, just put them aside when you make mutual decisions about your child. Many relationships fall apart due to lack of communication. Your relationship as parents does not have to suffer the same fate.

As parents, we are role models for our children, so we must conduct ourselves in the best way possible every chance we get. Sure, it is a challenging task to achieve, whether you are married, separated, or divorced. But if you can keep the focus on the best interests of your child, then hopefully you will be inspired to think before you react, to make decisions that are in the best interest of your child. Above all else, be consistent in your parenting and your communication.

Epilogue

As I write these final pages, I am listening to the sounds of water flowing in the background from my neighbor's fountain. It is early evening, the weather is warm, and our backyard is festive, adorned with glowing white lights. There is no other sound to distract my attention but the trickling of the fountain. For years, I longed for this exact moment: to sit in peace and listen to nothing but silence in my own backyard. My mind travels back to numerous memories of pushing my children on the swing set and the laughter from Juliette as she slid down the slide. I can still picture Kai throwing tennis balls to our golden retriever, Ashley, in our backyard when he was three. Cole loved climbing into the tree house with chalk in hand to transform the walls into a Pollack masterpiece at the early age of two. It seems like just yesterday, but it's been over five years.

It's fascinating to experience how quickly time goes by based on the age of your child alone. I find myself in a new chapter of parenting. Gone are the days of diapers and wipes and my kids learning how to tie their shoes and/or say the ABC's. Today it's, "Dad, can I have some money for gas?" or "Dad, I'll take out the trash later, I'm playing *GTA*." The one that really hits me in the heart is when I get no response but the eye roll and hair flip from my daughter Juliette. Welcome to parenting a teenager! But I must confess these moments are few and far between. More often than not I hear, "Dad, can we go hit the ball in the backyard now, please?" or "Hey, Dad, can we go for a hike today?" My favorite from Juliette is, "Dad, can we go to brunch?" I'm always down for a good meal.

I love that my children have grown into the bright, caring, artistic, and loving young adults they are today. But it didn't just happen on its own. I realize Meredith and I did a ton of work to ensure they were all guided down the right path. I feel fortunate to have put in the hard work both in my marriage and my parenting journey and have experienced the payoff firsthand.

It's an amazing time in this country to experience becoming a dad. Companies are acknowledging dad's role and offering paternity leave so dad can assist and bond with his newborn and partner. How cool is that? There are more men choosing to be stay-at-home dads as their partners advance in the professional world, and the need for two incomes is not as necessary as it used to be. It just goes to show how the role of a dad is changing with the times, constantly evolving. Our culture is evolving to reflect the love and attention fathers need and want to give to their children, just as our mothers have been doing for generations.

Of course, I can't imagine our family running any other way than with both Meredith and me contributing. Parenting is a big responsibility and you need all the help you can get, financially and emotionally. We are responsible for teaching our children the tools to survive in today's world amid all the chaos in our society. That's a huge obligation. But how amazing is it to think that we can be the change that we wish to see in the world by the way we choose to parent our children? No matter whether you live in New York, Topeka, or Portland, we can all stake a claim in the type of children we hope to raise today. You can be the first step in creating a different world for our children, and it all goes back to the simple question: What type of dad do you want to be?

Whether you view the role of a father as being a hero figure, a breadwinner, a leader, a partner, a guidepost, or a supporter, choose to be a dad who is involved and loving. Put in the hard work, because it will pay off later in life. There is no finer thing than to have raised a healthy, happy, well-rounded child.

A concert in our neighborhood just started in the distance, and I hear cheers from the crowd. My peace and quiet are now filled with the sounds of celebration and music. Life is good.

Cheers to this new phase in your life! May you have many years of a heart filled with love and affection for your new role as a dad, dude.

Acknowledgments

One of the many themes I discuss in this book is support. For a new parent it's virtually impossible to survive without it. I feel the same in regard to writing.

Thank you to my amazing wife, Meredith. You put up with me during my first book and still stuck with me when I committed to a second. I love you so much!

Kai, Juliette, and Cole. Because of you three, I've found a second love next to being a dad, and I thank you for your patience while I completed this project.

A huge thank-you to my friends and preschool teachers at Circle of Children. Your guidance gave me the wisdom to be the best father possible and share my experience with the world.

To my sister Nicole, I love you and am so grateful you shared your story with me. You're the best.

To Scout, Bill, Amara, Daphna, Charlie, Chris, Brian the Birth Guy, and Adam. Thanks for your willingness to open up and share your experiences with me.

Akiko, you are wonderful for keeping Diaper Dude running smoothly while I am off writing. Your talent and support are so appreciated!

Scott, you never disappoint. Even when you are out of the country, I can rely on your input. Love you, dude!

Thank you, Dr. Holly Lucille, ND, RN, for your medical expertise in this project.

Shab, thanks for your continued support.

Stacey, once again you rock, and I love working with you.

Marian, I am so excited for your belief in this project and am thankful for the chance to continue my story.

Frank Meyer, you are the man! Love working with you. I could not have completed this book without you.

Finally, I want to thank all fathers today who are putting in the time to make a difference and reverse the stereotype of what it means to be a dad.

Resources

Cohen, Michael, *The New Basics: A-to-Z Baby & Child Care for the Modern Parent* (New York: HarperCollins Publishers, 2005).

DeSouza, Luiza, *Eat, Play, Sleep: The Essential Guide to Your Baby's First Three Months* (New York: Atria Books, 2015).

Frankel, Alona, *Once upon a Potty*, Girl and Boy Editions (Richmond Hill, Ontario, Canada: 2007).

Gaffigan, Heidi Murkoff, Sharon Mazel, *What to Expect the First Year* (Workman Publishing Company, 2014).

Gaffigan, Jim, *Dad Is Fat* (New York: Three Rivers Press, 2014).

Gavigan, Christopher, *Healthy Child, Healthy World: Creating a Cleaner, Greener, Safer Home* (New York: Plume, 2009).

Gomi, Taro, *Everyone Poops* (St. Louis, MO: Turtleback School & Library, 2001).

Halsey, Dr. Claire, *Baby Development: Everything You Need to Know* (New York: DK Publishing, First American Edition, 2012).

Herald, Andy and Charlie Capen, *The Guide to Baby Sleep Positions: Survival Tips for Co-Sleeping Parents* (New York: Potter Style, 2013).

Karp, Harvey, *The Happiest Baby on the Block: The New Way to Calm Crying and Help Your Newborn Baby Sleep Longer*, 2nd ed. (New York: Bantam Books, 2015).

Levs, Josh, *All In: How Our Work-First Culture Fails Dads, Families, and Businesses—And How We Can Fix It Together* (New York: HarperOne, 2015).

Lewis, Michael, *Home Game: An Accidental Guide to Fatherhood* (New York: W. W. Norton & Company, 2010).

McGhee, Christina, *Parenting Apart: How Separated and Divorced Parents Can Raise Happy and Secure Kids* (New York: Berkley, 2010).

Murkoff, Heidi, *What to Expect the First Year* (New York: Workman Publishing Company, 2004).

Natterson, Cara Familian, *Your Newborn: Head to Toe* (New York: Little, Brown and Company, 2004).

Nelsen, Jane, *Positive Discipline* (New York: Ballantine Books, 2006).

Rockwell, Lizzy, *Good Enough to Eat: A Kid's Guide to Food and Nutrition* (New York: HarperCollins, 2009).

Sears, William and Martha Sears, *The Attachment Parenting Book: A Commonsense Guide to Understanding and Nurturing Your Baby* (Boston: Little, Brown and Company, 2001).

Sears, William, Martha Sears, Robert Sears, and James Sears, *The Baby Book: Everything You Need to Know About Your Baby from Birth to Age Two revised edition* (New York: Little, Brown and Company, 2013).

Siegel, Daniel J., MD, and Mary Hartzell, *Parenting from the Inside Out: How a Deeper Self-Understanding Can Help You Raise Children Who Thrive 10th Anniversary Edition* (New York: Tarcher/Penguin, 2013).

Zevin, Dan, *Dan Gets a Minivan: Life at the Intersection of Dude and Dad* (New York: Scribner, 2013).

Notes

1. Josh Levs, *How Our Work-Force Culture Fails Dads, Families and Businesses—And How We Can Fix It Together* (Harper Collins, 2015), 5.

2. *The Economist*, "More Hands to Rock the Cradle," May 16, 2015, http://www.economist.com/news/leaders/21651215-both -parents-should-be-paid-spend-time-home-their-babies- more-hands-rock-cradle.

3. Sarah Cools, John H. Fiva, and Lars Johannessen Kirke-bøen, May 4, 4; "Causal Effects of Paternity Leave on Children and Parents," *Scandinavian Journal of Economics*, July 2011, Vol. 117, Issue 3, pp. 801–828. DOI: 10.1111/sjoe.12113.

4. Chris Isadore, "San Francisco Paid Parental Leave," CNN .com, April 6, 2016, Money.cnn.com/2016/04/06/news/ economy/san-francisco-paid-parental-leave/.

5. NBC News Staff, "Measles Outbreak Linked to Disneyland Is Declared Over" (News.com, April 17, 2015), http://www

.nbcnews.com/storyline/measles-outbreak/measles
-outbreak-traced-disneyland-declared-over-n343686.

6. Katharine Stone, "How Many Women Get Postpartum Depression? The Statistics on PPD" (Postpartumprogress.com, 2010), http://www.postpartumprogress.com/how-many-women-get-postpartum-depression-the-statistics-on-ppd.

7. Heidi Murkoff and Sharon Mazel, *What to Expect When You're Expecting* (New York: Workman, 2008).

8. William Sears, MD, and Martha Sears, RN, *The Baby Book* (New York: Little, Brown, 1993).

Index

About the Authors

Photo by Jeff Linett

Actor-turned-father-turned-designer Chris Pegula is the creator of Diaper Dude, America's most high-profile line of hip gear for dads. After the birth of the first of his three children, Pegula noticed that most diaper bags and accessories were designed with a woman's sense of style in mind and created the Diaper Dude brand for dads. In addressing a simple need, Pegula revolutionized an industry. Since the launch of Diaper Dude, Pegula has emerged as a lifestyle expert on all things family-, parenting-, and partner-related, appearing on *Rachael Ray*, *The Nate Berkus Show*, *E! News*, *The Oprah Winfrey Show*, and numerous other TV and radio spots. Most recently Pegula founded True Dude, a lifestyle brand designed to honor today's men who leave a positive impact on the next generation. By aligning with the nonprofit Futures Without Violence (5 percent of all True Dude sales support the extraordinary efforts of Futures Without Violence), True Dude hopes to redefine what it means to be a dude in the twenty-first century. Pegula currently resides in Los Angeles with his wife and three children.

Photo by Scott Evansky

Frank Meyer wrote *From the Wall of Sound to the New York Underground,* cowrote *From Dude to Dad: The Diaper Dude Guide to Pregnancy* and *On the Road with the Ramones,* and has written for *Variety, LA Weekly,* and *New Times.* Meyer serves as content producer at Fender Musical Instruments by day, the front man for the notorious rock band the Streetwalkin' Cheetahs by night, and single father to a thirteen-year-old all the time.